Statistical Postprocessing of Ensemble Forecasts

Statistical Postprocessing of Ensemble Forecasts

Edited by

Stéphane Vannitsem
Royal Meteorological Institute of Belgium, Brussels, Belgium

Daniel S. Wilks
Cornell University, Ithaca, NY, United States

Jakob W. Messner
Technical University of Denmark, Kongens Lyngby, Denmark

ELSEVIER

Elsevier
Radarweg 29, PO Box 211, 1000 AE Amsterdam, Netherlands
The Boulevard, Langford Lane, Kidlington, Oxford OX5 1GB, United Kingdom
50 Hampshire Street, 5th Floor, Cambridge, MA 02139, United States

Library of Congress Cataloging-in-Publication Data
A catalog record for this book is available from the Library of Congress

British Library Cataloguing-in-Publication Data
A catalogue record for this book is available from the British Library

ISBN: 978-0-12-812372-0

For information on all Elsevier publications visit our
website at https://www.elsevier.com/books-and-journals

  Working together
to grow libraries in
developing countries

www.elsevier.com • www.bookaid.org

Publisher: Candice Janco
Acquisition Editor: Laura S Kelleher
Editorial Project Manager: Tasha Frank
Production Project Manager: Prem Kumar Kaliamoorthi
Cover Designer: Victoria Pearson

Typeset by SPi Global, India

Contents

CHAPTER 10 Postprocessing of Long-Range Forecasts 267

Bert Van Schaeybroeck, Stéphane Vannitsem

CHAPTER 11 Ensemble Postprocessing With R 291

Jakob W. Messner

Contributors

Roberto Buizza
European Centre for Medium-Range Weather Forecasts (ECMWF), Reading, United Kingdom

Sebastian Buschow
University of Bonn, Bonn, Germany

Petra Friederichs
University of Bonn, Bonn, Germany

Thomas M. Hamill
NOAA Earth System Research Lab, Physical Sciences Division, Boulder, CO, United States

Stephan Hemri
Heidelberg Institute for Theoretical Studies, Heidelberg, Germany; Presently at: Federal Office of Meteorology and Climatology MeteoSwiss, Zurich-Airport, Switzerland

Jakob W. Messner
Technical University of Denmark, Kongens Lyngby, Denmark

Annette Möller
Clausthal University of Technology, Clausthal-Zellerfeld, Germany

Pierre Pinson
Technical University of Denmark, Kongens Lyngby, Denmark

Roman Schefzik
German Cancer Research Center (DKFZ), Heidelberg, Germany

Nina Schuhen
Norwegian Computing Center, Oslo, Norway

Thordis L. Thorarinsdottir
Norwegian Computing Center, Oslo, Norway

Bert Van Schaeybroeck
Royal Meteorological Institute of Belgium, Brussels, Belgium

Stéphane Vannitsem
Royal Meteorological Institute of Belgium, Brussels, Belgium

Sabrina Wahl
University of Bonn; Hans-Ertel Center for Weather Research, Bonn, Germany

Daniel S. Wilks
Cornell University, Ithaca, NY, United States

Preface

A revolution in dynamical weather forecasting has occurred over the past 20 years, as the fundamental uncertainties regarding future atmospheric behavior, stemming from sensitivity to initial conditions ("chaotic" dynamics) and structural errors in the dynamical formulations, have come to be represented by ensemble forecasts. These multiple integrations of the governing physical equations are intended to represent probability distributions over future atmospheric states, thus allowing the formulation of probabilistic forecasts, which maximize utility and value to decision makers and other forecast users. Unfortunately, the collections of raw dynamical forecasts in the ensembles exhibit systematic errors in magnitude (bias) and dispersion (calibration) that must be corrected before the results can be interpreted probabilistically. Although such statistical corrections have a long history of use for postprocessing traditional deterministic forecasts, their extension for use with ensemble forecasts is a relatively new and still rapidly developing field, as witnessed by the wide variety of techniques developed and applied in various fields of geosciences.

The need for a general overview of these techniques was realized while the editors were organizing sessions on postprocessing of ensemble forecasts at meetings of the American Meteorological Society (AMS) and of the European Geophysical Union (EGU), which successfully attracted many people from diverse fields of geosciences. These sessions revealed, in particular, the wide variety of methodologies effectively used for specific applications and the potential of many new statistical postprocessing approaches. This book is an attempt to bring together and review the different methodologies that have been proposed in this context, and to illustrate their applications in a series of selected fields. It is comprised of a set of contributed chapters by international subject-matter experts describing the current state of the art in statistical postprocessing of ensemble forecasts, and illustrating use of these methods in several important applications including weather, hydrological and climate forecasts, and renewable energy forecasting.

We would like to acknowledge the excellent work of all these experts and express our gratitude to them for having made this compilation possible, namely Roberto Buizza, Sebastian Buschow, Petra Friederichs, Tom Hamill, Stephan Hemri, Annette Möller, Pierre Pinson, Roman Schefzik, Nina Schuhen, Thordis Thorarinsdottir, Bert Van Schaeybroeck, and Sabrina Wahl. We would like also to thank the reviewers of the different chapters for their dedicated work that improved the quality of the content of the book. Finally, we would like to acknowledge the constant support of the production team of Elsevier.

We hope that this book will help scientists of different horizons to find their way in this rapidly growing field of ensemble postprocessing.

Stéphane Vannitsem
Royal Meteorological Institute of Belgium, Brussels, Belgium

Daniel S. Wilks
Cornell University, Ithaca, NY, United States

Jakob W. Messner
Technical University of Denmark, Kongens Lyngby, Denmark

UNCERTAIN FORECASTS FROM DETERMINISTIC DYNAMICS

Daniel S. Wilks*, Stéphane Vannitsem[†]

Cornell University, Ithaca, NY, United States* Royal Meteorological Institute of Belgium, Brussels, Belgium[†]

CHAPTER OUTLINE

1.1 SENSITIVITY TO INITIAL CONDITIONS, OR "CHAOS"

In a startling paper that was published more than a half-century ago, Lorenz (1963) demonstrated that solutions to systems of deterministic nonlinear differential equations can exhibit sensitive dependence on initial conditions. That is, even though deterministic equations yield unique and repeatable solutions when integrated forward from a given initial condition, integrating systems exhibiting sensitive dependence from very slightly different initial conditions eventually yields computed states that diverge strongly from each other. Twelve years later, Li and Yorke (1975) coined the name "chaotic" dynamics, although this label is somewhat unfortunate in that it is not descriptive of the sensitive-dependence phenomenon.

The system of three coupled ordinary differential equations used by Lorenz (1963), and originally derived by Saltzman (1962), is deceptively simple:

$$\frac{dX}{dt} = -10X + 10Y \tag{1.1a}$$

$$\frac{dY}{dt} = -XZ + 28X - Y \tag{1.1b}$$

$$\frac{dZ}{dt} = XY - \frac{8}{3}Z \tag{1.1c}$$

Statistical Postprocessing of Ensemble Forecasts. https://doi.org/10.1016/B978-0-12-812372-0.00001-7

This system is a highly abstracted representation of thermal convection in a fluid, where X represents the intensity of the convective motion, Y represents the temperature difference between the ascending and descending branches of the convection, and Z represents departure from linearity of the vertical temperature profile. Despite its low dimensionality and apparent simplicity, the system composed of Eqs. (1.1a)–(1.1c) shares some key properties with the equations governing atmospheric flow, in particular an apparently erratic behavior whose characterization is at the heart of dynamical weather prediction. Accordingly, Lorenz (1963, p. 141) concluded that his results "indicate that prediction of the sufficiently distant future is impossible by any method, unless the present conditions are known exactly. In view of the inevitable inaccuracy and incompleteness of weather observations, precise very-long-range forecasting would seem to be nonexistent."

Palmer (1993) points out that in addition to sensitive dependence, the simple Lorenz system and the equations governing atmospheric motion also exhibit regime structure, multiple distinct time scales, and state-dependent variations in predictability. Because the Lorenz system has only three prognostic variables, these three properties, as well as their sensitive dependence on initial conditions, can be visualized in terms of trajectories on the system's phase-space attractor. A phase space is an abstract geometrical space, each of the coordinate axes of which corresponds to one of the prognostic variables in a dynamical system. The phase space for the Lorenz system (Eqs. 1.1a–1.1c) is therefore a three-dimensional volume. The attractor of a dynamical system is a geometrical object within the phase space toward which trajectories are attracted in the course of time, each point on which represents a dynamically self-consistent state, jointly for all of the prognostic variables. The understanding of the specific geometry and the dynamical properties of this type of attractor is the subject of the ergodic theory of chaos and of strange attractors (e.g., Eckmann & Ruelle, 1985).

Fig. 1.1, from Palmer (1993), shows an approximate rendering of the Lorenz attractor, projected onto the X–Z plane. The figure has been constructed by numerically integrating the Lorenz system

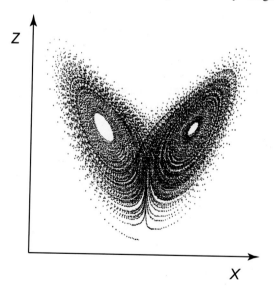

FIG. 1.1

Projection of a finite rendering of the Lorenz attractor onto the X–Z plane, yielding the Lorenz "butterfly."

From Palmer, T. N. (1993). Extended-range atmospheric prediction and the Lorenz model.
Bulletin of the American Meteorological Society, 74, 49–65.

forward for an extended time, with each dot representing the system state at a discrete time increment. The characteristic shape of this projection of the Lorenz attractor has come to be known as the Lorenz "butterfly." In a sense, the attractor can be thought of as representing the "climate" of its dynamical system, on which each point represents a possible instantaneous "weather" state. A sequence of these states then traces out a trajectory in the phase space, along the attractor.

Each wing of the attractor in Fig. 1.1 represents a regime of the Lorenz system. Trajectories in the phase space consist of some number of clockwise "laps" around an unstable fixed point of the dynamical equations on the left-hand $(X < 0)$ wing of the attractor, followed by a shift to the right-hand $(X > 0)$ wing of the attractor where some number of counterclockwise laps are executed around a second unstable fixed point, until the trajectory shifts again to the left wing, and so on. The transition from one wing to the other is performed in the vicinity of the third unstable fixed point of the dynamical equations. Circuits around one or the other of the wings occur on a faster time scale than residence times on each wing. The traces in Fig. 1.2, which are example time series of the X variable, illustrate that the fast

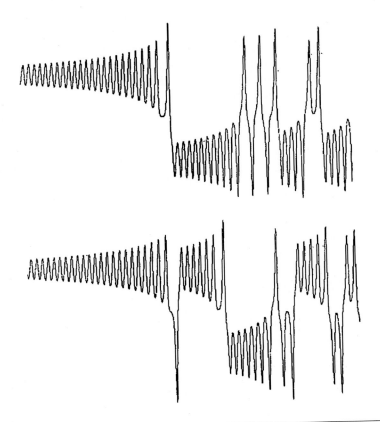

FIG. 1.2

Example time series for the X variable in the Lorenz system. The two time series have been initialized at nearly identical values.

From Palmer, T. N. (1993). Extended-range atmospheric prediction and the Lorenz model.
Bulletin of the American Meteorological Society, 74, 49–65.

oscillations around one or the other wings are variable in number, and that transitions between the two wing regimes occur suddenly. The two traces in Fig. 1.2 have been initialized at very similar points, and the sudden difference between them that begins after the first regime transition illustrates the sensitive-dependence phenomenon.

An especially interesting property shared by the Lorenz system and the real atmosphere is their state-dependent variations in predictability. That is, forecasts initialized in some regions of the phase space (corresponding to particular subsets of the dynamically self-consistent weather states) may yield better predictions than others. Fig. 1.3 illustrates this idea for the Lorenz system by tracing the trajectories of loops of initial conditions initialized at different parts of the attractor. The initial loop in Fig. 1.3a, on the upper part of the left wing, illustrates extremely favorable forecast evolution. These initial points remain close together throughout the 10-stage forecast (although of course they would eventually diverge if the integration were to be carried further into the future). The result is that the forecast from any one of these initial states would produce a good forecast of the trajectory from the (unknown) true initial condition, which might be located near the center of the initial loop. In contrast, Fig. 1.3b shows forecasts for the same set of future times when the initial conditions are taken as the points on the loop that is a little lower on the left wing of the attractor. Here, the dynamical predictions are reasonably good through the first half of the forecast period, but they diverge strongly toward the end of the period as some of the trajectories remain on the left-hand wing of the attractor while others undergo the regime transition to the right-hand wing. The result is that a broad range of the prognostic variables might be forecast from initial conditions near the unknown true initial condition, and there is no way to tell in advance which of the trajectories might represent good or poor forecasts.

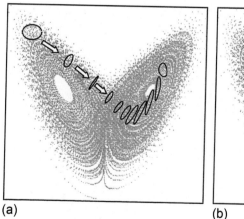 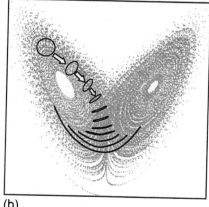

(a)   (b)

FIG. 1.3

Collections of forecast trajectories for the Lorenz system, initialized at (a) a high-predictability portion of the attractor, and (b) a moderate-predictability portion of the attractor. Any of the forecasts in panel (a) would likely represent the unknown true future state well, whereas many of the results in panel (b) would correspond to poor forecasts.

From Palmer, T. N. (1993). Extended-range atmospheric prediction and the Lorenz model.
Bulletin of the American Meteorological Society, 74, 49–65.

This high sensitivity is related to the divergence along the unstable direction on both sides of the saddle node present at the center of the attractor. The inhomogeneity of the predictability of the flow on the attractor is a property shared by many low-order systems such as the Lorenz model, as well as by higher complexity models up to and including operational forecasting systems, as discussed in two recent reviews (Vannitsem, 2017; Yoden, 2007).

1.2 UNCERTAINTY AND PROBABILITY IN "DETERMINISTIC" PREDICTIONS

In the middle of the past century, when dynamical weather prediction was not yet an operational tool but rather a research curiosity, Eady (1951) wrote:

> [T]he initial state of motion is never given precisely and we never know what small perturbations may exist below a certain margin of error. Since the perturbations may grow at an exponential rate, the margin of error in the forecast (final) state will grow exponentially as the period of forecast is increased, and this possible error is unavoidable whatever our method of forecasting… [T]he set of all possible future developments consistent with our initial data is a divergent set and any direct computation will simply pick out, arbitrarily, one member of the set. Clearly, if we are to glean any information at all about developments beyond the limited time interval, we must extend our analysis and consider the properties of the set or "ensemble" (corresponding to the Gibbs-ensemble of statistical mechanics) of all possible developments. Thus, long-range forecasting is necessarily a branch of statistical physics in its widest sense: both our questions and answers must be expressed in terms of *probabilities*.

Of course Eady could not have been aware of what today is called chaotic dynamics, but he realized that amplification of initial-condition errors would inevitably lead to uncertainty in dynamical forecasts, and that those uncertainties should be expressed in the language of probability.

The connection between uncertainty, probability, and dynamical forecasting can be approached using the phase space of the Lorenz attractor as a low-dimensional and comprehensible metaphor for the millions-dimensional phase spaces of realistic modern dynamical weather prediction models. Consider again the forecast trajectories portrayed in Fig. 1.3. Rather than regarding the upper-left loops as collections of initial states, imagine that they represent boundaries containing most of the probability, perhaps the 99% probability ellipsoids, for probability density functions defined on the attractor. When initializing a dynamical forecast model we can never be certain of the true initial state, but we may be able to quantify that initial-condition uncertainty in terms of a probability distribution, and that distribution must be defined on the system's attractor if the initial state is to be dynamically consistent with the governing equations. In effect, those governing equations will operate on the probability distribution of initial-condition uncertainty, advecting it across the attractor and distorting its initial shape in the process. If the initial probability distribution is a correct representation of the initial-condition uncertainty, and if the model's equations are a correct representation of the dynamics of the true system, then the subsequent advected and distorted probability distributions will correctly quantify the forecast uncertainty at future times. This uncertainty may be larger (as represented by Fig. 1.3b) or smaller (Fig. 1.3a), depending on the intrinsic predictability of the states in the initial region of the attractor. (To the extent that the forecast model equations are not complete and correct representations of the true dynamics, which is inevitable in atmospheric modeling, then additional uncertainty will be introduced.)

Using this concept of a probability distribution that quantifies the initial-condition uncertainty, Epstein (1969) proposed the method of stochastic-dynamic prediction. The historical and biographical background leading to this important paper has been reviewed by Lewis (2014). Denoting the (multivariate) uncertainty distribution as φ and the vector $\dot{\mathbf{X}}$ as containing the total derivatives with respect to time of the prognostic variables defining the coordinate axes of the phase space, Epstein (1969) begins with the conservation equation for total probability, φ,

$$\frac{\partial \varphi}{\partial t} + \nabla \cdot (\dot{\mathbf{X}} \varphi) = 0 \tag{1.2}$$

Eq. (1.2), also known as the Liouville equation (Ehrendorfer, 1994a; Gleeson, 1970), is analogous to the more familiar continuity (i.e., conservation) equation for mass. As noted by Epstein (1969),

It is possible to visualize the probability density in phase space, as analogous to mass density (usually ρ) in three-dimensional physical space. Note that $\rho \geq 0$ for all space and time, and $\iiint (\rho/M)dxdydz = 1$ if M is the total mass of the system. The "total probability" of any system is, by definition, one.

Eq. (1.2) states that any change in the probability contained within a small (hyper-) volume surrounding a point in phase space must be balanced by an equal net flux of probability through the boundaries of that volume. The governing physical dynamics of the system (e.g., Eqs. 1.1a–1.1c for the Lorenz system) are contained in the time derivatives $\dot{\mathbf{X}}$ in Eq. (1.2), also known as tendencies. Note that the integration of Eq. (1.2) is deterministic, in the sense that there are no random terms introduced on the right-hand sides of the dynamical tendencies. The Liouville equation is, in fact, the limiting case (drift-only case) of a more general approach in which stochastic diffusion forcings and jump processes are incorporated, known as the Chapman-Kolmogorov equation (e.g., Gardiner, 2009). Thus the terminology used by Epstein (1969) should not be confused with the current notion of a stochastic system.

Epstein (1969) considered that direct integration of Eq. (1.2) on a set of gridpoints within the phase space was computationally impractical, even for the idealized 3-dimensional dynamical system he used as an example. Instead he derived time-tendency equations for the elements of the mean vector and covariance matrix of φ (in effect, assuming multivariate normality for this distribution initially and at all forecast times) yielding a system of nine coupled differential equations (three each for the means, variances, and covariances), by assuming that the third and higher moments of the forecast distributions vanished. In addition to providing a (vector) mean forecast, the procedure characterizes state-dependent forecast uncertainty through the forecast variances and covariances that populate the forecast covariance matrix, the increasing determinant ("size") of which at increasing lead times can be used to characterize the increasing forecast uncertainty.

Stochastic-dynamic prediction in the phase-space in terms of the first and second moments of the uncertainty distribution, or related approaches to integration of Eq. (1.2) (Ehrendorfer, 1994b; Thompson, 1985), are even today computationally impractical when applied to realistic forecast models. Furthermore, forecasts of forecast uncertainty based on Eq. (1.2) assume that the system dynamics encoded in the elements of $\dot{\mathbf{X}}$ are correct and complete, whereas the violation of this assumption in realistic weather forecast models adds uncertainty to any forecast.

1.3 ENSEMBLE FORECASTING

Even though the stochastic-dynamic approach to forecasting as proposed by Epstein (1969) is out of reach computationally, it is theoretically sound and conceptually appealing. It provides the philosophical basis for addressing the problem of sensitivity to initial conditions in dynamical weather and climate models, which is currently best achieved through ensemble forecasting. Rather than computing the effects of the governing dynamics on the full continuous probability distribution of initial-condition uncertainty, ensemble forecasting proceeds by constructing a discrete approximation to this process. That is, a collection of individual initial conditions (each represented by a single point in the phase space) is chosen, and each is integrated forward in time according to the governing equations of the dynamical system. Ideally, the distribution of these states in the phase space at future times, which can be mapped to physical space, will then represent a sample from the statistical distribution of forecast uncertainty.

Ensemble forecasting is an instance of Monte-Carlo integration, (Metropolis & Ulam, 1949), the use of which in meteorology was foreshadowed by the quotation from Eady (1951) reproduced at the beginning of Section 1.2. Ensemble forecasting in meteorology appears to have been first proposed explicitly in a conference paper by Lorenz (1965):

> The proposed procedure chooses a finite ensemble of initial states, rather than the single observed initial state. Each state within the ensemble resembles the observed state closely enough so that the differences might be ascribed to errors or inadequacies in observation. A system of dynamic equations previously deemed to be suitable for forecasting is then applied to each member of the ensemble, leading to an ensemble of states at any future time. From an ensemble of future states, the probability of occurrence of any event, or such statistics as the ensemble mean and ensemble standard deviation of any quantity, may be evaluated.

Ensemble forecasting was first implemented in a meteorological context by Epstein (1969) as a means to provide representations of the true forecast distributions to which his (truncated) stochastic-dynamic calculations could be compared. He explicitly chose initial ensemble members as independent random draws from the initial-condition uncertainty distribution:

> Discrete initial points in phase space are chosen by a random process such that the likelihood of selecting any given point is proportional to the given initial probability density. For each of these initial points (i.e. for each of the sample selected from the ensemble) deterministic trajectories in phase space are calculated by numerical integration... Means and variances are determined, corresponding to specific times, by averaging the appropriate quantities over the sample.

Forecasts entailing more or less uncertainty are then characterized by larger or smaller ensemble variances. A more detailed exposition of the procedure was provided in the influential paper by Leith (1974).

In addition to computational tractability, an advantage of ensemble forecasting is that it permits bi- or multi-modal forecast distributions as ensemble members diverge, allowing representation of possible regime shifts. Lorenz (1965) specifically considered this attribute in his proposal for the method. In contrast, Epstein's (1969) truncated stochastic-dynamic formulation is limited in the allowed mathematical form of its predictive distributions because of its formulation in terms of distribution moments, so that only unimodal predictive distributions can be computed. This problem of multimodality was nicely addressed in the context of the atmospheric Lorenz 3-variable model through the development of a stochastic equation for the error growth by Nicolis (1992).

Both the stochastic-dynamic and ensemble approaches to representing the effects of initial-condition uncertainty initially assumed that the equations governing the physical dynamics were complete and correct. Of course, in practice dynamical weather forecast models are not perfect, and errors are introduced through spatial and temporal discretization, and through empirical formulations for unresolved processes. Pitcher (1974, 1977) represented the effects of these structural model errors through addition of random forcing terms to the prognostic equations following approaches developed in the context of stochastic modeling (e.g., Gardiner, 2009), and Leith (1974) suggested applying the same approach to ensemble forecasts.

This "stochastic parameterization" approach was first introduced into operational ensemble weather forecasting practice at the European Centre for Medium Range Weather Forecasts (ECMWF, Buizza, Miller, & Palmer, 1999), although the issue is not considered solved and research in this area is ongoing both from a practical forecasting side (e.g., Christensen, Lock, Moroz, & Palmer, 2017), and from a more theoretical perspective through the development of techniques deduced from first principles (e.g., Demaeyer & Vannitsem, 2017; Majda, Timofeyev, & Vanden Eijnden, 2001; Wouters & Lucarini, 2012).

Stochastic approaches for the representation of uncertainties are also very popular in the context of climate (e.g., Hasselman, 1976) and hydrological modeling (Bras & Rodriguez-Iturbe, 1984), due to a larger number of sources of uncertainties than in atmospheric modeling for weather forecasting. For climate modeling, many forcings influence the atmosphere that are either not fully understood or too expensive to incorporate at the current stage of development of climate models. In hydrology, both external forcings essentially coming from the atmosphere and the description of (small-scale) surface processes display important uncertainties. In both cases, these uncertainties are often best described with stochastic forcings.

1.4 POSTPROCESSING INDIVIDUAL DYNAMICAL FORECASTS

Statistical postprocessing of dynamical weather forecasts has a history that is almost as long as the history of dynamical weather forecasting itself. Operational dynamical forecasting began in 1956 in the U.S. (Fawcett, 1962), and dissemination to the public of products derived from statistically post-processed dynamical forecasts (Klein & Lewis, 1970) was initiated in 1968 (Carter, Dallavalle, & Glahn, 1989). These early forecasts were based on a technique known as "perfect prog" (e.g., Wilks, 2011), which required no training data from the dynamical model. Shortly thereafter, the preferred model output statistics (MOS, Glahn & Lowry, 1972) method began to be used when sufficient dynamical-model training data became available.

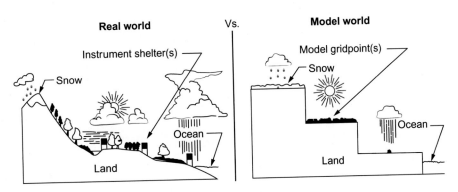

FIG. 1.4

Cartoon illustration of representativeness error inherent in making forecasts for small-scale variations in the real-world (left) on the basis of coarse grid-cell dynamical forecasts (right).

From Karl, T. R., Schlesinger, M. E., & Wang, W. C. (1989). A method of relating general circulation model simulated climate to the observed local climate. I. Central tendencies and dispersion. In: Preprints, sixth conference on Applied Climatology, American Meteorological Society (pp. 188–196).

The MOS approach continues to be preferred because it relates past forecasts from a particular dynamical forecast model to the subsequent weather elements of interest, and so is able to correct biases deriving from structural errors that are specific to that particular dynamical model. MOS methods also adjust for "representativeness errors," notably mismatches between grid-cell-scale dynamical forecast output and the local instrumental observations that are of primary interest to many forecast users, the correction of which is known in the climate-change literature as "downscaling" (e.g., Wilby & Wigley, 1997). These mismatches are illustrated by the cartoon in Fig. 1.4.

The original MOS forecast systems operated on the single-integration dynamical forecasts available at the time, and were nearly all structured as multiple linear regressions:

$$y_t = a + b_1 x_{t,1} + b_2 x_{t,2} + \cdots + b_m x_{t,m} + \varepsilon_t \tag{1.3}$$

where y_t, $t = 1, \ldots, n$, are the values to be predicted in a set of training data, the $x_{t,k}$ are any relevant predictor variables, and the regression coefficients b_k, $k = 1, \ldots, m$, are estimated by minimizing the sum of the squared residuals ε_t^2 over the n training samples. Often one of the predictor variables corresponds to the quantity y of interest if it is available as a prognostic variable in the dynamical model. However, due to the lower quality of the early dynamical models relative to those of the present day, these equations sometimes included 10 or more additional predictors (Glahn, 2014), such as other dynamical prognostic variables, recent surface observations, climatological values, and (trigonometric transformations of) the day of the year in order to represent some aspects of seasonality (e.g., Jacks et al., 1990).

Nearly all MOS forecasts based on single-integration dynamical forecasts were, and continue to be disseminated in nonprobabilistic formats, although some of the computations underlying these forecasts are probabilistic. For others, issued forecasts correspond to Eq. (1.3), operating on new predictor data $x_{t,k}$ with $\varepsilon_t = 0$, with no expression of uncertainty and so yielding an estimate of the conditional expectation for y given current values of the $x_{t,k}$. Although a probabilistic forecast can be constructed using Eq. (1.3), by assuming a Gaussian predictive distribution centered on y_t, with variance related to the regression mean-squared error (e.g., Glahn et al., 2009; Wilks, 2011), case-to-case (i.e., state-dependent) variations in forecast uncertainty are not represented. However, extension of this MOS concept to postprocessing of ensemble forecasts allows both correction of biases due to model errors as well as representation of variations in uncertainty based on variations in ensemble spread.

1.5 POSTPROCESSING ENSEMBLE FORECASTS: OVERVIEW OF THIS BOOK

Operational ensemble forecasting began in 1992 at both ECMWF and the U.S. National Meteorological Center (Molteni, Buizza, Palmer, & Petroliagis, 1996; Toth & Kalnay, 1993). As expected from prior research, ensemble-mean forecasts outperformed traditional high-resolution single-integration dynamical forecasts in terms of such metrics as mean-squared error, but the primary aim was to characterize and forecast the uncertainty on the basis of ensemble spread. Initially, relative frequency within each forecast ensemble was regarded as a rough estimate of the corresponding outcome probability, but it became quickly evident that these probability estimates were typically biased. In particular, the raw dynamical ensembles exhibited insufficient dispersion (e.g., Buizza, 1997; Hamill, 2001), which imparted overconfidence to their uncertainty forecasts (e.g., Wilks, 2011).

Evidently, ensemble forecasts require the same kinds of statistical postprocessing for bias correction as their traditional single-integration counterparts. Indeed, the same computer code is executed for both. But in addition, forecast ensembles require statistical postprocessing to adjust their dispersion to yield properly calibrated forecast probabilities. Ensemble-MOS methods thus aim to correct forecast errors deriving from both structural deficiencies in the dynamical models and forecast sensitivity to uncertain initial conditions. These methods began to be developed early in the present century, and a comparison among the first approaches to be proposed is provided in Wilks (2006). The past decade has seen an explosion of interest in the statistical postprocessing of ensemble forecasts, and the purpose of this book is to document the progress to date in this rapidly expanding field.

In Chapter 2, Buizza (2018) concludes the introductory section of this book by reviewing the construction of ensemble prediction systems, with a particular focus on operations at ECMWF, and underscores their need for postprocessing.

The second section of the book is devoted to exposition of the methods available for statistical postprocessing of ensemble forecasts. In Chapter 3, Wilks (2018) reviews univariate ensemble postprocessing, where forecasts for a single weather element, at one location and for one time in the future, are considered. Chapter 4, by Schefzik and Möller (2018), extends these methods for multivariate forecasts, where the postprocessed forecasts for multiple weather elements are meant to be statistically consistent with each other. Such methods are important where spatial and/or temporal coherence of the forecasts are important to the management of weather-sensitive enterprises. In Chapter 5, Friederichs, Wahl and Buschow (2018) consider the specialized perspective necessary for postprocessing forecasts for extreme, and therefore rare, events. The section concludes with the discussion in Chapter 6, by Thorarinsdottir and Schuhen (2018), of the methods of forecast verification devised specifically for evaluation of postprocessed ensemble forecasts.

Section three of this book is devoted to applications of ensemble postprocessing. Practical aspects of ensemble postprocessing are detailed by Hamill (2018) in Chapter 7, including an extended and illustrative case study. In Chapter 8, Hemri (2018) discusses ensemble postprocessing specifically for hydrological applications, where the spatial correlations among the forecast elements must be represented correctly if the forecasts are to have real-world utility. Pinson and Messner (2018) treat postprocessing in support of renewable energy applications, where the conversion of meteorological variables into power generation imposes additional challenges, in Chapter 9. Chapter 10, by Van Schaeybroeck and Vannitsem (2018), discusses postprocessing of monthly, seasonal, and interannual forecasts, which is especially difficult because for these lead times the predictable signal is typically small relative to the intrinsic uncertainty. Finally, in Chapter 11 Messner (2018) provides a guide to the ensemble-postprocessing software available in the R programming language, that should greatly help readers implement many of the ideas presented in this book.

REFERENCES

Bras, R., & Rodriguez-Iturbe, I. (1984). *Random Functions and Hydrology* (pp. 559). Reading, Massachusetts: Addison-Wesley Publishing Company.

Buizza, R. (1997). Potential forecast skill of ensemble prediction and spread and skill distributions of the ECMWF ensemble prediction system. *Monthly Weather Review, 125*, 99–119.

Buizza, R. (2018). Ensemble forecasting and the need for calibration. In S. Vannitsem, D. S. Wilks, & J. W. Messner (Eds.), *Statistical Postprocessing of Ensemble Forecasts*. Elsevier.

Buizza, R., Miller, M., & Palmer, T. N. (1999). Stochastic representation of model uncertainties in the ECMWF ensemble prediction system. *Quarterly Journal of the Royal Meteorological Society, 125*, 2887–2908.

Carter, G. M., Dallavalle, J. P., & Glahn, H. R. (1989). Statistical forecasts based on the National Meteorological Center's numerical weather prediction system. *Weather and Forecasting, 4*, 401–412.

Christensen, H. M., Lock, S. -J., Moroz, I. M., & Palmer, T. N. (2017). Introducing independent patterns into the stochastically perturbed parameterisation tendencies (SPPT) scheme. *Quarterly Journal of the Royal Meteorological Society, 143*, 2168–2181.

Demaeyer, J., & Vannitsem, S. (2017). Stochastic parameterization of subgrid-scale processes in coupled ocean-atmosphere systems: Benefits and limitations of response theory. *Quarterly Journal of the Royal Meteorological Society, 143*, 881–896.

Eckmann, J. -P., & Ruelle, D. (1985). Ergodic theory of chaos and strange attractors. *Reviews of Modern Physics, 57*, 617–656.

Eady, E. (1951). The quantitative theory of cyclone development. In T. Malone (Ed.), *Compendium of Meteorology* (pp. 464–469). Boston, Massachusetts: American Meteorological Society.

Ehrendorfer, M. (1994a). The Liouville equation and its potential usefulness for the prediction of forecast skill. Part I. Theory. *Monthly Weather Review, 122*, 703–713.

Ehrendorfer, M. (1994b). The Liouville equation and its potential usefulness for the prediction of forecast skill. Part II. Applications. *Monthly Weather Review, 122*, 714–728.

Epstein, E. S. (1969). Stochastic dynamic prediction. *Tellus, 21*, 739–759.

Fawcett, E. B. (1962). Six years of operational numerical weather prediction. *Journal of Applied Meteorology, 1*, 318–332.

Friederichs, P., Wahl, S., & Buschow, S. (2018). Postprocessing for extreme events. In S. Vannitsem, D. S. Wilks, & J. W. Messner (Eds.), *Statistical Postprocessing of Ensemble Forecasts*. Elsevier.

Gardiner, C. W. (2009). *Handbook of Stochastic Methods* (4th ed.). Berlin: Springer.

Glahn, H. R. (2014). A nonsymmetric logit model and grouped predictand category development. *Monthly Weather Review, 142*, 2991–3002.

Glahn, H. R., & Lowry, D. A. (1972). The use of model output statistics (MOS) in objective weather forecasting. *Journal of Applied Meteorology, 11*, 1203–1211.

Glahn, H. R., Peroutka, M., Wiedenfeld, J., Wagner, J., Zylstra, G., Schuknecht, B., et al. (2009). MOS uncertainty estimates in an ensemble framework. *Monthly Weather Review, 137*, 246–268.

Gleeson, T. A. (1970). Statistical-dynamical predictions. *Journal of Applied Meteorology, 9*, 333–344.

Hamill, T. M. (2001). Interpretation of rank histograms for verifying ensemble forecasts. *Monthly Weather Review, 129*, 550–560.

Hamill, T. M. (2018). Practical aspects of statistical postprocessing. In S. Vannitsem, D. S. Wilks, & J. W. Messner (Eds.), *Statistical Postprocessing of Ensemble Forecasts*. Elsevier.

Hasselman, K. (1976). Stochastic climate models part I. Theory. *Tellus, 28*, 473–485.

Hemri, S. (2018). Applications of postprocessing for hydrological forecasts. In S. Vannitsem, D. S. Wilks, & J. W. Messner (Eds.), *Statistical Postprocessing of Ensemble Forecasts*. Elsevier.

Jacks, E., Bower, J. B., Dagostaro, V. J., Dallavalle, J. P., Erickson, M. C., & Su, J. C. (1990). New NGM-based MOS guidance for maximum/minimum temperature, probability of precipitation, cloud amount, and surface wind. *Weather and Forecasting, 5*, 128–138.

Karl, T. R., Schlesinger, M. E., & Wang, W. C. (1989). A method of relating general circulationmodel simulated climate to the observed local climate. I. Central tendencies and dispersion. Preprints. In: *Proceedings of the Sixth Conference on Applied Climatology*, (pp. 188–196). American Meteorological Society.

Klein, W. H., & Lewis, B. M. (1970). Computer forecasts of maximum and minimum temperatures. *Journal of Applied Meteorology, 9*, 350–359.

Leith, C. E. (1974). Theoretical skill of Monte Carlo forecasts. *Monthly Weather Review, 102*, 409–418.

Lewis, J. M. (2014). Edward Epstein's stochastic-dynamic approach to ensemble weather prediction. *Bulletin of the American Meteorological Society, 95*, 99–116.

Li, T. Y., & Yorke, J. A. (1975). Period three implies chaos. *American Mathematical Monthly, 82*, 985–992.

Lorenz, E. N. (1963). Deterministic nonperiodic flow. *Journal of the Atmospheric Sciences, 20*, 130–141.

Lorenz, E. N. (1965). On the possible reasons for long-period fluctuations of the general circulation. In: *Proceedings of the WMO-IUGG Symposium on Research and Development Aspects of Long-Range Forecasting* (pp. 203–211). Boulder, CO: World Meteorological Organization, WMO Tech. Note 66.

Majda, A. J., Timofeyev, I., & Vanden Eijnden, E. (2001). A mathematical framework for stochastic climate models. *Communications on Pure and Applied Mathematics, 54*, 891–974.

Messner, J. W. (2018). Ensemble postprocessing with R. In S. Vannitsem, D. S. Wilks, & J. W. Messner (Eds.), *Statistical Postprocessing of Ensemble Forecasts*. Elsevier.

Metropolis, N., & Ulam, S. (1949). The Monte-Carlo method. *Journal of the American Statistical Association, 44*, 335–341.

Molteni, F., Buizza, R., Palmer, T. N., & Petroliagis, T. (1996). The ECMWF ensemble prediction system: Methodology and validation. *Quarterly Journal of the Royal Meteorological Society, 122*, 73–119.

Nicolis, C. (1992). Probabilistic aspects of error growth in atmospheric dynamics. *Quarterly Journal of the Royal Meteorological Society, 118*, 553–568.

Palmer, T. N. (1993). Extended-range atmospheric prediction and the Lorenz model. *Bulletin of the American Meteorological Society, 74*, 49–65.

Pinson, P., & Messner, J. W. (2018). Application of postprocessing for renewable energy. In S. Vannitsem, D. S. Wilks, & J. W. Messner (Eds.), *Statistical Postprocessing of Ensemble Forecasts*. Elsevier.

Pitcher, E. J. (1974). *Stochastic-dynamic prediction using atmospheric data* (Ph.D. dissertation) (pp. 151). University of Michigan. https://deepblue.lib.umich.edu/bitstream/handle/2027.42/7101/bad1099.0001.001.pdf.

Pitcher, E. J. (1977). Application of stochastic dynamic prediction to real data. *Journal of the Atmospheric Sciences, 34*, 3–21.

Saltzman, B. (1962). Finite amplitude free convection as an initial value problem—I. *Journal of the Atmospheric Sciences, 19*, 329–341.

Schefzik, R., & Möller, A. (2018). Ensemble postprocessing methods incorporating dependence structures. In S. Vannitsem, D. S. Wilks, & J. W. Messner (Eds.), *Statistical Postprocessing of Ensemble Forecasts*. Elsevier.

Thompson, P. D. (1985). Prediction of the probable errors of prediction. *Monthly Weather Review, 113*, 248–259.

Thorarinsdottir, T. L., & Schuhen, N. (2018). Verification: Assessment of calibration and accuracy. In S. Vannitsem, D. S. Wilks, & J. W. Messner (Eds.), *Statistical Postprocessing of Ensemble Forecasts*. Elsevier.

Toth, Z., & Kalnay, E. (1993). Ensemble forecasting at NMC: The generation of perturbations. *Bulletin of the American Meteorological Society, 74*, 2317–2330.

Van Schaeybroeck, B., & Vannitsem, S. (2018). Postprocessing of Long-Range Forecasts. In S. Vannitsem, D. S. Wilks, & J. W. Messner (Eds.), *Statistical Postprocessing of Ensemble Forecasts*. Elsevier.

Vannitsem, S. (2017). Predictability of large-scale atmospheric motions: Lyapunov exponents and error dynamics. *Chaos, 27*, 032101.

Wilby, R. L., & Wigley, T. M. L. (1997). Downscaling general circulation model output: A review of methods and limitations. *Progress in Physical Geography, 21*, 530–548.

Wilks, D. S. (2006). Comparison of ensemble-MOS methods in the Lorenz '96 setting. *Meteorological Applications, 13*, 243–256.

Wilks, D. S. (2011). *Statistical Methods in the Atmospheric Sciences* (3rd ed., pp. 676). Amsterdam: Elsevier.

Wilks, D. S. (2018). Univariate ensemble postprocessing. In S. Vannitsem, D. S. Wilks, & J. W. Messner (Eds.), *Statistical Postprocessing of Ensemble Forecasts*. Elsevier.

Wouters, J., & Lucarini, V. (2012). Disentangling multi-level systems: Averaging, correlations and memory. *Journal of Statistical Mechanics: Theory and Experiment, 2012*, P03003.

Yoden, S. (2007). Atmospheric predictability. *Journal of the Meteorological Society of Japan, B85*, 77–102.

ENSEMBLE FORECASTING AND THE NEED FOR CALIBRATION

2

Roberto Buizza

European Centre for Medium-Range Weather Forecasts (ECMWF), Reading, United Kingdom

CHAPTER OUTLINE

2.1 THE DYNAMICAL WEATHER PREDICTION PROBLEM

2.1.1 HISTORICAL BACKGROUND

Weather affects many human activities, and this is why people have studied and tried to predict the weather for thousands of years.

The word "meteorology" comes from ancient Greece: Aristotle described weather phenomena in Μετεωρολογικά, a treatise written at around 340 BCE that is considered the first attempt to address a broad range of meteorological topics. This was the first time that the word "meteor" was used to describe the clouds from which precipitation falls, which originate from the Greek word μετεωρος (meteoros), meaning "high in the sky." From this word comes the modern term "meteorology," the study of clouds and weather.

Statistical Postprocessing of Ensemble Forecasts. https://doi.org/10.1016/B978-0-12-812372-0.00002-9

In 1922, Lewis Fry Richardson organized the first dynamical weather prediction experiment when he tried to produce a hindcast of the situation of May 20, 1910. He used a simple mathematical model of the principal features of the atmosphere, and used data taken at a specific time (7 AM) to calculate the weather 6 h later. As discussed in Lynch (2006), Richardson's forecast failed dramatically, predicting a huge 145 hectopascal rise in pressure over 6 h when the pressure actually was more or less static. However, detailed analysis by Lynch has shown that the cause was a failure to apply smoothing techniques to the data, which rule out unphysical surges in pressure. When these are applied, Richardson's forecast is revealed to be essentially accurate.

Dynamical weather prediction as we know it today starts in 1950, when the group led by Jules Charney completed the first successful dynamical weather prediction experiment at Princeton University (see e.g., Charney, Fjörtoft, & von Neumann, 1950). From 1948 to 1956, Charney was at the Institute for Advanced Study at Princeton, where he headed the Meteorological Research Group. There he worked with John von Neumann and tried to use computers and numerical techniques to improve weather forecasting. From 1959 to 1965, he was a member of the Committee on Atmospheric Sciences of the National Academy of Sciences and the chairperson of the academy's Committee on International Meteorological Cooperation. In those roles, he conceived and helped organize GARP, the Global Atmospheric Research Program, which was the first international effort in weather research.

Before we move to discuss in more detail how a forecast is produced, let us first define the meaning of few terms that are going to be used in this chapter, and second, briefly review the dynamical forecasting process.

List of important terms:

- *Dynamical system*: a system, of which the time evolution can be described by a set of physics equations by integrating them in time.
- *Atmosphere*: the free part of the atmosphere, described in terms of its state variables, that is, the wind components (and/or vorticity and divergence), temperature, pressure, and micro-physics variables (including water vapor).
- *Earth-system*: the land, ocean (waves and currents), sea-ice and atmosphere, described in terms of the state variables of all components.
- *Forecast state (or prediction, or simply forecast)*: the future state of a system starting from specific initial conditions.
- *Initial conditions*: the state of a system at the initial time.
- *Weather*: the state of the atmosphere at a specific point and time, and/or averaged over a 3-dimensional volume and over a short time period (say, up to a few days), expressed in terms of the state variables.
- *Climate*: the state of the Earth-system averaged over a long time period (say, longer than a few days), mostly also over a 3-dimensional volume, and expressed in terms of the state variables.
- *Short-range forecast*: a prediction valid for few days (say, up to 2–3 days).
- *Medium-range forecast*: a prediction valid for up to 2 weeks.
- *Extended-range forecast*: a prediction valid for a period longer than 2 weeks.
- *Monthly and sub-seasonal forecast*: prediction valid for up to 1–2 months.
- *Seasonal forecast*: prediction valid for a period longer than 2 months.
- *Dynamical forecast*: a prediction generated by solving numerically the physics equations that describe the evolution of the Earth-system, or of some of its components.

- *Single forecast*: one single forecast that describes the possible future state of a system.
- *Ensemble of forecasts*: an ensemble of *m* forecasts (also called ensemble members) that describe the possible distribution of future states of a system.
- *Probabilistic forecast*: a forecast expressed in terms of a probability, for example, computed by counting the number of ensemble members predicting a specific phenomenon.

In the rest of this chapter, we will be discussing how ensemble methods can be used to provide users with forecasts of the state of the *Earth-system*, which is a *dynamical system*, and we will illustrate how ensembles have been designed to be able to provide probabilistic forecasts for the *whole globe* for *up to 1 year*. The reader should keep in mind that the *words in italics* in this paragraph have important implications:

- *Earth-system*: this means that we will be talking about ensemble methods used to predict the state of the Earth-system, and not just the atmosphere.
- *Dynamical system*: the fact that the Earth-system is a dynamical system means that to address the prediction problem we can use a set of dynamical equations that describe how it evolves in time from an initial state.
- *Whole globe*: although, generally speaking, ensemble methods applied to global or regional predictions are very similar, there could be some important differences in the way initial and model uncertainties are simulated; in this chapter we will focus on the global ensembles.
- *Up to 1 year*: the fact that the time range up to which the predictions are issued is 1 year means that the initial conditions are very important, and the prediction problem is essentially an initial-value problem.

Let us now briefly review how we produce a global dynamical forecast at ECMWF. Fig. 2.1 is a schematic of the process followed at the European Centre for Medium-Range Weather Forecasts (ECMWF) to generate a global forecast:

i. As many observations as possible are collected every day, and are exchanged using a global telecommunication network, so that weather prediction centers have timely access to them.

ii. A few times a day (for global prediction, today this happens every 6 h, at times coinciding with what are called the synoptic times: 00, 06, 12, and 18 UTC, where UTC stands for Coordinated Universal Time), a data assimilation procedure (see Daley, 1993; Kalnay, 2002) is performed to estimate the state of the system at a specific time T: this procedure merges observations collected a few hours before or centered on time T with a short-range forecast that provides an estimate of the state of the atmosphere.

iii. At the end of the data assimilation procedure, initial conditions are available to launch the next forecast; in weather prediction, these initial conditions are also called the atmospheric analysis.

iv. From the initial conditions, the next forecast is launched: at ECMWF, for example, forecasts with different forecast lengths are launched at the four synoptic times every day.

v. Short-range forecasts are then used as input to the next data assimilation procedure, and to generate forecast products.

vi. All analysis and forecast data (and reforecast data as well if generated, see Section 2.6) are also copied to an archive system, so that users can go back and revisit each case for diagnostic and verification purposes.

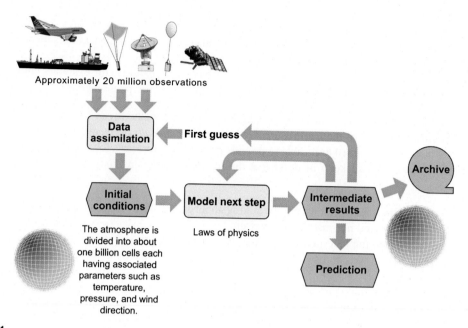

FIG. 2.1

Schematic of the dynamical weather prediction process followed at the European Centre for Medium-Range Weather Forecasts to generate a global forecast.

2.1.2 OBSERVATIONS

The past 50 years have seen a huge increase in the number of Earth-system observations (Fig. 2.2), and of their quality. Since the end of the 1930s, surface and subsurface (for the ocean) data have been complemented by upper-air data (e.g., from soundings). The 1970s saw the start of satellite soundings. Before the satellite era, the number of observations taken in the atmosphere on a daily basis was in the hundreds of thousands. Today, ECMWF receives daily a few hundreds of millions of observations, of which about 10% are used to estimate the state of the atmosphere (the initial conditions). Of these data, about 95% of the atmosphere observations come from satellites (Fig. 2.3).

One of the reasons why only 10% of the observations are used is that many quantities are observed by different observing systems, and only a subset of the observed quantities is used in the data assimilation procedure. Furthermore, some observations are rejected after a quality control procedure.

Although we can say that today the atmosphere is well observed, this cannot be said for the other components of the Earth-system that we need to initialize, especially if we want to produce subseasonal and seasonal forecasts. For example, the land state (soil moisture and temperature, snow cover) is poorly observed. Furthermore, very few observations are taken in the ocean: considering the sub-surface ocean, for example, today we only receive a few thousand observations a day, compared with a few million a day for the atmosphere.

Observational spatial coverage depends on the observation type (see Figs. 2.4 and 2.5). Land surfaces remain the regions with the largest number of observations, coming both from conventional and

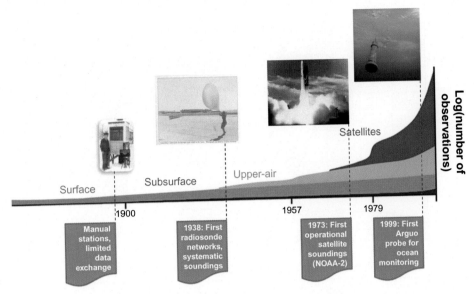

FIG. 2.2

Schematic of the time evolution of the number of observations of the Earth-system (land, ocean, and atmosphere) from 1900 to date. Note that the vertical scale is logarithmic.

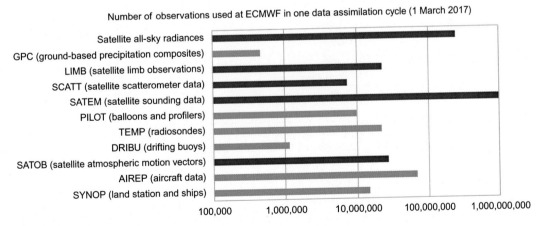

FIG. 2.3

Number of observations, classified in terms of 11 categories (red bars indicate satellite observations), used in one data assimilation cycle at ECMWF in March 2017; note that the x-axis is logarithmic (from Alan Geer, ECMWF; personal communication).

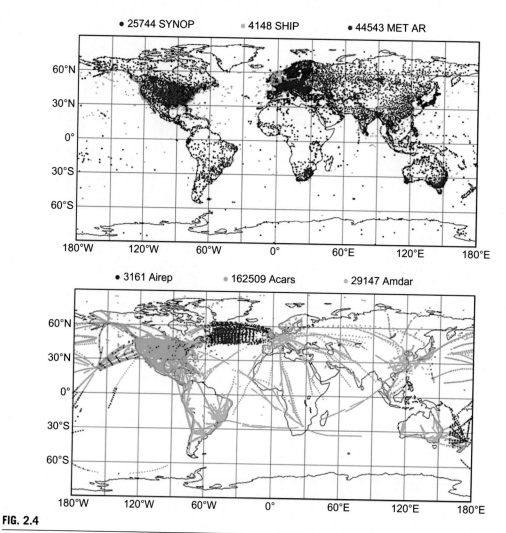

FIG. 2.4

Spatial coverage of the observations from synop stations and buoys (top panel) and aircraft (bottom panel) that were available for use at ECMWF to generate the 00 UTC analysis of the 2nd of March 2017.

satellite platforms. Oceans' surfaces are today well observed, especially at low latitudes, but the Polar Regions are still rather poorly observed, with very little high-quality surface data.

Observation quality is also very important, because it influences the accuracy of the best estimate of the true state of the atmosphere that we generate using data assimilation procedures. In general, observation quality depends on the instrument, and for satellites, also on the instrument position with respect to the area under observation. Furthermore, satellite observation quality is influenced by the state of the atmosphere, with lower (higher) observation errors in cases of clear-sky (cloudy) conditions.

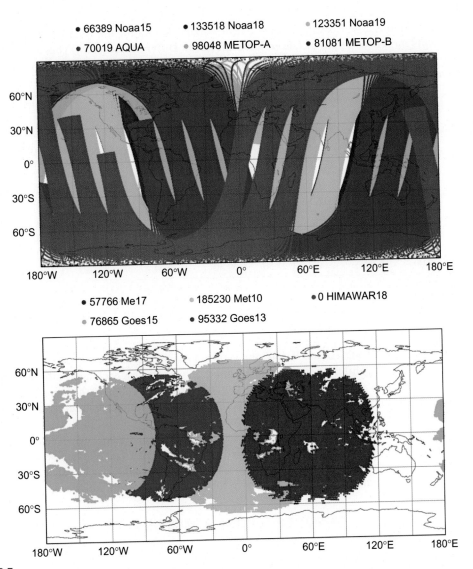

FIG. 2.5

Spatial coverage of the satellite observations from orbiting AMSU-A instruments (top panel) and from geo-stationary satellites (bottom panel) that were available for use at ECMWF to generate the 00 UTC analysis of the 2nd of March 2017.

For example, because water vapor concentration is higher close to the surface, satellite observations are less accurate closer to the Earth's surface.

The fact that observations are affected by errors that depend on the type of observation and on the region of the atmosphere that is observed is taken into account in the data assimilation procedures. Table 2.1 lists, for example, the prescribed root-mean-square errors for observations of the two

Table 2.1 Columns 2–4: Prescribed Root-Mean-Square Errors for the U and V Wind Components for Three Types of Observations at Seven Different Heights (From the Surface at 1000 to 50 hPa), Used in the ECMWF Data Assimilation Procedure (See ECMWF IFS Documentation for Cycle 41R3, 2017)

V Level (hPa)	U and V Wind Comp. (m/s)			HEIGHT Obs. (m)			Temperature Obs. (K)		
	TEMP/ PILOT	SATOB	SYNOP	TEMP/ PILOT	SYNOP (Manual Land)	SYNOP (Auto Land)	TEMP	AIREP	SYNOP (Land)
1000	1.80	2.00	3.00	4.30	5.60	4.20	1.40	1.40	2.00
850	1.80	2.00	3.00	4.40	7.20	5.40	1.25	1.18	1.50
700	1.90	2.00	3.00	5.20	8.60	6.45	1.10	1.00	1.30
500	2.10	3.50	3.40	8.40	12.10	9.07	0.95	0.98	1.20
250	2.50	5.00	3.20	11.80	25.40	19.05	1.15	0.95	1.80
100	2.20	5.00	2.20	18.10	39.40	29.55	1.30	1.30	2.00
50	2.00	5.00	2.00	22.50	59.30	44.47	1.40	1.50	2.40

Columns 5–7: as columns 2–4 but for height observations. Columns 8–10: as columns 2–4 but for temperature.

horizontal wind components (U and V), and the geopotential height and the temperature, at different vertical levels. By using different root-mean-square error estimates for each observation, the data assimilation can take into account their accuracy and give more (less) weight to the more (less) accurate observations. Complex techniques can also take into account the fact that the observations taken close to each other can have correlated observation errors.

The key message to take from this section is that observations are affected by errors, which depend on their type, pressure level, variable observed, method used to derive the variable, location, and state of the atmosphere. Observation errors can (and must) be taken into account when estimating the state of the atmosphere, which are the initial conditions from which forecasts are initialized.

2.1.3 THE EQUATIONS OF MOTION FOR THE ATMOSPHERE

In the introduction of this chapter we said that the Earth-system, and in particular both the ocean and the atmosphere, are dynamical systems governed by Newton's laws of physics, as applied to a fluid. To understand the complexity of the dynamical prediction problem, and how forecasts are generated, we will briefly review hereafter the equations used to simulate the atmosphere, and we will discuss how they are solved numerically to generate weather forecasts.

The starting point from which the equations are deduced is:

$$\vec{F} = m\,\vec{a}$$

$$(2.1)$$

where F stands for the 3-dimensional force (a vector), m for mass and a for the 3-dimensional acceleration (a vector), and the energy and mass conservation equations for a fluid (see Holton, 2012; Hoskins & James, 2014).

Following Holton (2012), the hydrostatic form of the dynamic equations, written for a unitary fluid volume of the atmosphere on a rotating sphere (the Earth), takes the following form:

(momentum conservation)

$$\frac{d\vec{v}}{dt} = -2\vec{\Omega} \times \vec{v} - \frac{1}{\rho}\vec{\nabla}p + \vec{g} + \vec{P}_v \tag{2.2}$$

(thermodynamic energy equation)

$$c_v\frac{dT}{dt} + p\frac{d\alpha}{dt} = P_T \tag{2.3}$$

(water vapor conservation)

$$\frac{dq}{dt} = P_q \tag{2.4}$$

(continuity equation)

$$\frac{1}{\rho}\frac{d\rho}{dt} + \vec{\nabla} \cdot \vec{v} = 0 \tag{2.5}$$

(hydrostatic balance)

$$\frac{dp}{dz} = -\rho g \tag{2.6}$$

where this is the meaning of the different symbols:

- $\vec{v}$ is the two dimensional horizontal wind vector;
- $\vec{\Omega}$ is the Earth angular velocity vector, directed along the Earth rotation axis;
- ρ is the atmosphere density;
- p is the pressure;
- $\vec{g}$ is the gravity vector, with magnitude g and directed toward the Earth's center;
- T is the temperature;
- R is the gas constant for dry air (=287 J kg^{-1} K^{-1});
- ω is the vertical wind component;
- c_v is the specific heat at constant volume (=717 J kg^{-1} K^{-1});
- $\alpha = \frac{1}{\rho}$ is the specific volume (i.e., volume per unit mass);
- q is the specific humidity;
- $\vec{P}_v, P_T, P_q$ are the tendencies of, respectively, the horizontal wind components, temperature and specific humidity, due to physical processes (such as convection, the interaction of clouds with radiation, etc.).

The P-terms on the right-hand-sides ($\vec{P}_v, P_T, P_q$) represent the effects of physical processes such as convection, the effect of clouds on solar radiation, the effect of mountains on the wind, and the effect of turbulence on the energy and momentum transport. They are the most difficult terms to be modeled, and they are the causes of most of the model errors.

These equations are a simplified form (in the sense that, e.g., they do not include an equation for the cloud liquid water or the cloud ice concentration, which are often also considered) of the equations that are solved numerically to generate a weather forecast. They are solved on a 3-dimensional grid that

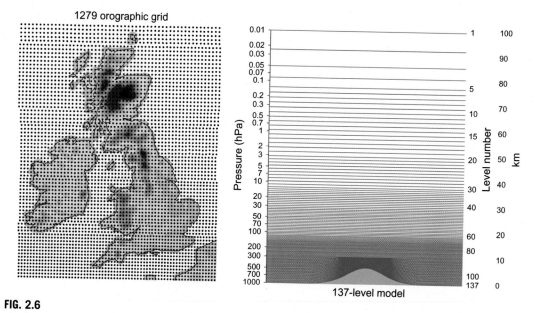

FIG. 2.6

Horizontal grid (left) and vertical levels (right) used by the ECMWF ensemble model version, which has a Tco639L91 resolution: a cubic-octahedral representation in the horizontal with a 639 total wave number and a grid spacing of about 18 km, and 91 vertical levels up to 0.01 hPa (approximately 80 km).

covers the whole globe, with a horizontal spacing that varies, for global models, from about 9 km (for the ECMWF high-resolution model version) to about 200 km (for models used for seasonal prediction and climate projections). Vertically, these models use between 20 and 137 vertical levels (this is the value for the ECMWF high-resolution model version) and cover the lowest 40-to-80 km of the atmosphere, (i.e., they have a top of the atmosphere up to about 0.01 hPa). For the ECMWF medium-range/monthly ensemble (ENS), which is used to produce global ensemble forecasts four times a day (at 00, 06, 12. and 18 UTC), the horizontal resolution is about 18 km and the number of vertical levels is 91 (up to 0.01 hPa) (see Fig. 2.6).

Table 2.2 lists the characteristics of the ECMWF Integrated Forecasting System (IFS) components used to produce the single, high-resolution and the ensemble analyses and forecasts at the time of writing (March 2017). Note that ECMWF uses a spectral model, with a corresponding cubic-octahedral grid in geographical space (Wedi et al., 2015). If we consider, for example, the two main ensembles, the Ensemble of Data Assimilation (EDA) and the medium-range/monthly ensemble (ENS), they use a Tco639L137 resolution. Tco639L137 means that the model uses a spectral resolution with up to 639 wave numbers, with a cubic-octahedral grid, and in the vertical 137 levels up to 0.01 hPa (i.e., TOA, the Top Of the Atmosphere). Since summer 2016, it should be pointed out that ENS is also run at 06 and 18 UTC as part of a special project: the data from these ensembles are available only to the ECMWF Member States who have been funding this project.

Table 2.2 Key Configuration of the Eight Components of the ECMWF Operational Suite Run at 00 and 12 UTC at the Time of Writing (March 2017; See Main Text for More Details): Description (Column 2), Number of Members (Column 3), Resolution (Column 4), Forecast Length (Column 5), Name of the Coupled Ocean/Sea-Ice Model and Resolution of the Ocean Model (Column 6) and Method Used to Simulate Initial and/or Model Uncertainties

IFS Component	Description	#	Horizontal and Vertical Resolution	Forecast Length	Ocean/Sea-Ice Model	Uncertainty Simulation
4DVar analysis	Atm/land/wave High-resolution analysis	1	Tco1279 (9 km) L137 (TOA 0.01 hPa)	–	No	No
EDA[25] analyses	Atm/land/wave Ensemble of data assimilation	25	Tco639 (18 km) L137 (TOA 0.01 hPa)	–	No	Yes: - Observations - Model: SPPT(1)
ORAS4[5] analyses	Ocean Ensemble of analyses	5	1 degree 42 layers	–	NEMO	Yes: - Observations
ORAS5[5] analyses	Ocean Ensemble of analyses	5	0.25 degree 75 layers	–	NEMO/ LIM2	Yes: - Observations
ERA-Interim analysis	Atm/land/wave Reanalysis	1	TL255 (80 km) L60 (TOA 0.1 hPa)	–	No	No
HRES forecast	Atm/land/wave High-resolution	1	Tco1279 (9 km) L137 (TOA 0.01 hPa)	10 d	No	No
ENS[51] forecast	Atm/land/wave/ ocean Medium-range and monthly ensemble	51	Tco639 (18 km) L91 (TOA 0.01 hPa) Tco319 (36 km) L91 (TOA 0.01 hPa)	15 d 15–46 d	NEMO/ LIM2 0.25 degree 75 layers	Yes: - ICs: EDA, SVs, ORAS5; - Model: SPPT(3), SKEB;
S4[51] forecast	Atm/land/wave/ ocean Seasonal ensemble	51	TL255 (80 km) L91 (TOA 0.01 hPa)	7 m 13 m	NEMO 1 degree 42 layers	Yes: - ICs: EDA, SVs, ORAS4; - Model: SPPT(3), SKEB;

In column 1, EDA[25] denotes the 25-member Ensemble of Data Assimilation, ORAS4[5] denotes the 5-member Ocean Reanalysis System-4, HRES the high-resolution model, ENS the medium-range/monthly ensemble, S4 the Seasonal System4. In column 4, Tco indicates a spectral triangular truncation with a cubic-octahedral grid, and TOA means "Top Of the Atmosphere." In column 6, NEMO is the ocean model and LIM is the sea-ice model (see text). In column 7, SPPT is the Stochastically Perturbed Parameterized Tendency model error scheme (Buizza, Miller, & Palmer, 1999). The reader is referred to the text for more details.

As discussed later in the chapter, these IFS components are used to provide users with estimates of the probability distribution function (PDF) of analyses and forecast states:

- EDA[25]: the 25-member, 18-km, L137 (137 vertical levels) Ensemble of Data Assimilation, which provides flow-dependent statistics and estimates of the analysis PDF.
- 4DVar: the single, 9-km single L137 analysis.
- ORAS4[5]: the 5-member ensemble of ocean analyses, version S4 with a 1.00° resolution and 42 vertical layers.
- ORAS5[5]: the 5-member ensemble of ocean analyses, version S5 with a 0.25° resolution and 75 vertical layers.
- LIM2: the Louvain-la-Neuve sea-ice model (Rousset et al., 2015).
- ERA—I: the 80-km, L60 ERA-Interim reanalysis, which is used to generate the ICs for the ENS and S4 reforecast suites.
- HRES: the single, 9-km resolution, L137, 10-day forecast.
- ENS[51]: the 51-member, L91 coupled ensemble, which provides forecasts at 18-km resolution up to day 15, and at 36-km resolution from day 15 to 46 (only at 00 UTC, on Mondays and Thursdays).
- S4[51]: the 51-member, L91, 80-km coupled seasonal ensemble System-4 (S4), which provides forecasts and estimates of the forecast PDF for the seasonal time scale.

If we denote with $\vec{x}(t)$ the state vector of the atmosphere at time t, which includes the horizontal wind components, the temperature, the water vapor concentration, and the logarithm of the surface pressure:

$$\vec{x}(t) \equiv \left(\overrightarrow{v(t)}, T(t), q(t), \ln(p_s(t)) \right) \tag{2.7}$$

we can write schematically the dynamical equations as:

$$\frac{\partial \vec{x}}{\partial t} = \vec{F}\left(\vec{x}, t\right) \tag{2.8}$$

Eq. (2.8) states that the time variation of the state variable x at each grid point depends on the state of the system itself, and on the time.

Note that the dimension of the state vector (i.e., the number of degrees of freedom of the problem that we are trying to solve) is huge. If we take, for example, one member of the ECMWF medium-range ensemble ENS (see Table 2.2), and we assume for simplicity that the state vector is defined in spectral space (with $N = 639$), and includes the four variables (two horizontal wind components, temperature and specific humidity) on all vertical levels (91) plus the surface pressure, then:

$$N_{tot} = (91 \cdot 4 + 1) \cdot \frac{(N+1) \cdot (N+2)}{2} = 74,868,800 \tag{2.9}$$

Formally, we can denote the solution of Eq. (2.8), that is, a forecast, as:

$$\vec{x}(t) = \vec{x}(0) + \int_0^t \vec{F}\left(\vec{x}, \tau\right) d\tau \tag{2.10}$$

Eq. (2.10) is solved numerically to generate a forecast by computing, at each time step, the right-hand-side tendency term $\vec{F}\left(\vec{x}, t\right)$, and integrating the equation in time numerically. Note that this term is known only in an approximate way, partly because some of the terms (such as the $\vec{P_v}, P_T, P_q$ terms) simulate in an approximate way the real processes, and partly because the equations are solved numerically, by approximating derivatives with finite differences and integrals by areas numerically computed. These approximations contribute to forecast errors, and they are also called model uncertainties: they are actually one of the main sources of forecast errors.

Forecast models have limitations in the way they describe the dynamical and physical evolution of the Earth system, arising both from numerical approximations and from the assumptions involved in the parametrizations of sub-grid physical processes. A properly designed, reliable ensemble forecasting system should aim at representing the random errors in the tendencies in order to predict reliable uncertainty estimates. This can be achieved by using alternative numerical and physical formulations in each integration, and/or by including a stochastic component designed to represent the difference between alternative, physically plausible representations of a given process. Today most of the operational ensembles include one or a combination of schemes and approaches to simulate the effect of model uncertainties, either based on a multimodel approach, a perturbed parameter approach, a perturbed-tendency approach, a stochastic back-scatter approach, or a combination of them.

Eq. (2.10) clearly shows that the solution (i.e., the forecast) depends on the initial state $\vec{x}(0)$. Clearly, if the initial state is affected by errors, for example, because observations only cover part of the globe and/or are affected by observational errors, as we discussed herein, then the forecast can also be wrong. Errors linked to the initial state of the atmosphere are also called initial uncertainties: they are another main source of forecast errors.

2.1.4 COMPUTATION OF THE INITIAL CONDITIONS (ANALYSIS)

Determining the initial state of the Earth-system is very complex, given that the system has many degrees of freedom, and that many, if not all of them, need to be initialized properly. Considering the ECMWF IFS, for example, as discussed herein (see Table 2.2), the atmosphere has about 100 million degrees of freedom if simulated at the ECMWF ENS resolution (about 18 km in the horizontal and with 91 vertical levels in the vertical). This number increases to about 450 million for a simulation performed with the ECMWF high-resolution version, which has a 2-times finer horizontal resolution than ENS and 137 instead of 91 vertical levels.

The initial states of all these components define the analysis, that is, the state of the system at $t = 0$ in Eq. (2.10). The analysis is computed by comparing a short-term forecast, started from the last available initial condition, with all the available observations within a time window (order 10 millions), using a procedure called "data assimilation."

At ECMWF, a 4-dimensional variational assimilation procedure (4d-Var) is used to compute the analysis (Rabier, Järvinen, Klinker, Mahfouf, & Simmons, 1999), by comparing a 12-h forecast started from the previously available analysis with all the observations collected inside this window (see Fig. 2.7). 4d-Var computes the analysis by finding the minimum of a cost function that measures the distance between the short-range forecast trajectory and the observations.

Let us denote by $\vec{x}_b(t)$ the short-range forecast started from the previously available analysis, also called the first guess, and by $\vec{x}_a(t)$ the analysis we are looking for.

Define:

$$\delta\vec{x} = \vec{x}_a(t) - \vec{x}_b(t) \qquad (2.11)$$

to be the correction that needs to be added to the short-range forecast to compute the analysis.

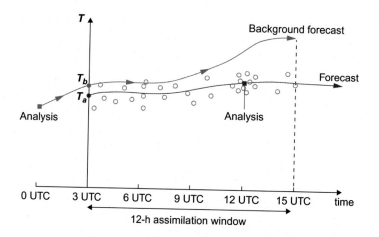

FIG. 2.7

Schematic of the 4-dimentional variational scheme used at ECMWF to compute the analysis at 3 UTC, that is, the initial conditions for a forecast starting at 3 UTC. The figure illustrates the case of temperature at one specific location, generated by assimilating all the observations collected between 3 and 15 UTC. At the beginning of the assimilation window, Tb represents the "background" temperature given by a forecast started from the previous analysis, and Ta represents the "analysis" temperature at the end of the assimilation process. The difference between Ta and Tb is the correction, introduced by the assimilation procedure. The circles between 3 and 15 UTC illustrate observations used to generate the analysis at 3 UTC.

Let us define the cost function of 4d-Var in terms of the correction $\vec{\delta x}$ as:

$$J\left(\vec{\delta x}\right) = \frac{1}{2}\vec{\delta x}^T \cdot \overline{\overline{B^{-1}}} \cdot \vec{\delta x} + \frac{1}{2}\left(\overline{\overline{H}} \cdot \vec{\delta x} - \vec{d}\right)^T \cdot \overline{\overline{R^{-1}}} \cdot \left(\overline{\overline{H}} \cdot \vec{\delta x} - \vec{d}\right) \tag{2.12}$$

where

- $\overline{\overline{B}}$ is a matrix defined by the forecast error statistics;
- $\overline{\overline{R}}$ is a matrix defined by the observation error statistics;
- $\overline{\overline{H}}$ is the observation operator, that map the state vector $\vec{x}$ from the model phase space onto the observation space (e.g., it is the operator that would compute the temperature at a station location from the model temperature values at the closest grid points);
- $\vec{d} = \vec{o} - \overline{\overline{H}} \cdot \vec{x}_b$ is the distance between the observation and the first guess, also called the innovation vector.

The first term of the cost function measures the distance of the solution from the first guess, while the second term measures the distance of the solution from the observations.

The (inverses of the) two matrices $\overline{\overline{B}}$ and $\overline{\overline{R}}$ define the relative weight given to the two terms, in other words, the confidence that one would give to the first guess with respect to the observations. The matrix $\overline{\overline{B}}$ also determines how much an observation at a specific location can influence the analysis at a nearby point. Because, as we discussed herein, both the first guess (which is a short-range forecast) and the observations are affected by errors, and these errors depend, for example, on the state of the system and on the location, the relative value of the two matrices is not constant.

4d-Var aims to compute the correction $\vec{\delta x}$ that minimizes the cost function J in Eq. (2.12). This minimum is computed by applying complex minimization procedures that involve the definition of a tangent forward and its adjoint operator. The reader is referred to Daley (1993) and Kalnay (2002) for an introduction to data assimilation.

Other data assimilation procedures would solve a different minimization problem, using different methods, but the essence of the problem remains the same: merge all the available information (observations and first-guess estimates of the state of the system) to compute the analysis.

From Eq. (2.12) it is clear that the quality of the analysis depends on the observations (more precisely, on their quality, quantity, and coverage), on the fidelity of the model (because it is used to define the first guess and to define the observation operators) and on the data assimilation assumptions, such as the assumption that errors are distributed normally. This latter dependency is linked to the choice of the assimilation methods, and, for 4d-Var, on the cost function definition and on the way the weight matrices are defined and computed. Uncertainties in any of these three areas will affect the knowledge of the initial conditions, and will propagate in time during the time integration of Eq. (2.10) to generate a forecast.

4d-Var is one of the methods currently used to generate the initial conditions of a weather forecast. Generally speaking, the methods used to initialize the ensembles can be grouped in three main categories: variational methods (3-dimensional or 4-dimensional), ensemble methods (EnKF, ETKF) or hybrid methods, which combine an ensemble component, used to provide flow-dependent statistics, and a variational component. For more information on this subject, the reader is referred to more topical books, for example, Daley (1993), Ghil et al. (1997), and Kalnay (2002).

2.2 THE CHAOTIC NATURE OF THE ATMOSPHERE

In the previous sections, we have discussed the main steps involved in solving the dynamical prediction problem and issuing a forecast. In particular, we have discussed the fact that observation and model uncertainties, and data assimilation assumptions can impact the estimation of the actual state (the analysis, the initial conditions) of the atmosphere, and in general of the Earth-system. Initial errors can then propagate in time and affect a forecast.

If the error propagation and its growth rate were slow, then we would be able to issue accurate forecasts despite these uncertainties, because forecast errors would remain small, say, as large as the initial uncertainties. Unfortunately, this is not the case. If the error propagation were the same every single day, one could think of using some sophisticated statistical technique to estimate the error characteristics, and then correct a forecast *a posteriori*. Unfortunately, this is not the case either: errors' propagation and growth rates depend on the atmospheric flow itself (Lorenz, 1993).

Lorenz (1963) highlighted this sensitivity of the error growth to the atmospheric flow very clearly with his 3-dimensional simplified model of a chaotic atmosphere. He showed that trajectories starting from very close initial conditions would diverge as they evolved, with the rate of divergence depending on the state of the systems. In some cases, trajectories starting close by would remain close for a long time, in other cases they would diverge after some time, while in others they would start diverging very quickly.

This behavior is typical of chaotic systems, which exhibit a strong sensitivity to initial conditions (see also Lorenz, 1993; Palmer & Hagedorn, 2006).

The combination of initial and model uncertainties, data assimilation assumptions, and the chaotic nature of the atmosphere makes weather forecasting extremely difficult, with forecast errors varying considerably from day to day.

This predictability variation is detected routinely in operational weather forecasting. Fig. 2.8 shows an example of the variation of a measure of forecast accuracy, for the ECMWF single ENS control forecasts over Europe of the 500 hPa geopotential height field. The "control" is the ensemble member that starts from the unperturbed initial conditions, and is integrated without any simulation of model uncertainties (see also Section 2.7). The accuracy measure is the anomaly correlation coefficient (ACC, with the anomaly computed with respect to climatology; see Wilks, 2011), which has a value of 1.00 for a perfect forecast. Fig. 2.8 shows that from one day to the next, the ACC can vary substantially, especially for the longer lead times (note the differing vertical scales in Fig. 2.8). From the graph, it is clear that some flow patterns are more difficult to predict than others, not only a few days in advance, but also in the short forecast range. For example, there is a clear dip in the ACC of the +24 h forecasts issued between the 29th and the 31st of January. Similarly, the +72 h forecasts issued between the 27th and the 29th of January show a dip, as the +120 h forecasts issued between the 25th and the 27th of January, and the +168 h forecasts issued between the 23rd and the 25th of January.

2.3 FROM SINGLE TO ENSEMBLE FORECASTS

Looking at the time evolution of accuracy measures such as those shown in Fig. 2.8, in the 1980s meteorologists started wondering whether there could be a way to identify in advance whether the flow of a certain period would be more, or less, predictable than, say, on average throughout a season. In other words, they started wondering whether there could be a way to forecast the forecast skill when a forecast was generated, and to identify whether there were regions where the forecast would be more, or less, accurate than average. In other words, they started wondering whether one could devise a method to provide a reliable and accurate estimate of the forecast skill, either expressed in terms of a range of forecast values, or of probabilities that certain specific events would occur.

The reader is referred to other chapters in this book for a more detailed discussion of the problem of forecast verification, of the metrics that can be used, and of the meaning of reliable, accurate, and valuable forecasts. Hereafter, we will discuss two key questions, and link the concepts of reliable and accurate probabilistic forecast to their generation.

2.3.1 FORECAST RELIABILITY AND ACCURACY

A forecast is reliable if, on average over many cases, when I say that I am 90% confident that the forecast is accurate, or when I say that there is a 90% probability that the temperature will be above 30°, then 90% of the times this event actually happens. An unreliable forecast is one for which, on average over many cases, there is no correspondence between the forecast probability and the probability occurrence. One way to measure forecast reliability is to compare forecast probabilities and relative frequencies of occurrence, and compute the squared difference between the two.

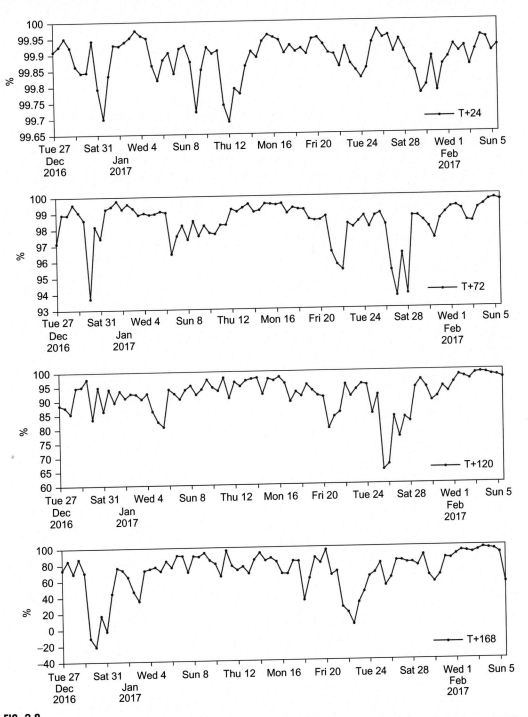

FIG. 2.8

Anomaly correlation coefficient of the ECMWF single ENS control forecast (the ensemble member starting from the unperturbed initial conditions) issued at 00 and 12 UTC of each day of January 2017, for the prediction of the 500 hPa geopotential height over Europe: starting from the top to the bottom, the four panels show the ACC for the 24, 72, 120, and 168 h forecasts.

Another way of measuring reliability is to compare the average root-mean-square error of the ensemble-mean with the average spread of the ensemble, measured by its standard deviation. In fact, in a reliable ensemble, the verification or analysis should be included in the range spanned by the ensemble most of the time: if this is the case, on average, the error of the ensemble-mean, that is, the average distance of the ensemble-mean from the analysis, should be similar to the average distance of the ensemble-mean from the ensemble members (i.e., the ensemble standard deviation). As pointed out by Hamill (2001), care should be taken when applying only this method to assess reliability, because it is possible to have systematic over-spread in one area compensated for by under-spread in another.

A forecast is accurate if there is a very good match between the forecast field and the true state of the system. If one considers a single forecast, then forecast accuracy could be measured by the anomaly correlation coefficient, as was shown in Fig. 2.8, or by the root-mean-square distance between the forecast and the verification (say the analysis) fields. If instead one considers a probabilistic forecast, then forecast accuracy could be measured by computing the distance between the forecast PDF and the observed PDF (see Wilks, 2011 and/or Jolliffe & Stephenson, 2011 for more information about forecast verification measures).

The second question that we want to consider in this section was already one of the most important questions that forecasters and scientists were asking themselves in the 1980s: how can we provide a reliable and accurate estimate of the forecast skill?

As we mentioned already, this could not be done by applying statistical methods based on past cases, because the atmosphere does not repeat itself and it does not show cyclical behaviors. Although it appears as if there is some similarity between past and current events, if one takes into account all the millions of degrees of freedom of the system and compares past and current events, there is a very poor similarity among them. Furthermore, data for the past might not include events that occur in the future. This is particularly true for the rare and extreme events that cause the most severe damage, which means that other methods must be devised.

Leith (1974) was one of the first to suggest a Monte Carlo technique to provide a range of possible forecast outcomes, whereby by using an ensemble of forecasts instead of only one, he could provide an estimate of the forecast skill and the range of possible future scenarios. He estimated that about 10 members would be enough to get a good estimate of the mean of the distribution of possible forecast states and some indication of the variations around the mean.

Fig. 2.9 is a schematic of the idea behind the ensemble approach he was suggesting, which is to move from a single-forecast approach to one that complements it with a number of forecasts, designed to sample all possible sources of forecast error. With a single-forecast approach to dynamical prediction, it is not possible to estimate the range of possible forecast outcomes, or to estimate whether the forecast accuracy could be higher or lower than average. By contrast, with an ensemble of reliable forecasts, one could estimate the range of possible outcomes and compute an estimate of the forecast accuracy.

The key issue that then everyone in the field had to start addressing in ensemble prediction was how to design a reliable and accurate ensemble, given that the system had order few million degrees of freedom, and they could afford to generate only order 10 ensemble members.

In the 1980s, for example, Hoffman and Kalnay (1983) proposed a lagged average approach, whereby single forecasts issued on consecutive days could be combined to provide an ensemble of forecasts. Because these ensembles started at different times, with the youngest forecast being characterized by the lowest errors, they were given different weights when combined to provide an estimate

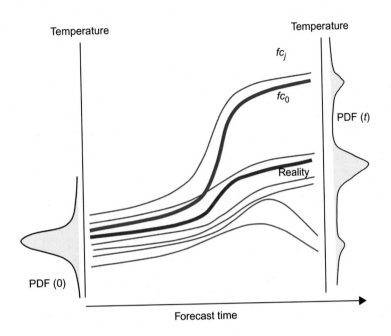

FIG. 2.9

Schematic of an ensemble approach to dynamical prediction, whereby many ensemble members are used to estimate the probability density function (PDF) of initial and forecast states. "fc_0" represents the control forecast, "fc_j" represents a single perturbed forecast, PDF(0) and PDF(t) denotes the PDF at initial and forecast time t.

of the forecast PDF. One of the problems with this approach was that they could only combine a few ensembles, to avoid having too large a difference between the forecast initial dates. The second problem was that, especially in the short-range, there were quite large differences between the errors of the youngest and oldest forecasts, and thus only very few members were given a nonnegligible weight. These ensembles were providing some useful indications, especially for the medium and long forecast range, but they were not very valuable for the short forecast range.

Hollingsworth (1980) tried to develop a "burst" ensemble, with all members starting at the same initial time. He reported results based on Monte Carlo experiments whereby forecasts were started from the same time but slightly different initial conditions, defined by adding random perturbations to the analysis. Results were inconclusive because the initial perturbations did not grow as forecast error does, and all the forecasts ended up being very similar. These ensembles did not diverge as they should, and were producing too-narrow PDFs that often did not include the true state of the atmosphere as one possible realization. This result is not surprising, if we contrast the number of forecasts that were tested, order 10, to the number of degrees of freedom of the system, which even at the low resolution that was used in the 1980s is order few millions. It is indeed difficult to imagine that a few random perturbations could simulate the evolution of the forecast errors.

These earlier experiments indicated that although the idea of a Monte Carlo approach is simple (just run a *number m* of forecasts to build an ensemble), designing a reliable and accurate ensemble is not. Simply using random perturbations of the initial conditions, or using lagged ensembles, did not work, because these methods were not able to sample and represent the sources of forecast errors.

2.3.2 ARE ENSEMBLE FORECASTS MORE VALUABLE THAN A SINGLE FORECAST?

One way to address this question and measure the difference between single and ensemble-based probabilistic forecasts is to use the concept of the Potential Economic Value (PEV; Buizza, 2001; Richardson, 2000; Zhu, Toth, Wobus, Richardson, & Mylne, 2002) of a forecasting system. The PEV is defined by considering a simple cost-loss model, whereby a user can decide to incur a cost C to protect against a loss L, linked to a specific weather event. Forecasts can then be assessed by considering users with different C/L ratios, and by constructing the curve, that shows the savings that users can make if they used the forecasts. Clearly, PEV is a function of the reliability and accuracy of the forecasts.

Fig. 2.10 shows the average PEV for both the ECMWF single high-resolution forecast and the ENS probabilistic forecast of four events:

- 2-m temperature cold anomaly (with respect to climatology) lower than 4°;
- 2-m temperature warm anomaly (with respect to climatology) greater than 4°;
- 10-m wind-speed stronger than 10 m/s;
- Total precipitation larger than 1 mm.

The $t + 144$ h forecasts have been verified against SYNOP observations over Europe and Northern Africa, and the PEV have been computed for the 3-month period October to December 2016. Fig. 2.10 shows that the ENS-based probabilistic forecasts have a higher PEV for all ranges of users.

2.4 SOURCES OF FORECAST ERRORS

The 1980s saw a lot of work being done to try to understand perturbation and error growth in the atmosphere, and to identify the sources of forecast errors that needed to be simulated in an ensemble to be able to provide reliable and accurate forecasts. It was clear that error sources could be linked to the observations, the dynamical model, and the data assimilation procedure, but initially attention was focused on the simulation of forecast errors due to initial uncertainties. The reason for this choice was that results indicated that initial uncertainties dominated perturbation growth during the short forecast range (see, e.g., the work of Harrison, Palmer, Richardson, & Buizza, 1999).

At ECMWF, Buizza and Palmer (1995) investigated the use of singular vectors (SVs), which are the perturbations with the fastest growth over a finite time interval, to simulate initial uncertainties. SVs provided a very good basis to define the initial perturbations of the ECMWF ensemble: compared with random initial perturbations, they were characterized by a very good growth rate, similar to the forecast error growth rate. SVs were indeed used in the first version of the ECMWF ensemble that started producing operational forecasts on the 24th of November 1992. They remained the only type of initial perturbations used in the ECMWF ensemble until 2008, when the ensemble of data assimilations (EDA) started being used, together with singular vectors (Buizza, Leutbecher, & Isaksen, 2008).

At the National Centers for Environmental Prediction (NCEP), Toth and Kalnay (1993) used bred-vectors (BVs) to define the initial perturbations of the NCEP ensemble. The ECMWF and the NCEP ensembles were the first ones used in operational dynamical weather prediction. These two ensembles were followed, in 1995, by the Canadian ensemble, which was developed following a Monte Carlo approach that was designed to simulate both initial uncertainties due to observation errors and data assimilation assumptions, and model uncertainties (Houtekamer, Lefraive, & Derome, 1996).

It is interesting to contrast the rationale behind the decision by ECMWF and NCEP to use a different set of perturbations to initialize their ensembles. The ECMWF SV approach is based on the notion that

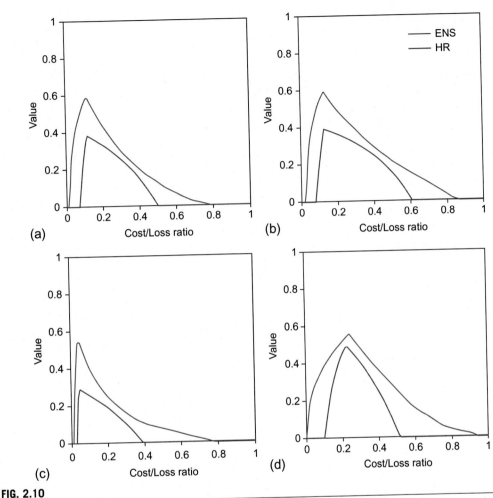

FIG. 2.10

Potential Economic Value (PEV) of ECMWF single high-resolution forecasts (red lines) and ENS-based probabilistic forecasts (blue lines), for cost loss ratios C/L ranging from 0 to 1, for four different forecasts: 2-m temperature cold anomaly lower than 4 degrees (a), 2-m temperature warm anomaly greater than 4 degrees (b), 10-m wind-speed stronger than 10 m/s (c) and total precipitation larger than 1 mm (d). PEV average values have been computed considering the ECMWF operational forecasts for October-November-December 2016, verified against SYNOP observations.

given an initial uncertainty, perturbations along the directions of maximum growth amplify more than those along other directions. If the forecast error evolves linearly and a proper initial norm is used (Barkmeijer, Buizza, & Palmer, 1999; Buizza & Palmer, 1995), the resulting ensemble captures the largest amount of forecast error variance at optimization time (Ehrendorfer & Tribbia, 1997). The BV cycle aims to emulate the data-assimilation cycle, and it is based on the notion that analyses generated by data assimilation will accumulate growing errors by the virtue of perturbation dynamics (Toth & Kalnay, 1993, 1997). This is due to the fact that neutral or decaying errors detected by an assimilation scheme in the early part of the assimilation window will be reduced, and what remains of them will decay due to the dynamics of such perturbations by the end of the assimilation window. In contrast, even if growing errors are reduced by the assimilation system, what remains of them will amplify by the end of the assimilation window.

Following the Canadian example, the simulation of model uncertainties was introduced in the ECMWF ensemble in 1999, using a stochastic approach to simulate the effect of model errors linked to the physical parameterization schemes (Buizza et al., 1999). At present, four main approaches are followed in ensemble prediction to represent model uncertainties:

- A multimodel approach, where different models are used for each ensemble member; models can differ entirely or only in some components (e.g., in the convection scheme).
- A perturbed parameter approach, where all ensemble integrations are made with the same dynamical model but with different parameters defining the settings of the model components; one example is the Canadian ensemble (Houtekamer et al., 1996).
- A perturbed-tendency approach, where stochastic schemes designed to simulate the random model error component are used to simulate the fact that tendencies are known only approximately: one example is the ECMWF Stochastically Perturbed Parametrization Tendency scheme (SPPT, Buizza et al., 1999; Palmer et al., 2009).
- A stochastic back-scatter approach, where a Stochastic Kinetic Energy Backscatter scheme (SKEB) is used to simulate processes that the model cannot resolve, for example, the upscale energy transfer from the scales below the model resolution to the resolved scales: an example is the SKEB scheme used in the ECMWF-ENS (Berner, Shutts, Leutbecher, & Palmer, 2008; Palmer et al., 2009; Shutts, 2005).

Given that forecast models will never be perfect due to the necessary numerical approximations and due to approximations in the representation of physical processes, model uncertainty has to be taken into account and simulated in ensemble prediction systems (Hamill & Swinbank, 2015).

ECMWF (http://www.ecmwf.int/), NCEP (http://www.emc.ncep.noaa.gov/gmb/ens/) and the Meteorological Service of Canadian (MSC, https://weather.gc.ca/ensemble/index_e.html) were the first three meteorological centers to use ensembles to produce operational forecasts. They were followed by many other centers, which developed and implemented global ensembles for the medium-range, sub-seasonal and seasonal time scales, and for short-range, regional predictions. Following the example of MSC, ensembles are also increasingly used to provide estimates of the PDF at analyses time, and to define the initial conditions of ensemble forecasts.

2.5 CHARACTERISTICS OF THE OPERATIONAL GLOBAL ENSEMBLE SYSTEMS

Meteorological centers producing global ensemble forecasts have been exchanging data since 2007 as part of TIGGE, the THORPEX Interactive Grand Global Ensemble (Bougeault et al., 2009; Swinbank et al., 2016), a project of the World Meteorological Organization (https://www.wmo.int/pages/prog/arep/wwrp/new/thorpex_gifs_tigge_index.html). Every day, as part of TIGGE, more than 500 global forecasts are exchanged: these forecasts can be accessed in quasi-real-time (with a 48-h delay) from the TIGGE data portal.

Table 2.3 shows a summary of the operational global ensembles available from the TIGGE website hosted at ECMWF (http://www.ecmwf.int/en/research/projects/tigge):

Table 2.3 Key Characteristics of the Nine Operational Global Medium-Range Ensembles (See Buizza, 2014 for More Details): Blue (Brown) Color Identifies the Ensemble With the Finest (Coarsest) Characteristics, Listed in Alphabetical Order (Column 1): Initial Uncertainty Method (Column 2), Model Uncertainty Simulation (Y/N, Column 3), Truncation and Approximate Horizontal Resolution (Column 4), Number of Vertical Levels and Top of the Atmosphere in hPa (Column 5), Forecast Length in Days (Column 6), Number of Perturbed Members for Each Run (Column 7), and Number of Runs Per Day

Center	Initial Unc. Method (Area)	Model Unc.	Truncation (Degrees, km)	# Vert Lev (TOA, hPa)	Fcst Length (d)	# Pert Mem	#Runs Per Day (UTC)
BMRC	SV (NH,SH)	NO	TL119 (1.5°; 210 km)	19 (10.0)	10	32	2 (00/12)
CMA	BV (globe)	NO	T213 (0.56°; 70 km)	31 (10.0)	10	14	2 (00/12)
CPTEC	EOF (40S:30N)	NO	T126 (0.94°, 120 km)	28 (0.1)	15	14	2 (00/12)
ECMWF	SV (NH, SH, TC) + EDA (globe)	YES	Tco639 (0.14°; 16 km)	91 (0.01)	0–15	50	4 (00/06/ 12/18)
			Tco319 (0.28°; 32 km)		15/46		
JMA	SV (NH, TR, SH)	YES	TL479 (0.38°; 50 km)	60 (0.1)	11	25	2 (00/12)
KMA	ETKF (globe)	YES	N320 (0.35°; 40 km)	70 (0.1)	10	23	4 (00/06/ 12/18)
MSC	EnKF (globe)	YES	600 × 300 (0.6°, 66 km)	40 (2.0)	16/32	20	2 (00/12)
NCEP	EnKF (globe)	YES	T254 (0.70°; 90 km)	28 (2.7)	0–8	20	4 (00/06/ 12/18)
			T190 (0.95°; 120 km)		8–16		
UKMO	ETKF (globe)	YES	N216 (0.45°; 60 km)	70 (0.1)	15	23	2 (00/12)

Note that the BMRC ensembles stopped production in July 2010. In column 2, SV stands for Singular Vector, BV stands for Bred Vector, EOF stands for Empirical Orthogonal Function, EDA for Ensemble of Data Assimilation, ETKF stands for Ensemble Transformed Kalman Filter, EnKF stands for Ensemble Kalman Filter, ETR stands for Ensemble Transformed with Rotation (see text for details).

1. BMRC, the Australian Bureau of Meteorology (only up to July 2010);
2. CMA, the China Meteorological Administration;
3. CPTEC, the Brazilian Center for Weather Prediction and Climate Studies (Centro de Previsao de Tempo e Estudos Climatico);
4. ECMWF, the European Centre for Medium-Range Weather Forecasts;
5. JMA, the Japanese Meteorological Administration;
6. KMA, the Korean Meteorological Administration;
7. MSC, the Meteorological Service of Canada;
8. NCEP, the National Centers for Environmental Prediction;
9. UKMO, The UK met Office.

Each TIGGE ensemble member is defined by the numerical integration of the model equations adopted by its production center to simulate the Earth-system. In other words, each single T-hour forecast starting at day d is given by the time integration of a set of model equations from initial time 0 to time T:

$$e_j(d;T) = e_j(d;0) + \int_0^T \left[A_j(t) + P_j(t) + dP_j(t) \right] dt \tag{2.13}$$

where A_j is the tendency due to the adiabatic processes (say advection, Coriolis force, pressure gradient force), P_j is the tendency due to the parameterized physical processes (say convection, radiation, turbulence) and dP_j represents the tendency due to stochastic model errors and unresolved processes.

In the MSC ensemble, each numerical integration starts from initial conditions defined by an independent data assimilation procedure:

$$e_j(d,0) = F\left[e_j(d - T_A, T_A), o_j(d - T_A, d) \right] \tag{2.14}$$

where $F[..,..]$ denotes the data assimilation process of merging the model first guess $e_j(d - T_A, T_A)$ and the observations spanning the time period T_A (the window covered by the data assimilation process), from $(d - T_A)$ to d. The first guess $e_j(d - T_A; T_A)$ is given by the T_A-hour integration of the model equations from $(d - T_A)$ to d. The data assimilation process uses the observations $o_j(d - T_A, d)$, taken between $(d - T_A)$ and d, and available at the time when the data assimilation procedure starts. More precisely, each member's initial conditions are selected among the 192 members of their Ensemble Kalman Filter.

In all the other eight ensembles, each numerical integration starts from initial conditions defined by adding a perturbation to an unperturbed initial state:

$$e_j(d,0) = e_0(d,0) + de_j(d,0) \tag{2.15}$$

$$e_0(d,0) = F[e_0(d - T_A, T_A), o_0(d - T_A, d)] \tag{2.16}$$

where the unperturbed initial conditions are defined by the data assimilation process spanning the time period T_A. The initial perturbations $de_j(d,0)$ are defined in different ways in each ensemble.

Eqs. (2.13)–(2.16) provide a simple, unified framework to describe how the jth member of the 502 TIGGE forecasts is produced every day. Buizza (2014) provides a summary of the way initial and model uncertainties are simulated in each ensemble, and lists the main references that describe each ensemble.

All these ensembles have been designed to provide reliable and accurate forecasts: how are they actually performing? Let us focus on the ECMWF medium-range/monthly ensemble (ENS), which is considered to provide the most reliable and accurate forecasts in the medium-range.

Fig. 2.11a shows a measure of the ECMWF ENS reliability for winter 2016–17 (November and December 2016, and January 2017), for the prediction of the 500 hPa geopotential height (i.e., for the prediction of synoptic scale features in the free atmosphere, at about 5000 m from the Earth's surface). Fig. 2.11b shows the difference between the seasonal average ensemble spread and ensemble-mean error for the past 4 years: positive (negative) values indicate that the ensemble spread is larger (smaller) than the error, thus indicating an over-dispersive (under-dispersive) ensemble. For this variable (and considering the caveat mentioned in Section 2.3 on using the match between

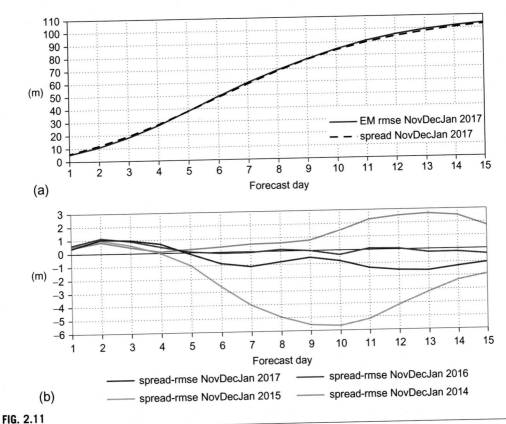

(a)

(b)

- spread-rmse NovDecJan 2017   — spread-rmse NovDecJan 2016
- spread-rmse NovDecJan 2015   — spread-rmse NovDecJan 2014

FIG. 2.11

(a) ECMWF ensemble average spread measured by the standard deviation (red dashed line) and average root-mean-square error of the ensemble-mean (red solid line) for the prediction of the 500 hPa geopotential height over the Northern Hemisphere in winter 2016/17 (November and December 2016, and January 2017).

(b) Difference (spread-error) for winter 2016/17 (red line), winter 2015/16 (blue line), winter 2014/15 (black line) and winter 2013/14 (cyan line). The vertical axis is in meters, the horizontal axis is in forecast days. Forecasts have been verified against analysis.

ensemble-spread and error as a measure of reliability) the ECMWF ensemble is very reliable, with average over/under dispersions of only few percentage points (note that the vertical scale in the bottom panel is 10-times smaller than the one for the top panel).

Fig. 2.12 shows a measure of the ECMWF ENS accuracy, the continuous ranked probability skill score (CRPSS), also for the prediction of the 500 hPa geopotential height over the Northern Hemisphere for the past four winters. The CRPS is a conventional accuracy metric for probabilistic forecasts of continuous predictands. The CRPSS is the skill-score associated to the CRPS: it is 1.00 for a perfect forecast, and 0 for a forecast with is as good as a statistical forecast based on climatology (see Wilks, 2011 for the definition of CRPS and CRPSS). Fig. 2.12 shows that the ECMWF ENS probabilistic forecasts are more skilful than a statistical probabilistic forecast based on the past events for the whole ENS forecast range. Values are higher in the short forecast range, and then gradually decrease as forecast errors start affecting each ensemble member.

Ensemble forecast reliability and accuracy are not so good for other variables, especially for variables close to the surface such as precipitation and wind speed, verified at specific locations. These variables are strongly affected by local characteristics, for example, by the interaction of the atmospheric flow with the terrain and the vegetation type, because the models include only an approximate representation of processes such as the interaction of the surface flow with the orography. Furthermore,

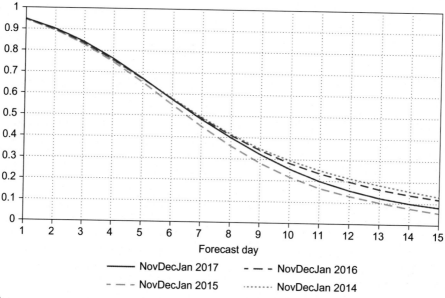

FIG. 2.12

Continuous ranked probability skill score (CRPSS, see text for more details) for the ECMWF ensemble probabilistic prediction of the 500 hPa geopotential height over the Northern Hemisphere in winter 2016/17 (November and December 2016, and January 2017; solid red line), winter 2015/16 (dashed blue line), winter 2014/15 (dashed green line) and winter 2013/14 (dashed cyan line). Forecasts have been verified against analyses.

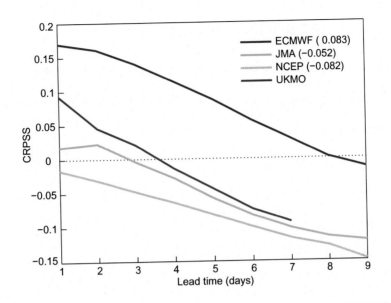

FIG. 2.13

Continuous ranked probability skill score (CRPSS, see text for more details) for the ECMWF ensemble probabilistic prediction of precipitation (solid red line) over the globe in winter 2016/17 (November and December 2016, and January 2017; solid red line); the blue, green and yellow lines show the corresponding lines for the UKMO, JMA and NCEP ensembles. Forecasts have been verified against SYNOP observations.

one should take into account that predicting surface variables at specific locations where weather is affected substantially by local characteristics (e.g., the orography, the terrain characteristics, the land-sea contrast) is particularly challenging. This can be seen very clearly in Fig. 2.13, which shows the CRPSS of the ECMWF ENS for the prediction of precipitation in winter 2016/17. Compared with the prediction of the 500 hPa geopotential height (Fig. 2.12), precipitation CRPSS values not only reach the zero, no-skill value at forecast day 8, but they also have much lower values in the short forecast range (around 0.15 at forecast day 1 compared with about 0.95 for the 500 hPa geopotential height). The same applies to the other ensembles shown in Fig. 2.12, which cross the zero, no-skill line even earlier.

The results shown in Figs. 2.11–2.13 were based on raw ensemble forecasts: in other words, the ensemble forecast data were used as they were produced by the model, without applying any correction. They indicate that even the best global ensemble system in the world, the ECMWF ENS, does not provide reliable and accurate forecasts of surface variables beyond a few days.

It is worth mentioning that a TIGGE follow-on project, the sub-seasonal-to-seasonal (S2S) WMO project (http://s2sprediction.net/), started in 2016. It aims to promote the understanding of predictability over the sub-seasonal and seasonal time scale, and help the development of ensembles for this time range. S2S aims to build a database (Vitart et al., 2017) similar to the TIGGE one, whereby all available S2S ensembles are shared and can be freely accessed with a 3-week delay mode.

2.6 **THE VALUE OF A REFORECAST SUITE**

The fact that ensembles are not able to provide reliable and accurate raw forecasts of some variables, unfortunately of the ones that interest most the forecast users (i.e., variables that describe the state of the Earth system close to the surface where people live), has been known for some time. Despite continuous advances, performance improvements for these variables have been much slower than for variables that describe the free atmospheric flow (e.g., the 500 hPa geopotential height, or the 850 hPa geopotential temperature).

One way to address this problem is to correct forecasts by using the statistics of past forecast errors. For example, one could estimate the bias of ensemble forecasts considering the forecasts done for past years, and remove it from the most recent forecast, or estimate whether the ensemble suffers from any over- or under-dispersion, and correct the ensemble spread accordingly. Similarly, if one is interested in probabilistic forecasts for specific events, one could verify quantile forecasts, build so-called quantile-to-quantile scatter diagrams, and if there are discrepancies between the forecast and observed statistics, correct for them, as discussed extensively in the following chapters of the book.

All these methods require a training dataset. With this set of past forecast-observation pairs, correction coefficients for a regression-based calibration scheme can be determined. This has been shown to work well when a reforecast training dataset is available (Hagedorn, Hamill, & Whitaker, 2008; Hagedorn, Buizza, Hamill, Leutbecher, & Palmer, 2012; Hamill, Whitaker, & Mullen, 2006; Hamill, Hagedorn, & Whitaker, 2008; Hamill et al., 2013). A reforecast dataset is a collection of forecasts performed with a very similar if not identical system from the past, usually going back for a considerable number of years or decades. Consistency between the forecast and reforecast systems is very important for the calibration technique to be effective.

From an operational point of view, ECMWF has been the first center complementing its ensemble forecast suite with a reforecast suite, with the two being as close as possible. Table 2.4 lists the key characteristics of the forecast and reforecast suites of the ECMWF medium-range/monthly (ENS) and seasonal (S4; see Molteni et al., 2011) ensembles.

For ENS, reforecasts are run for 20 years, twice a week, with an 11-member instead of a 51-member ensemble. They are initialized in a slightly different way: the control, unperturbed analysis is different (ERA-Interim instead of the operational HRES) because the latest version of the high-resolution analyses and the EDA analyses are available only for very recent years. For example, past analyses at the current resolution are available only since the implementation of the Tco1279L137 model version in March 2015. Furthermore, EDA-based initial perturbations are generated in a slightly different way (see Buizza et al., 2008 for more details): this is because the EDA operational production started in 2010 with 10 members, and the EDA members were increased to 25 in November 2013: there is no availability of the EDA analyses before 2008. These reforecasts are run "on the fly" (i.e., every week), as we progress with the forecasts (more precisely, reforecasts for the past 20 years are generated so that each week, the reforecasts of the past 20 years for the 5 weeks centered on the current one are all available). This is because the ENS model cycle changes about twice a year, and the best way to have the reforecasts available when they are needed is to run them "on the fly." Running them beforehand, when a new model cycle is implemented, would require the availability of huge computing resources in a brief amount of time, something that is not feasible.

Table 2.4 Key Characteristics of the Forecast and Reforecast Suites of the ECMWF Medium-Range/Monthly Ensemble (ENS) and the Seasonal Ensemble (S4)

Ensemble		Initial Uncertainties	Model Unc.	Truncation (Degrees, km)	# Vert Lev (TOA, hPa)	Fcst Length	# Pert Mem	# Years	Atm Unpert ICs
ENS	Forecast	SV + EDA(d) + ORAS5	YES	Tco639 (0.14°; 16 km)	91 (0.01)	0–15	50 twice a day	n/a	HRES an
				Tco319 (0.28°; 32 km)		15/46			
	Reforecast	SV + EDA (const) + ORAS5	YES	Tco639 (0.14°; 16 km)	91 (0.01)	0–15	11 twice a week	20	ERA-Interim
				Tco319 (0.28°; 32 km)		15/46			
S4	Forecast	SV + SST pert + ORAS5	YES	TL255L91 (0.75°; 80 km)	91 (0.01)	0–7/13 m	50 once a month	n/a	HRES an
	Reforecast	SV + SST pert + ORAS5	YES	TL255L91 (0.75°; 80 km)	91 (0.01)	0–7/13 m	15 once a month	30	ERA-Interim

Yellow boxes identify the parameters for which the forecast and reforecast suites differ. Column 3 (initial uncertainties) indicates the methodology used to generate the initial perturbations; column 4 whether model uncertainties are simulated; column 5 the horizontal resolution; column 6 the vertical resolution; column 7 the forecast length (in days or months); column 8 the number of years included in the reforecast suite; column 9 the "unperturbed" analysis, used as the center of the ensemble initial conditions.

For Seasonal System-4 (S4), reforecasts are run for 30 years, with the same initial date (i.e., the first of each month), with a 15-member instead of a 51-member ensemble. Furthermore, the control, un-perturbed analysis is different (ERA-Interim instead of the operational HRES), for the same reason as ENS. Because S4 uses a frozen model cycle (i.e., there are no changes in the model for about 5 years, see Molteni et al., 2011), the seasonal reforecasts are generated upfront, in the preoperational phase, and not continuously as the days progress.

To give an idea of the rather large amount of computing resources required to generate a reforecast dataset, let us compute the number of forecasts that are run as part of the forecast and reforecast suites.

For the medium-range/monthly ensemble, given that the reforecasts are run "on the fly," let us compute the number of forecasts run in 1 year:

- ENS forecasts: 1 year of ENS forecasts means running 37,128 (52*7*2*51) forecasts up to 15 days, of which 5304 (52*2*51) are extended to 46 days.
- ENS reforecasts: 20 years of ENS reforecasts mean running 22,880 (20*52*2*11) forecasts up to 46 days.

Thus for ENS over 1 year, the cost of the reforecasts is about 60% of the cost of the forecasts.

By contrast, because the seasonal system is frozen for about 5 years (this has been the case for all the ECMWF operational seasonal systems up to now), let us compute the number of forecasts and reforecasts run over a 5-year period:

- S4 forecasts: 5 years of S4 forecasts includes running 6060 (5*12*51) forecasts up to 7 months, of which 2020 are extended to 13 months.
- S4 reforecasts: 30 years of S4 reforecasts mean running 5400 (30*12*15) ensemble forecasts, of which 1800 are extended to 13 months.

Thus, for S4 over the 5-year operational activity of S4, the cost of the reforecasts is about 90% the cost of the forecasts.

The reforecasts are used not only to compute error statistics (e.g., biases, quantile-to-quantile correspondence) or assess the ensemble reliability (spread/error ratio, reliability diagrams) so that methods could be devised to correct the forecasts. The reforecasts are also used to provide an estimate of the expected skill of the forecast ensembles. This is extremely important especially if one considers extreme and rare events, such as intense extra-tropical cyclonic developments and wind storms, extreme precipitation events and heat/cold waves, and El Niño and La Niña conditions. Having a reforecast suite that covers many years of the same season allows us to have a better, more statistically sound data set of cases with similar characteristics, that can be analysed to understand how the forecast system behaves in such cases.

Looking at Table 2.4, one could raise the question whether there is an optimal configuration in terms of ensemble size (number of members) for both the forecast and reforecast suites, and in terms of the number of years spanned by the reforecasts. The issue of ensemble size was already looked at in the early days of ensemble prediction: Leith (1974) concluded that about 10 members are enough to have a good estimate of the first moment (the ensemble-mean) of the PDF of forecast states.

Concerning the ensemble size for the forecast suite, ECMWF started in 1992 with 32 members, and then in 1996 increased the number to 51: since then it has not been increased. Today, all operational ensembles have few tens of members (see Table 2.3). Throughout the years, when more computer power has become available, all meteorological centers have used it to increase the resolution rather than the size of their ensembles. The main reason behind this choice has been that if an ensemble is not capable of simulating a certain phenomenon, there is no chance it will provide a probability for this phenomenon to occur. This is particularly true for the extreme, severe events that require a fine resolution to be properly simulated.

Concerning the size of the reforecast ensemble, priority has been given to use the same resolution as in the forecast ensemble, and to span at least 20 years for the medium-range/monthly ensemble and 30 years for the seasonal ensemble. As a consequence, to keep the cost of the reforecast suite lower than that of the forecast suite, we had to compromise and keep the ensemble size lower than that of the forecast ensemble. At the same time, we have tried to keep it as close as possible, so that the properties of the two suites would be very similar (e.g., in terms of reliability).

2.7 A LOOK INTO THE FUTURE

As mentioned earlier in this chapter, ensembles are increasingly used not only to estimate the forecast PDF, but also to provide a good estimate of the initial time (analysis) PDF. They are also used to generate climate projections, and in the production of climate reanalyses, so that users can use the ensemble of reanalyses to estimate the analysis uncertainty of past, compared with current states. This is very important, for example, if one wants to extract climate trends by looking at many decades of data. Because past observations are more sparse and of a lower quality, clearly the past estimates of the state of the Earth-system are less accurate than the current one. This is, for example, very relevant for the ocean, for which we have had access to good quality data only in the past 15 years (say since the 2000s, since the deployment of the argo floats).

The paradigm shift from single to ensemble-based, probabilistic forecasts that started in 1992, with the operational implementation of the ECMWF and the NCEP ensembles, will continue. At ECMWF, the 10-year strategy for 2016–25 (http://www.ecmwf.int/en/about/what-we-do/strategy) has the following goals:

"To provide forecast information needed to help save lives, protect infrastructure and promote economic development in Member and Co-operating States through:

- *Research at the frontiers of knowledge to develop an integrated global model of the Earth system to produce forecasts with increasing fidelity on time ranges up to 1 year ahead. This will tackle the most difficult problems in numerical weather prediction such as the currently low level of predictive skill of European weather for a month ahead.*
- *Operational ensemble-based analyses and predictions that describe the range of possible scenarios and their likelihood of occurrence and that raise the international bar for quality and operational reliability. Skill in medium-range weather predictions in 2016, on average, extends to about 1 week ahead. By 2025 the goal is to make skillful ensemble predictions of high-impact weather up to 2 weeks ahead. By developing a seamless approach, we also aim to predict large-scale patterns and regime transitions up to 4 weeks ahead, and global-scale anomalies up to a year ahead."*

Thus, the future will see an even more widespread use of ensembles of Earth-system models, to provide reliable and accurate forecasts of the state of the ocean, land, sea-ice, and atmosphere.

2.8 SUMMARY: THE KEY MESSAGES OF THIS CHAPTER

To conclude, let us summarize in few bullets the key points that were discussed in this chapter:

- **i.** Atmosphere and Earth-system (land, ocean, sea-ice, and atmosphere) forecasts are produced using dynamical systems with two key ingredients: observations and models designed to represent reality as closely as possible.
- **ii.** The chaotic nature of the atmosphere (and in general of some of the Earth-system components) makes the prediction problem very difficult, because error growth is flow dependent.
- **iii.** The two key sources of forecast errors are observation errors and model uncertainties and approximations; a third important source of forecast errors are the data-assimilation assumptions made when computing the analysis (i.e., the forecast initial conditions).
- **iv.** Although we know the sources of forecast error and the role of the chaotic nature of the Earth-system, it is difficult to design an ensemble that simulates all of the sources of forecast error in a reliable and accurate way, especially for variables close to the surface: postprocessing and calibration techniques can help us to generate more accurate and reliable products, than can be generated by raw ensemble data alone.

v. In 1992, ensembles started being part of the operational suites at ECMWF and NCEP, introducing a paradigm shift from single to ensemble-based probabilistic forecasts.

vi. Ensembles should have two key properties, reliability and accuracy: their performance should be judged considering both of them.

vii. Today, almost all meteorological centers use ensembles: about 500 forecasts generated as part of 10 global ensembles are shared daily as part of the TIGGE. Their performance, in terms of reliability and accuracy, varies a lot.

viii. Reforecast ensembles were introduced about 10 years ago to help correct raw, ensemble-based probabilistic forecasts, and to provide a more statistically-sound assessment of ensemble performance.

ix. At ECMWF, both the medium-range/monthly and the seasonal ensembles have a reforecast suite, which includes slightly fewer members than the forecast suites and span 20 and 30 years, respectively.

x. The future will see an increased use of ensembles.

xi. The future ensemble-based products will see the development of more sophisticated calibration and postprocessing techniques that can be applied to raw ensemble data to provide users with more reliable and accurate products.

REFERENCES

Barkmeijer, J., Buizza, R., & Palmer, T. N. (1999). 3D-Var Hessian singular vectors and their potential use in the ECMWF ensemble prediction system. *Quarterly Journal of the Royal Meteorological Society, 125,* 2333–2351.

Berner, J., Shutts, G., Leutbecher, M., & Palmer, T. N. (2008). A spectral stochastic kinetic energy backscatter scheme and its impact on flow-dependent predictability in the ECMWF ensemble prediction system. *Journal of the Atmospheric Sciences, 66,* 603–626.

Bougeault, P., et al. (2009). The THORPEX interactive grand global ensemble (TIGGE). *Bulletin of the American Meteorological Society, 91,* 1059–1072.

Buizza, R. (2001). Accuracy and economic value of categorical and probabilistic forecasts of discrete events. *Monthly Weather Review, 129,* 2329–2345.

Buizza, R. (2014). *The TIGGE medium-range, global ensembles. ECMWF research department technical memorandum n. 739, ECMWF, Shinfield Park, Reading RG2-9AX, UK* (pp. 53). http://www.ecmwf.int/sites/default/files/elibrary/2014/7529-tigge-global-medium-range-ensembles.pdf.

Buizza, R., & Palmer, T. N. (1995). The singular-vector structure of the atmospheric general circulation. *Journal of the Atmospheric Sciences, 52,* 1434–1456.

Buizza, R., Miller, M., & Palmer, T. N. (1999). Stochastic representation of model uncertainties in the ECMWF ensemble prediction system. *Quarterly Journal of the Royal Meteorological Society, 125,* 2887–2908.

Buizza, R., Leutbecher, M., & Isaksen, L. (2008). Potential use of an ensemble of analyses in the ECMWF ensemble prediction system. *Quarterly Journal of the Royal Meteorological Society, 134,* 2051–2066.

Charney, J. G., Fjörtoft, R., & von Neumann, J. (1950). Numerical integration of the Barotropic vorticity equation. *Tellus, 2,* 237–254.

Daley, R. (1993). *Atmospheric Data Analysis* (pp. 466). Cambridge University Press.

Ehrendorfer, M., & Tribbia, J. (1997). Optimal prediction of forecast error covariances through singular vectors. *Journal of the Atmospheric Sciences, 54,* 286–313.

ECMWF IFS Documentation for Cycle 41R3. (2017). *Part I: Observations: Freely available from the ECMWF.* Web site: http://www.ecmwf.int/sites/default/files/elibrary/2016/17114-part-i-observations.pdf.

ECMWF Strategy. (2015). *The strength of a common goal. Available from the ECMWF web site* http://www.ecmwf.int/sites/default/files/ECMWF_Strategy_2016-2025.pdf.

Ghil, M., Ide, K., Bennet, A., Courtier, P., Kimoto, M., & Nagata, M.et al. (1997). *Data Assimilation in Meteorology and Oceanography.* Tokyo, Japan: Meteor. Soc. Japan.

Hagedorn, R., Hamill, T. M., & Whitaker, J. S. (2008). Probabilistic forecast calibration using ECMWF and GFS ensemble reforecasts. Part I. Two-meter temperatures. *Monthly Weather Review, 136,* 2608–2619.

Hagedorn, R., Buizza, R., Hamill, T. M., Leutbecher, M., & Palmer, T. N. (2012). Comparing TIGGE multi-model forecasts with re-forecast calibrated ECMWF ensemble forecasts. *Quarterly Journal of the Royal Meteorological Society, 138,* 1814–1827.

Hamill, T. M. (2001). Interpretation of rank histograms for verifying ensemble forecasts. *Monthly Weather Review, 129,* 550–560.

Hamill, T. M., Bates, G. T., Whitaker, J. S., Murray, D. R., Fiorino, M., Galarneau, T. J., Jr., et al. (2013). NOAA's second-generation global medium-range ensemble reforecast data set. *Bulletin of the American Meteorological Society, 94,* 1553–1565.

Hamill, T. M., Hagedorn, R., & Whitaker, J. S. (2008). Probabilistic forecast calibration using ECMWF and GFS ensemble forecasts. Part II. Precipitation. *Monthly Weather Review, 136,* 2620–2632.

Hamill, T. M., & Swinbank, R. (2015). Chapter 11: Stochastic forcing, ensemble prediction systems and TIGGE. In *Vol. 1156. Seamless Prediction of the Earth System: From Minutes to Months* (pp. 187–212). World Meteorological Organization.

Hamill, T. M., Whitaker, J. S., & Mullen, S. L. (2006). Reforecasts, an important dataset for improving weather predictions. *Bulletin of the American Meteorological Society, 87,* 33–46.

Harrison, M. S. J., Palmer, T. N., Richardson, D. S., & Buizza, R. (1999). Analysis and model dependencies in medium-range ensembles: Two transplant case studies. *Quarterly Journal of the Royal Meteorological Society, 126,* 2487–2515.

Hoffman, R. N., & Kalnay, E. (1983). Lagged average forecasting, an alternative to Monte Carlo forecasting. *Tellus A, 35A,* 100–118.

Hollingsworth, A. (1980). *An experiment in Monte Carlo forecasting* (pp. 65–95). *Proceedings of the ECMWF workshop on stochastic-dynamic forecasting. Available from ECMWF, Shinfield Park, Reading RG2 9AX, UK.* http://www.ecmwf.int/sites/default/files/elibrary/1979/9945-experiment-monte-carlo-forecasting.pdf.

Holton, J. (2012). *An Introduction to Dynamic Meteorology Vol. 88* (5th ed., pp. 552). Oxford, UK: Academic Press.

Hoskins, B. J., & James, I. N. (2014). *Fluid Dynamics of the Middle Atmosphere* (pp. 432). Chichester, UK: Wiley-Blackwell.

Houtekamer, P. L., Lefraive, L., & Derome, J. (1996). A system simulation approach to ensemble prediction. *Monthly Weather Review, 124,* 1225–1242.

Jolliffe, I. T., & Stephenson, D. B. (2011). *Forecast Verification: A Practitioner's Guide in Atmospheric Science* (2nd ed.). Chichester, UK: Wiley.

Kalnay, E. (2002). *Atmospheric Modelling, Data Assimilation and Predictability* (pp. 368). Chichester, UK: Cambridge University Press.

Leith, C. E. (1974). Theoretical skill of Monte Carlo forecasts. *Monthly Weather Review, 102,* 409–418.

Lynch, P. (2006). *The Emergence of Numerical Weather Prediction* (pp. 290). Cambridge, UK: Cambridge University Press.

Lorenz, E. (1963). Deterministic nonperiodic flow. *Journal of the Atmospheric Sciences, 20,* 130–141.

Lorenz, E. (1993). *The Essence of Chaos* (pp. 240). Seattle, WA: University of Washington Press.

Molteni, F., Stockdale, T., Balmaseda, M., Balsamo, G., Buizza, R., Ferranti, L., et al. (2011). *The new ECMWF seasonal forecast system (system 4)* (pp. 51). *ECMWF research department technical memorandum n. 656, ECMWF, Shinfield Park, Reading RG2-9AX, UK.* Available from the ECMWF web site: http://www.ecmwf.int/en/elibrary/technical-memoranda.

NEMO: The Nucleus for European Modelling of the Ocean, a state-of-the-art modelling framework for oceanographic research, operational oceanography seasonal forecast and climate studies, see http://www.nemo-ocean.eu/About-NEMO and references therein

Palmer, T. N., Buizza, R., Doblas-Reyes, F., Jung, T., Leutbecher, M., Shutts, G. J., et al. (2009). *Stochastic parametrization and model uncertainty* (pp. 42). *ECMWF Research Department Technical Memorandum No. 598*, Shinfield Park, Reading RG2-9AX, UK: ECMWF.

Palmer, T. N., & Hagedorn, R. (2006). *Predictability of Weather and Climate* (pp. 702). Cambridge, UK: Cambridge University Press.

Rabier, F., Järvinen, H., Klinker, E., Mahfouf, J. F., & Simmons, A. (1999). *The ECMWF operational implementation of four dimensional variational assimilation. Part I: experimental results with simplified physics* (pp. 26). *ECMWF Research Department Technical Memorandum No. 271*. Available from the ECMWF web site: http://www.ecmwf.int/en/elibrary/technical-memoranda.

Richardson, D. S. (2000). Skill and economic value of the ECMWF ensemble prediction system. *Quarterly Journal of the Royal Meteorological Society, 126*, 649–668.

Rousset, C., Vancoppenolle, M., Madec, G., Fichefet, T., Flavoni, S., Barthélemy, A., et al. (2015). The Louvain-La-Neuve sea ice model LIM3.6: global and regional capabilities. *Geoscientific Model Development, 8*, 2991–3005.

S2S, The sub-seasonal to seasonal project: http://s2sprediction.net/.

Shutts, G. (2005). A kinetic energy backscatter algorithm for use in ensemble prediction systems. *Quarterly Journal of the Royal Meteorological Society, 131*, 3079–3100.

Swinbank, R., et al. (2016). The TIGGE project and its achievements. *Bulletin of the American Meteorological Society, 97*, 49–67.

TIGGE: Information and data can be accessed from the ECMWF web site: https://software.ecmwf.int/wiki/display/TIGGE, http://www.ecmwf.int/en/research/projects/tigge

Toth, Z., & Kalnay, E. (1993). Ensemble forecasting at NMC: the generation of initial perturbations. *Bulletin of the American Meteorological Society, 74*, 2317–2330.

Toth, Z., & Kalnay, E. (1997). Ensemble forecasting at NCEP and the breeding method. *Monthly Weather Review, 125*, 3297–3319.

Vitart, F., et al. (2017). The sub-seasonal to seasonal (S2S) project database. *Bulletin of the American Meteorological Society, 98*, 163–173.

Zhu, Y., Toth, Z., Wobus, R., Richardson, D., & Mylne, K. (2002). The economic value of ensemble-based weather forecasts. *Bulletin of the American Meteorological Society, 83*, 73–83.

Wedi, N. P., Bauer, P., Deconinck, W., Diamantakis, M., Hamrud, M., Kuehnlein, C., et al. (2015). *The modelling infrastructure of the integrated forecasting system: recent advances and future challenges* (pp. 48). *ECMWF Research Department Technical Memorandum No. 760*. Available from the ECMWF web site: http://www.ecmwf.int/en/elibrary/technical-memoranda.

Wilks, D. (2011). *Statistical Methods in the Atmospheric Sciences* (3rd ed., pp. 676). San Diego, CA: Academic Press.

UNIVARIATE ENSEMBLE POSTPROCESSING

Daniel S. Wilks

Cornell University, Ithaca, NY, United States

CHAPTER OUTLINE

3.1 INTRODUCTION

In principle, a forecast ensemble is a random sample from the forecast-uncertainty distribution, so that the proportion of ensemble members predicting a given event or condition should be a sample estimate of the probability for that event or condition. This proportion should correspond well to the subsequent observed relative frequency when evaluated over a large sample of forecast cases. If so, the ensembles are probabilistically calibrated. In addition to exhibiting calibration, the resulting probability forecasts preferably should be as sharp (i.e., specific) as possible (Brier, 1950; Gneiting, Balabdaoui, & Raftery, 2007; Murphy & Winkler, 1987).

Statistical Postprocessing of Ensemble Forecasts. https://doi.org/10.1016/B978-0-12-812372-0.00003-0
Copyright © 2018 Elsevier Inc. All rights reserved.

49

A forecast ensemble is meant to represent the inherent uncertainty of a particular forecast. Small ensemble dispersion (all ensemble members similar to each other) should indicate smaller uncertainty, and large ensemble dispersion (substantial differences among ensemble members) should correspond to larger forecast uncertainty. Buizza (2018, Chapter 2 of this book) has outlined the structural deficiencies of ensemble-forecast formulations, and why postprocessing is needed. Random samples from distributions of initial-condition uncertainty are impractical, if only because those distributions are unknown. In addition, the dimensionality of the forecast problem (millions) is many orders of magnitude larger than feasible ensemble sizes (tens), so that ensembles could not sample the relevant distributions adequately even if they were known. Furthermore, errors and inaccuracies in the formulation of the dynamical forecast models introduce biases into the forecasts. Therefore, ensembles require postprocessing for the same reasons as traditional single-integration dynamical forecasts, but in addition to bias correction they require adjustments to their dispersion in order to achieve probabilistic calibration.

Univariate methods for ensemble postprocessing extend the traditional model output statistics (MOS) approach that has been in use for correcting nonprobabilistic dynamical forecasts for nearly half a century (Glahn & Lowry, 1972; Wilks, 2011). The need for statistical postprocessing of ensemble forecasts was known before production of these ensembles became computationally feasible (Leith, 1974). The aim of any MOS procedure is to statistically characterize errors in a database of past forecasts made with a particular system, and then to use these error characteristics to correct future forecasts made by that system. Thus, Glahn et al. (2009) write that "MOS… is a synonym for model postprocessing." Rather than single point values, ensemble postprocessing will result either in a full predictive distribution, or a discrete collection of transformed ensemble members that are meant to represent a sample from a corrected predictive distribution. For conventional nonprobabilistic dynamical forecasts, MOS aims to correct errors in (conditional) mean. For ensembles, there are errors in dispersion also, so ensemble-MOS presents a more difficult problem.

This chapter outlines methods for univariate ensemble postprocessing (i.e., in settings where the forecast is a univariate probability distribution for a single prognostic variable). Schefzik and Möller (2018, Chapter 4 of this book) describe extensions of these methods to the joint probability distribution of two or more prognostic variables. Here the discussion is mainly focused on methods for constructing postprocessed ensemble forecasts, whereas Chapters 7–10 address practical implementation of the methods in particular settings.

By far the most commonly used univariate ensemble postprocessing methods have been nonhomogeneous regression methods, and Bayesian model averaging (BMA). These and closely allied methods are described in Sections 3.2 and 3.3, respectively. Fully Bayesian univariate ensemble postprocessing methods that have been proposed are reviewed in Section 3.4, and nonparametric approaches are described in Section 3.5. Section 3.6 concludes the chapter with a brief comparison of the more popular methods.

3.2 NONHOMOGENEOUS REGRESSIONS, AND OTHER REGRESSION METHODS

3.2.1 NONHOMOGENEOUS GAUSSIAN REGRESSION (NGR)

Nonhomogeneous Gaussian regression (NGR) is one of the two most frequently used ensemble-MOS methods. It was independently proposed by Jewson, Brix, and Ziehmann (2004) and Gneiting, Raftery, Westveld, and Goldman (2005), where it was named EMOS (for ensemble MOS). However, as will be clear from subsequent sections of this review, it is one of many ensemble-MOS methods, so that the more descriptive and specific name NGR was applied to it by Wilks (2006). It is a *regression*-based method in that the conditional mean of the predictive distribution is defined as an optimized linear combination of the ensemble members, analogously to Eq. (1.3). However, unlike ordinary regression models, in which the "error" and predictive variances are assumed to be constant, in NGR these variances are formulated to depend linearly on the ensemble variance: they are *nonhomogeneous*. This property allows NGR predictive distributions to exhibit more uncertainty when the ensemble dispersion is large, and less uncertainty when the ensemble dispersion is small. Finally, in the original (Gneiting et al., 2005; Jewson et al., 2004) formulations, and in many subsequent applications, the predictive distributions are specified as *Gaussian*.

Specifically, for each forecasting occasion t the Gaussian predictive distribution has mean and variance that are particular to the ensemble for that occasion,

$$y_t \sim N\left[\mu_t, \sigma_t^2\right] \tag{3.1}$$

where y_t is the quantity being predicted. The Gaussian predictive distribution has mean

$$\mu_t = a + b_1 x_{t,1} + b_2 x_{t,2} + \cdots + b_m x_{t,m} \tag{3.2}$$

and variance

$$\sigma_t^2 = c + d\, s_t^2 \tag{3.3}$$

where $x_{t,k}$ is the kth ensemble member for the tth forecast, s_t^2 is the ensemble variance,

$$s_t^2 = \frac{1}{m} \sum_{k=1}^{m} (x_{t,k} - \bar{x}_t)^2 \tag{3.4}$$

m is the ensemble size, and the ensemble mean is

$$\bar{x}_t = \frac{1}{m} \sum_{k=1}^{m} x_{t,k} \tag{3.5}$$

The $m + 3$ regression constants a, b_k, c, and d are the same for each forecast occasion t and need to be estimated using past training data. (Division by $m - 1$ in Eq. 3.4 will yield identical forecasts because the larger sample variance estimates will be compensated by a smaller estimate for the parameter d.) Usually the $x_{t,k}$ are dynamical-model predictions of the same quantity as y_t, but these can be any useful predictors (e.g., Messner, Mayr, & Zeileis, 2017).

Eq. (3.2) is a general specification for the predictive mean, which would be appropriate when the m ensemble members are nonexchangeable, meaning that they have distinct statistical characteristics, as is the case for example when a multimodel ensemble is composed of single integrations from each of its m constituent models. Very often a forecast ensemble comprises exchangeable members, having the same statistical characteristics, so that the regression coefficients b_k must be the same apart from estimation errors. In this common situation of statistically exchangeable ensemble members, Eq. (3.2) is replaced by

$$\mu_t = a + b\bar{x}_t \tag{3.6}$$

in which case there are four regression parameters a, b, c, and d to be estimated. Mean functions of intermediate complexity are also sometimes appropriate. For example, a two-model ensemble in which the members within each model are exchangeable, termed a "compound ensemble" by

Bröcker and Smith (2008), would require two b coefficients operating on the two within-model ensemble means. Sansom, Ferro, Stephenson, Goddard, and Mason (2016) extend NGR for seasonal forecasting, where time-dependent biases deriving from progressive climate warming are important (e.g., Wilks & Livezey, 2013), by specifying

$$\mu_t = a + b_1 \bar{x}_t + b_2 t \tag{3.7}$$

rather than using Eq. (3.6).

Probabilistic NGR forecasts for the predictand y_t are computed using

$$\Pr\{y_t \leq q\} = \Phi\left(\frac{q - \mu_t}{\sigma_t}\right) \tag{3.8}$$

where q is any quantile of interest in the predictive distribution, and $\Phi(\bullet)$ indicates the cumulative distribution function (CDF) of the standard Gaussian distribution. The predictive mean (Eq. 3.6 or Eq. 3.2) corrects unconditional forecast bias (consistent over- or under-forecasting) when $a \neq 0$, and corrects conditional forecast biases when the $b_k \neq 1/m$ in Eq. (3.2) or $b \neq 1$ in Eq. (3.6). The (square root of) the predictive variance (Eq. 3.3) corrects dispersion errors, and further allows incorporation of any spread-skill relationship (i.e., positive correlation between ensemble spread and ensemble-mean error; e.g., Delle Monache, Eckel, Rife, Nagarajan, & Searight, 2013) into the forecast formulation. In Eq. (3.3), ensembles exhibiting correct dispersion characteristics correspond to $c \approx 0$ and $d \approx 1$, larger values of d reflect stronger spread-skill relationships, and $d \approx 0$ indicates lack of a useful spread-skill relationship.

Unlike the situation for conventional linear regression, there is no analytic formula that can be used for NGR parameter fitting. An innovative idea proposed by Gneiting et al. (2005) is to estimate the NGR parameters a, b_k, c, and d by minimizing the average continuous ranked probability score (CRPS; Matheson & Winkler, 1976; Wilks, 2011) over the training data. For Gaussian predictive distributions this is

$$\overline{CRPS}_G = \frac{1}{n}\sum_{t=1}^{n}\sigma_t\left\{\frac{y_t - \mu_t}{\sigma_t}\left[2\Phi\left(\frac{y_t - \mu_t}{\sigma_t}\right) - 1\right] + 2\phi\left(\frac{y_t - \mu_t}{\sigma_t}\right) - \frac{1}{\sqrt{\pi}}\right\} \tag{3.9}$$

where $\phi(\bullet)$ denotes the probability density function of the standard Gaussian distribution, n is the number of training samples, μ_t is defined by Eq. (3.2) or Eq. (3.6) as appropriate, and σ_t is the square root of the quantity in Eq. (3.3). Both $c \geq 0$ and $d \geq 0$ are required mathematically, which can be achieved by setting $c = \gamma^2$ and $d = \delta^2$, and optimizing over γ and δ.

A more conventional parameter estimation approach, used by Jewson et al. (2004), is maximization of the Gaussian log-likelihood function

$$L_G = -\sum_{t=1}^{n}\left[\frac{(y_t - \mu_t)^2}{2\sigma_t^2} - \ln(\sigma_t)\right] \tag{3.10}$$

which formally assumes independence among the n training samples. Maximization of this objective function is equivalent to minimization of the average logarithmic score (Good, 1952), which is also known as the Ignorance score (Roulston & Smith, 2002). Maximizing the log-likelihood also requires an iterative solution, but is much less computationally demanding than CRPS minimization, although it is less robust to the influence of outlying extreme values (Gneiting & Raftery, 2007). Gneiting et al. (2005)

found that use of Eq. (3.10) for estimation of NGR parameters yielded somewhat overdispersive predictive distributions, although Baran and Lerch (2016a), Gebetsberger, Messner, Mayr, and Zeileis (2017a), Prokosch (2013), Williams, Ferro, and Kwasniok (2014), and Williams (2016) have reported little difference in forecast performance between use of Eqs. (3.9) and (3.10) for estimating the NGR parameters.

Möller and Groß (2016) suggest use of autoregressive time-series models to represent autocorrelation in NGR forecast errors, and found that the method improved the underdispersion exhibited by conventional NGR for the day-ahead temperature forecasts in their study. Siegert, Sansom, and Williams (2016a) propose use of the "predictive bootstrap" (Harris, 1989) to account for parameter-estimation uncertainty in NGR. The effect on the resulting predictive distributions is that they exhibit heavier tails, especially for smaller training-sample sizes, analogously to use of Student-t rather than Gaussian distributions in conventional regression for small to moderate sample sizes (e.g., Glahn et al., 2009).

3.2.2 NONHOMOGENEOUS REGRESSIONS WITH MORE FLEXIBLE PREDICTIVE DISTRIBUTIONS

By construction, the NGR formulation described herein can yield only Gaussian predictive distributions. Some meteorological and climatological predictands are not Gaussian or approximately Gaussian, so that modifications to NGR are required if the method is to describe them adequately. One possible approach is to subject the target predictand and its predictor counterparts in the ensemble to a Box-Cox ("power," or "normalizing") transformation (Box & Cox, 1964; Wilks, 2011),

$$\ddot{y}_t = \begin{cases} (y_t^\lambda - 1)/\lambda, & \lambda \neq 0 \\ \ln(y_t), & \lambda = 0 \end{cases} \tag{3.11}$$

before fitting the NGR model (Hemri, Lisniak, & Klein, 2015), which is intended to make the climatological distribution of y as nearly Gaussian as possible. Here λ is an additional parameter to be estimated. Eq. (3.11) can be applied for strictly positive y_t only, although Yeo and Johnson (2000) define an extension that is applicable to general real-valued data.

In a special case of this approach, Baran and Lerch (2015, 2016a) investigate use of nonhomogeneous lognormal regressions (equivalent to NGR where the predictand y has been log-transformed), so that predictive probabilities on the transformed scale are estimated using

$$\Pr\{y_t \leq q\} = \Phi\left[\left(\ln(q) - \ln\left(\frac{\mu_t^2}{\sqrt{\mu_t^2 + \sigma_t^2}} \right) \right) \bigg/ \sqrt{\ln\left(1 + \frac{\sigma_t^2}{\mu_t^2}\right)} \right] \tag{3.12}$$

where σ_t^2 and μ_t are parameterized according to Eqs. (3.3) and (3.6), respectively. The counterpart of Eq. (3.9) for average lognormal CRPS is given in Baran and Lerch (2015).

Another possible approach to nonhomogeneous regression is to specify non-Gaussian predictive distributions. Messner, Mayr, Wilks, and Zeileis (2014a) modeled square-root transformed windspeeds using nonhomogeneous regressions with logistic predictive distributions,

$$\Pr\{y_t \leq q\} = \frac{\exp[(q - \mu_t)/\sigma_t]}{1 + \exp[(q - \mu_t)/\sigma_t]} \tag{3.13}$$

Here the conditional means μ_t were modeled using Eq. (3.6), and the corresponding (strictly positive) scale parameters were specified using

$$\sigma_t = \exp(c + ds_t) \tag{3.14}$$

where s_t is the square root of the quantity in Eq. (3.4). Logistic distributions are similar in shape to Gaussian distributions, but with heavier tails. Messner et al. (2014a) estimate the regression parameters in Eq. (3.13) using maximum likelihood, and Taillardat, Mestre, Zamo, and Naveau (2016) provide the expression necessary for minimum-CRPS estimation. Scheuerer and Möller (2015) investigate nonhomogeneous regressions with gamma-distributed predictive distributions, and provide a closed-form CRPS equation.

In another approach to modeling positively skewed predictands in a nonhomogeneous ensemble regression setting, Lerch and Thorarinsdottir (2013) forecast probability distributions for maximum daily windspeed using generalized extreme-value (GEV) predictive distributions. In this case the prediction probabilities are given by

$$\Pr\{y_t \le q\} = \begin{cases} \exp\left\{-\left[1 + \xi\left(\dfrac{q - \mu_t}{\sigma_t}\right)\right]^{-1/\xi}\right\}, & \xi \neq 0 \\[2ex] \exp\left\{-\exp\left[-\left(\dfrac{q - \mu_t}{\sigma_t}\right)\right]\right\}, & \xi = 0 \end{cases} \tag{3.15}$$

Lerch and Thorarinsdottir (2013) used Eq. (3.6) to specify the location parameter μ_t but in their application found better results when the scale parameter depended on the ensemble mean according to

$$\sigma_t = \exp(c + d\bar{x}_t) \tag{3.16}$$

which ensures $\sigma_t > 0$ as required. Eq. (3.16) does not allow explicit dependence on the ensemble spread, but does reflect the heteroscedastic nature of variables such as wind speeds and precipitation amounts, which often exhibit greater variability for larger values. The GEV shape parameter ξ was held constant (independent of the ensemble statistics) in order to provide better stability for parameter estimation, as is often done (Coles, 2001). Lerch and Thorarinsdottir (2013) found that better numerical stability was achieved when fitting the parameters a, b, c, and d using the method of maximum likelihood, although minimum CRPS estimation can also be implemented using the result in Friederichs and Thorarinsdottir (2012). When $\xi > 0$ (found in the overwhelming majority of forecasts by Lerch & Thorarinsdottir, 2013) Eq. (3.15) allows $y_t > \mu_t - \sigma_t/\xi$, so that GEV predictive distributions may yield nonzero probability for negative values of the predictand. These probabilities were quite low when maximum daily wind speeds were being forecast, but are expected to be more substantial when the predictand is daily precipitation.

3.2.3 TRUNCATED NONHOMOGENEOUS REGRESSIONS

The original NGR approach (Section 3.2.1) will often be appropriate when the predictand of interest exhibits a reasonably symmetric distribution, for example temperature or sea-level pressure. When the distribution of the predictand is not symmetric but unimodal, and if any discontinuity in its distribution at zero is small, NGR for the transformed predictand or nonhomogeneous regressions with different predictive forms, as in Section 3.2.2, may work well. However, strictly nonnegative predictands often exhibit large discontinuities in their probability densities at zero, which may prevent simple transformations from achieving adequate approximation to Gaussian shape, and which may be poorly represented by alternative conventional predictive distributions. This problem is especially acute in the case of relatively short-duration (such as daily) precipitation accumulations, which often feature a large probability 'spike' at zero. This section describes nonhomogeneous regression approaches based on truncation and censoring, which can be successful at addressing this issue.

Thorarinsdottir and Gneiting (2010) proposed using zero-truncated Gaussian predictive distributions in a nonhomogeneous regression framework for forecasting positively skewed and nonnegative predictands. A zero-truncated Gaussian distribution is a Gaussian distribution having nonzero probability only for positive values of the random variable. Similar to the conventional Gaussian distribution it has two parameters, μ and σ, but the probability density function is

$$f_{TG}(y_t) = \phi\left(\frac{y_t - \mu_t}{\sigma_t}\right)\left[\sigma_t \Phi\left(\frac{\mu_t}{\sigma_t}\right)\right]^{-1}, \quad y_t > 0 \qquad (3.17)$$

Fig. 3.1 shows a hypothetical zero-truncated Gaussian distribution, for which the ratio $\mu_t/\sigma_t = 1.5$. Because the probability corresponding to negative values has been "cut off," the factor $\Phi(\mu_t/\sigma_t)$ is necessary in order for Eq. (3.17) to integrate to unity. In effect, the probability for $y_t \leq 0$ has been spread proportionally across the rest of the distribution. The mean of the distribution in Eq. (3.17) is necessarily larger than the location parameter μ_t, and the variance is necessarily smaller than σ_t^2, because Eq. (3.17) specifies zero probability for $y_t \leq 0$ (e.g., Baran, 2014). Expressions for the mean and variance in terms of these two distribution parameters can be found in Baran (2014) and Johnson, Kotz, and Balakrishnan (1994).

The average CRPS for the truncated Gaussian distribution (Taillardat et al., 2016; Thorarinsdottir & Gneiting, 2010), which is the counterpart to Eq. (3.9), can again be used to fit the regression parameters a, b_k, c, and d. Having obtained these estimates, either Eq. (3.2) or Eq. (3.6) is used to compute the location parameter μ_t for the current forecast, and Eq. (3.3) is used to compute the scale parameter σ_t. Forecast probabilities can then be calculated using

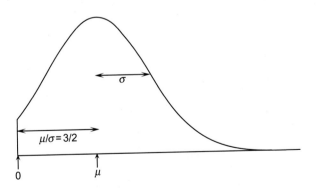

FIG. 3.1

A hypothetical truncated Gaussian distribution, where the ratio $\mu/\sigma = 3/2$, so that the leftmost $\Phi(-3/2) = 0.067$ portion of the probability in a full Gaussian distribution has been "cut off" and spread proportionally across the rest of the distribution.

$$\Pr\{y_t \leq q\} = \left[\Phi\left(\frac{q - \mu_t}{\sigma_t}\right) - \Phi\left(\frac{-\mu_t}{\sigma_t}\right) \right] / \Phi\left(\frac{\mu_t}{\sigma_t}\right) \tag{3.18}$$

which converges to Eq. (3.8) for $\mu_t \gg \sigma_t$.

Hemri, Scheuerer, Pappenberger, Bogner, and Haiden (2014) used zero-truncated Gaussian regressions following square-root transformation of the predictand. Scheuerer and Möller (2015) used nonhomogeneous regressions with truncated logistic predictive distributions (Eq. 3.13). Junk, Delle Monache, and Alessandrini (2015) fit zero-truncated Gaussian regressions using training data restricted to near-analogs of the current forecast, following an idea from Hamill and Whitaker (2006).

Observing that the heavier tails of the lognormal probability model better represent the distribution of stronger wind speeds whereas the truncated Gaussian model is better over the remainder of the distribution, Baran and Lerch (2016a) investigated probability mixtures of nonhomogeneous lognormal and zero-truncated Gaussian predictive distributions, so that predictive probabilities are computed using

$$\Pr\{y_t \leq q\} = w \Pr_{\mathrm{TG}}\{y_t \leq q\} + (1 - w) \Pr_{\mathrm{LN}}\{y_t \leq q\} \tag{3.19}$$

where the indicated truncated Gaussian (TG) and lognormal (LN) probabilities are specified by Eqs. (3.18) and (3.12), respectively. Baran and Lerch (2016a) estimated the mixing probability w simultaneously with the parameters for the two distributions, yielding a high-dimensional estimation problem. Baran and Lerch (2016b) proposed simplifying the parameter estimation in this setting, by first estimating optimal parameters for each of the constituent distributions individually via CRPS minimization, and then finding the optimal mixing probability for weighting these previously fitted distributions. They found similar predictive performance with substantially reduced computational cost for the latter distribution-combination method, and clearly better calibrated predictive distributions derived from the mixtures as compared with the individual lognormal or truncated Gaussian distributions.

Lerch and Thorarinsdottir (2013) introduced a regime-switching model, where GEV predictive distributions (Eq. 3.15) are used when the ensemble median is above a threshold θ, and truncated Gaussian predictive distributions (Eq. 3.18) are used otherwise, with θ being an additional parameter requiring estimation. Baran and Lerch (2015) investigate a similar regime-switching model, where the predictive distribution is lognormal rather than GEV if the ensemble median is larger than θ.

3.2.4 CENSORED NONHOMOGENEOUS REGRESSIONS

Censoring, in contrast to truncation, allows a probability distribution to represent values falling hypothetically below a censoring threshold, even though those values have not been observed. In the context of ensemble postprocessing, the censoring threshold is generally zero, and any probability corresponding to negative predictand values is assigned to exactly zero, yielding a probability spike there. Specifying the parameters for these censored predictive distributions using regression models yields what is known in the economics literature as Tobit regression (Tobin, 1958).

Scheuerer (2014) describes a nonhomogeneous regression model having zero-censored GEV predictive distributions, so that that any nonzero probability for negative y_t is assigned to zero precipitation. Assuming $\xi \neq 0$ in Eq. (3.15) (with an obvious modification for $\xi = 0$), predictive probabilities are computed using

$$\Pr(y_t \leq q) = \begin{cases} \exp\left\{-\left[1+\xi\left(\dfrac{q-\mu_t}{\sigma_t}\right)\right]^{-1/\xi}\right\}, & q \geq 0 \\ 0, & q < 0 \end{cases} \tag{3.20}$$

so that all probability for $y_t \leq 0$ is assigned to zero. Scheuerer (2014) links the censored GEV parameters to the ensemble statistics using the relationships

$$\mu_t = a + b_1 \bar{x}_t + b_2 \sum_{k=1}^{m} I(x_{t,k} = 0) - \frac{\sigma_t}{\xi}[\Gamma(1-\xi)-1] \tag{3.21}$$

where $I(\bullet)$ is the indicator function whose value is 1 if its argument is true, and is zero otherwise, $\Gamma(\bullet)$ denotes the gamma function, and

$$\sigma_t = c + \frac{d}{m^2} \sum_{k=1}^{m} \sum_{j=1}^{m} |x_{t,k} - x_{t,j}| \tag{3.22}$$

Division by $m(m-1)$ rather than m^2 might be preferred because the m terms in Eq. (3.22) for which $k=j$ will be identically zero (Ferro, Richardson, & Weigel, 2008; Wilks, 2018). The distribution parameters were estimated by minimizing average CRPS, with the shape parameter ξ assumed independent of the ensemble statistics as was also done by Lerch and Thorarinsdottir (2013).

Using a similar approach, Scheuerer and Hamill (2015) and Baran and Nemoda (2016) implement nonhomogeneous regressions using zero-censored shifted-gamma (i.e., Pearson Type III) predictive distributions for precipitation amounts. The probability density function for this distribution can be written as

$$f_{P-III}(y_t) = \frac{\left(\dfrac{y_t - \eta_t}{\beta_t}\right)^{\alpha_t - 1} \exp\left(-\dfrac{y_t - \eta_t}{\beta_t}\right)}{\beta_t \Gamma(\alpha_t)} \tag{3.23}$$

where both the shape α_t and scale β_t parameters are required to be positive, and the shift parameter $\eta_t < 0$ in this application. Another way of looking at Eq. (3.23) is that the quantity $y_t - \eta_t$ follows an ordinary 2-parameter gamma distribution (Eq. 3.38). As is also the case for the Scheuerer (2014) zero-censored GEV model, any nonzero probability for negative values of y_t is assigned to $y_t = 0$, so that

$$\Pr\{y_t \leq q\} = \begin{cases} F_{\gamma(\alpha_t)}\left(\dfrac{q-\eta_t}{\beta_t}\right), & q \geq 0 \\ 0, & q < 0 \end{cases} \tag{3.24}$$

where $F_{\gamma(\alpha)}$ denotes the CDF for the standard ($\beta = 1$) 2-parameter gamma distribution (evaluation of which requires numerical integration) with shape parameter α.

Fig. 3.2, from Scheuerer and Hamill (2015) illustrates the idea of attributing all probability for negative y_t values to zero, for two shifted gamma distributions. Conceptually, one can imagine a latent process having positive probability for negative outcomes (darker gray), which maps to the probability of exactly zero precipitation (i.e., a dry day). This approach is distinct from truncation, illustrated in

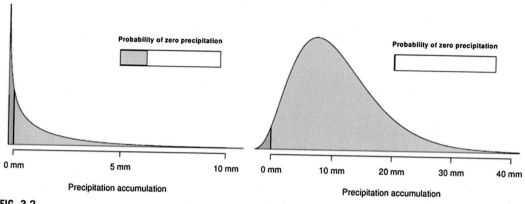

Probability of zero precipitation

0 mm   5 mm   10 mm

Precipitation accumulation

Probability of zero precipitation

0 mm   10 mm   20 mm   30 mm   40 mm

Precipitation accumulation

FIG. 3.2

Two examples of shifted-gamma (Pearson type III) probability density functions, showing that implied probabilities for negative precipitation are assigned to zero precipitation. These nominally negative precipitation amounts extend leftward to the value of the shift parameter, η.

From Scheuerer, M., & Hamill, T. M. (2015). Statistical postprocessing of ensemble precipitation forecasts by fitting censored, shifted gamma distributions. Monthly Weather Review, 143, 4578–4596.

Fig. 3.1, where the probability for below-zero values is distributed proportionally across the positive predictand values. Truncated formulations are likely more appropriate when the variable in question is essentially continuous from its lower limit, whereas censored formulations allow representation of a discontinuous probability spike at this limit.

Scheuerer and Hamill (2015) consider a variety of regression forms to relate the ensemble statistics to the parameters of the censored, shifted-gamma predictive distributions, including use of ensemble predictions for precipitable water as some of the x's in the equivalent of Eq. (3.2). Baran and Nemoda (2016) represent the gamma distribution mean $\mu_t = \alpha_t \, \beta_t$ using either Eq. (3.2) or Eq. (3.6) as appropriate, and represent the distribution variance $\sigma_t^2 = \alpha_t \beta_t^2$ using

$$\sigma_t^2 = c + d\bar{x}_t \tag{3.25}$$

Similarly to Eq. (3.16), this formulation does not include explicit dependence of the predictive variance on the ensemble spread, but does portray greater uncertainty for larger predicted values. Both Scheuerer and Hamill (2015) and Baran and Nemoda (2016), estimate the regression parameters by minimizing the average CRPS. Baran and Nemoda (2016) report that maximum likelihood estimation yields inferior performance in this setting.

Messner, Mayr, Zeileis, and Wilks (2014b) proposed nonhomogeneous regression for square-root transformed wind speeds, using censored logistic predictive distributions. Here the regression parameters were estimated using maximum likelihood, although the CRPS for this distribution is given in Taillardat et al. (2016). Probabilities on the transformed scale are then computed using

$$\Pr\{y_t \leq q\} = \begin{cases} \dfrac{\exp\left[(q - \mu_t)/\sigma_t\right]}{1 + \exp\left[(q - \mu_t)/\sigma_t\right]}, & q \geq 0 \\[2ex] 0, & \text{otherwise} \end{cases} \tag{3.26}$$

where the mean and scale parameters depend on the ensemble statistics according to Eqs. (3.6) and (3.14), respectively. Stauffer, Mayr, Messner, Umlauf, and Zeileis (2017) further extend this censored-logistic approach for precipitation forecasts, where in some cases all ensemble members may forecast zero, by defining the logistic distribution parameters as

$$\mu_t = a + b_1 I\left(\sum_{k=1}^{m} x_k = 0\right) + b_2 \bar{x}_t \left[1 - I\left(\sum_{k=1}^{m} x_k = 0\right)\right] \tag{3.27a}$$

and

$$\ln(\sigma_t) = c + d \ln(s_t)\left[1 - I\left(\sum_{k=1}^{m} x_k = 0\right)\right] \tag{3.27b}$$

Thus regression specifications for both mean and standard deviation parameters are invoked if at least one ensemble member forecasts nonzero precipitation, and fixed $\mu_t = a + b_1$ and $\sigma_t = \exp(c)$ are used if all ensemble members are dry. Stauffer et al. (2017) apply these equations to power-transformed precipitation, and estimate the regression parameters using maximum likelihood. The logarithms in Eq. (3.27b) ensure that the predictive standard deviation is always positive. Gebetsberger, Messner, Mayr, and Zeileis (2017b) point out that Eq. (3.27b) allows the logarithmic link for the standard deviation to be defined even when all ensemble members are zero, in which case $s_t = 0$.

An alternative to censoring for representing the finite probability of zero precipitation is to construct a mixture model where one element of the mixture is a probability p_t for zero precipitation (Bentzien & Friederichs, 2012). In that paper this probability was specified using logistic regression (Section 3.2.5), and then combined with either a gamma or lognormal distribution F_t for the nonzero precipitation amounts, yielding the probability specification

$$\Pr\{y_t \leq q\} = p_t + (1 - p_t)F_t(q| q > 0) \tag{3.28}$$

Bentzien and Friederichs (2012) also proposed specifying probabilities for large extremes using generalized Pareto distributions, which arise in extreme-value theory (Coles, 2001) for the distribution of values above a high threshold, appended to the right tails of the gamma or lognormal predictive distributions.

3.2.5 LOGISTIC REGRESSION

Hamill, Whitaker, and Wei (2004) first proposed use of logistic regression in the context of ensemble postprocessing, although the method has a much longer history within the framework of generalized linear modeling (Nelder & Wedderburn, 1972). Hamill et al. (2004) used the ensemble mean as the sole predictor, in an application where only the two climatological terciles were used as the prediction thresholds, q:

$$\Pr\{y_i \leq q\} = \frac{\exp\left[a_q + b_q \bar{x}_t\right]}{1 + \exp\left[a_q + b_q \bar{x}_t\right]} \tag{3.29}$$

When fitting logistic regression parameters, the training-data predictands are binary, being one if the condition in curly brackets on the left-hand side of Eq. (3.29) is true, and zero otherwise. Fig. 3.3 illustrates this idea, with the training data indicated by dots. This form of regression is well suited

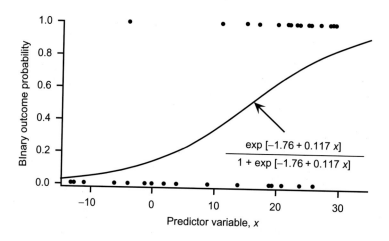

FIG. 3.3

Example logistic regression function, with parameters fit to the binary training data (*dots*).
Modified from Wilks, D. S. (2011). Statistical methods in the atmospheric sciences (3rd ed., 676 pp.). Academic Press.

for specifying probabilities that an observed value y_t will be above or below a selected quantile, because its S-shaped ($b_q > 0$) or reverse-S-shaped ($b_q < 0$) form is bounded on the unit interval. Logistic regression parameters are most frequently estimated using maximum likelihood, although Taillardat et al. (2016) provide the formula for minimum CRPS optimization.

As indicated in Eq. (3.29), separate regression coefficients must, in general, be estimated for each predictand threshold q in logistic regression. Especially when logistic regressions are estimated for large numbers of predictand thresholds, it becomes increasingly likely that probability specifications using Eq. (3.29) may be inconsistent, implying negative probabilities for some outcomes. Wilks (2009) proposed extending ordinary logistic regression in a way that unifies the regression functions for all quantiles of the distribution of the predictand, by assuming equal regression coefficients b_q, and specifying the regression intercepts as a nondecreasing function of the target quantile,

$$\Pr\{y_t \leq q\} = \frac{\exp[a\,g(q) + b\bar{x}_t]}{1 + \exp[a\,g(q) + b\bar{x}_t]} \tag{3.30}$$

The function $g(q) = \sqrt{q}$ was found to provide good results for the data used in that paper. This formulation ensures coherent probability specifications, and requires estimation of fewer parameters than does conventional logistic regression. The parameters are generally fit using maximum likelihood (Messner et al., 2014a) using a selected set of observed quantiles, but once fit Eq. (3.30) can be applied for any value of q. This approach has come to be known as extended logistic regression (XLR).

The mechanism of XLR can most easily be appreciated by realizing that the regression function in Eq. (3.29) is linear when expressed on the log-odds scale:

$$\ln\left(\frac{\Pr\{y_t \leq q\}}{1 - \Pr\{y_t \leq q\}}\right) = a\,g(q) + b\bar{x}_t \tag{3.31}$$

Fig. 3.4a shows example XLR probability specifications for selected predictand quantiles q, compared with corresponding results in when logistic regressions (Eq. 3.29) have been fit individually for the

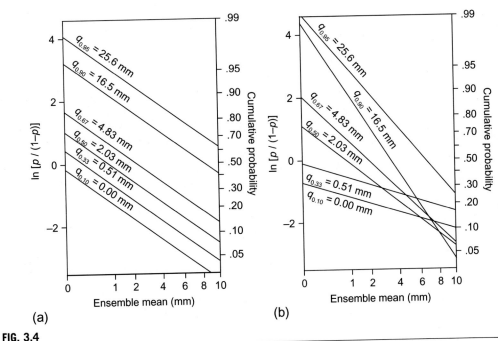

(a)   (b)

FIG. 3.4

Example logistic regressions for precipitation amounts, plotted on the log-odds scale. (a) Forecasts from Eq. (3.30) using $g(q) = \sqrt{q}$, evaluated at selected quantiles, are shown by the parallel lines, which cannot yield logically inconsistent sets of forecasts. (b) Regressions for the same quantiles, fitted separately using Eq. (3.29). Because these regressions are not constrained to be parallel, logically inconsistent forecasts are inevitable for sufficiently extreme values of the predictor.

From Wilks, D. S. (2009). Extending logistic regression to provide full-probability-distribution MOS forecasts. Meteorological Applications, 16, *361–368.*

same predictand quantiles in Fig. 3.4b. The common slope parameter b in Eqs. (3.30) and (3.31) force regressions for all quantiles to be parallel in log-odds in Fig. 3.4a; whereas the individual logistic regressions in Fig. 3.4b cross, leading in some cases to cumulative probability specifications for smaller precipitation amounts being larger than specifications for larger amounts.

Of course the logistic regression function inside the exponentials of Eq. (3.29) or (3.30) can include multiple predictors, in the form $b_1 x_{t,1} + b_2 x_{t,2} + \cdots + b_m x_{t,m}$ analogously to Eq. (3.2), where the various x's may be nonexchangeable ensemble members or other covariates. Messner et al. (2014a) point out that even if one or more of these involves ensemble spread, these forms do not explicitly represent any spread-skill relationship exhibited by the forecast ensemble, but that the logistic regression framework can be further modified to do so (Eq. 3.13).

Eqs. (3.30) and (3.31) can be seen as a continuous extension of the proportional-odds logistic regression approach (McCullagh, 1980) for specifying cumulative probabilities for an ordered set of discrete outcomes. Messner et al. (2014a) applied proportional-odds logistic regression to prediction of the climatological deciles of wind speed and precipitation, using both ordinary (homoscedastic) and nonconstant-variance (heteroscedastic) formulations. Hemri, Haiden, and Pappenberger (2016) applied the more conventional homoscedastic proportional-odds logistic regression to postprocessing ensemble forecasts of cloud cover. The cloud-cover predictand is measured in "octas," or discrete eighths of the sky hemisphere, so that the nine ordered predictand values are $y_t = 0/8, 1/8, \ldots, 8/8$. The (homoscedastic) proportional-odds logistic regression forecasts are then formulated as

$$\ln\left(\frac{\Pr\{y_t \le q\}}{1 - \Pr\{y_t \le q\}}\right) = a_q + b_1 x_{t,1} + b_2 x_{t,2} + \cdots + b_m x_{t,m} \tag{3.32}$$

where $q = 0/8, 1/8, \ldots, 8/8$, as before the predictors $x_{t,k}$ might be nonexchangeable ensemble members and/or other statistics derived from the ensemble, and the intercepts are strictly ordered so that $a_{0/8} < a_{1/8} < \cdots < a_{8/8}$. The overall result is very much like that shown in Fig. 3.4a, with regression functions that are parallel in the log-odd space and that have monotonically increasing intercepts a_q, but it differs in that intermediate functions between the plotted lines are not defined because of the discreteness of the predictand.

3.3 BAYESIAN MODEL AVERAGING, AND OTHER "ENSEMBLE DRESSING" METHODS

3.3.1 BAYESIAN MODEL AVERAGING (BMA)

BMA, introduced as an ensemble postprocessing tool by Raftery, Gneiting, Balabdaoui, and Polakowski (2005), is the second of the two most commonly used ensemble-MOS methods. In common with the regression methods described in Section 3.2, BMA yields a continuous predictive distribution for the forecast variable y. However, rather than imposing a particular parametric form, BMA predictive distributions are mixture distributions, or weighted sums of m component probability distributions, each centered at the corrected value of one of the m ensemble members being postprocessed. The BMA procedure is an example of the process sometimes referred to as "ensemble dressing," because it is the aggregate result of m probability distributions being metaphorically draped around each corrected ensemble member.

Construction of a BMA predictive distribution can be expressed in general as

$$f_{BMA}(y_t) = \sum_{k=1}^{m} w_k f_k(y_t) \tag{3.33}$$

where each w_k is the nonnegative weight associated with the component probability density $f_k(y_t)$ pertaining to the kth ensemble member $x_{t,k}$, and $\Sigma_k w_k = 1$. Fig. 3.5 illustrates the process for a five-member ensemble of nonexchangeable members, where the component distributions are Gaussian, have been constrained to have equal variance, and the weights can be interpreted as probabilities that the respective members will provide the best forecast (Raftery et al., 2005). Fig. 3.5 emphasizes that, even when the component densities are Gaussian, their weighted sum can take on a wide variety of shapes. In the case of Fig. 3.5, the BMA density is bimodal, reflecting the bifurcation of this small ensemble into two groups.

In order for BMA and other ensemble dressing procedures to work correctly, the raw ensemble members must first be debiased in order to correct systematic forecast errors exhibited in the available training data. Usually this initial debiasing step is accomplished using linear regressions, although

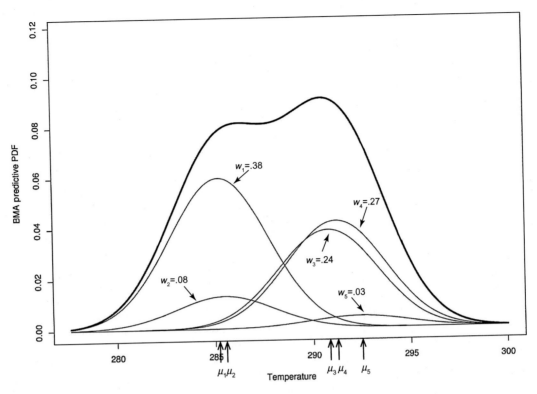

FIG. 3.5

Illustration of a BMA predictive distribution *(heavy curve)* constructed from a bias-corrected ensemble of size $m = 5$. The magnitudes of the bias-corrected ensemble members are indicated by the arrows on the horizontal axis, and the indicated weights for the five Gaussian components *(light curves)* correspond to their areas. The BMA predictive distribution is the weighted sum.

Modified from Raftery, A. E., Gneiting, T., Balabdaoui, F., & Polakowski, M. (2005). Using Bayesian model averaging to calibrate forecast ensembles. Monthly Weather Review, 133, 1155–1174.

more sophisticated methods could be used if appropriate. For nonexchangeable ensembles, separate regressions are fit for each ensemble member to reflect their different error statistics (Raftery et al., 2005), so that the bias correction is accomplished by centering each component distribution at mean

$$\mu_{t,k} = a_k + b_k x_{t,k}, \quad k = 1, \ldots, m \tag{3.34}$$

where as usual the regression coefficients minimize the average squared difference over the training period between the observed values y_t and the conditional regression specifications $\mu_{t,k}$. When the ensemble members are exchangeable, then these correction equations should be the same for each member, which can be accomplished by constraining the regression parameters in Eq. (3.34) to be equal for all ensemble members (Wilson, Beauregard, Raftery, & Verret, 2007), or with a regression involving the ensemble mean as the sole predictor (Hamill, 2007; Williams et al., 2014),

$$\mu_{t,k} = (x_{t,k} - \bar{x}_t) + (a + b\bar{x}_t) \tag{3.35}$$

Very often Gaussian distributions are adopted for the component probability densities $f_k(y_t)$ in Eq. (3.33). The corresponding component standard deviations σ_k and weights w_k are then usually estimated by maximizing the log-likelihood function

$$L_{\text{BMA}} = \sum_{t=1}^{n} \sum_{k=1}^{m} \left[\ln(w_k) - \frac{(y_t - \mu_{t,k})^2}{2\sigma_k^2} - \ln(\sigma_k) \right] \tag{3.36}$$

over the n available training samples. In this context, maximum likelihood estimation is preferably computed using the expectation-maximization (EM) algorithm (McLachlan & Krishnan, 1997), which is particularly convenient for fitting parameters of mixture distributions such as Eq. (3.33) (e.g., Wilks, 2011). The Gaussian BMA predictive probabilities are then computed as

$$\Pr\{y_t \le q\} = \sum_{k=1}^{m} w_k \Phi\left(\frac{q - \mu_{t,k}}{\sigma_k}\right) \tag{3.37}$$

This equation reflects the fact that case-to-case differences in the spread of BMA predictive distributions derive from the dispersion of the corrected ensemble members, $\mu_{t,k}$, because the standard deviations σ_k of the component ("dressing") distributions are fixed.

When all ensemble members are exchangeable, then the standard deviations σ_k and weights w_k would be constrained to be equal (Wilks, 2006; Wilson et al., 2007). Similarly if there are groups of exchangeable members within the ensemble (e.g., exchangeable members from each of several dynamical models), then these parameters (and also the debiasing means in Eq. 3.34) would be equal within each group (Fraley, Raftery, & Gneiting, 2010).

As was also the case for the nonhomogeneous regression methods (Section 3.2), basing BMA postprocessing on Gaussian component distributions may be inappropriate for predictands having skewed distributions, and/or those that can take on only nonnegative values. Duan, Ajami, Gao, and Sorooshian (2007) approach the first of these problems simply by forecasting Box-Cox (i.e., "power") transformed predictands (Eq. 3.11).

The problem of nonnegative predictands, which is relevant especially for wind speed and precipitation amount forecasting, is somewhat more difficult. Sloughter, Gneiting, and Raftery (2010) describe BMA forecasts for wind speed using gamma distributions for the component probability densities,

$$f_k(y_t) = \frac{(y_t/\beta_{t,k})^{\alpha_{t,k}-1} \exp(-y_t/\beta_{t,k})}{\beta_{t,k}\Gamma(\alpha_{t,k})} \tag{3.38}$$

which are combined according to Eq. (3.33). Eq. (3.38) is equivalent to Eq. (3.23) when $\eta_t = 0$. Sloughter et al. (2010) link the ensemble statistics to the parameters of these component gamma densities using Eq. (3.34) for the mean, and

$$\sigma_{t,k} = c + dx_{t,k} \tag{3.39}$$

for the standard deviation, where $\mu_{t,k} = \alpha_{t,k}\beta_{t,k}$ and $\sigma_{t,k} = \beta_{t,k}\sqrt{\alpha_{t,k}}$ relate these regressions to the two gamma distribution parameters $\alpha_{t,k}$ and $\beta_{t,k}$. They report that estimating member-specific c_k and d_k parameters did not appreciably improve the resulting forecasts; and that nearly equivalent results are obtained when the debiasing parameters a and b are also not member-specific, even though the ensemble members in that study were not exchangeable. The basic parameter fitting procedure in Sloughter et al. (2010) was to estimate the a_k and b_k using conventional linear regressions, and then to find the maximum-likelihood values of the parameters c, d, and the weights w_k using the EM algorithm. Forecast probabilities are then computed using

$$\Pr\{y_t \leq q\} = \sum_{k=1}^{m} w_k F_{\gamma(\alpha_{t,k})}(q/\beta_{t,k}) \tag{3.40}$$

where again $F_{\gamma(\alpha)}$ denotes the CDF for the standard ($\beta = 1$) gamma distribution with shape parameter α.

Baran (2014) proposed using truncated Gaussian component distributions in BMA for forecasting wind speeds, analogously to the nonhomogeneous regression approach of Thorarinsdottir and Gneiting (2010). The component probability density functions to be weighted in Eq. (3.33) have the same form as Eq. (3.17):

$$f_k(y_t) = \frac{\phi[(y_t - \mu_{t,k})/\sigma]}{\sigma \Phi(\mu_{t,k}/\sigma)} \tag{3.41}$$

with the location parameters $\mu_{t,k}$ defined using Eq. (3.34), and the scale parameter σ assumed to be the same for each ensemble member. Forecast probabilities are then computed using

$$\Pr\{y_t \leq q\} = \sum_{k=1}^{m} w_k \left[\Phi\left(\frac{q - \mu_{t,k}}{\sigma}\right) - \Phi\left(\frac{-\mu_{t,k}}{\sigma}\right) \right] \bigg/ \Phi\left(\frac{\mu_{t,k}}{\sigma}\right) \quad q > 0 \tag{3.42}$$

Baran (2014) compared the more traditional approach of fitting Eq. (3.34) using linear regression followed by maximum likelihood estimation of the BMA weights w_k in Eq. (3.33) and the common scale parameter σ, to full maximum likelihood estimation simultaneously for all parameters a_k, b_k, w_k, and σ; finding better performance for the full maximum likelihood approach, at a cost of approximately 40% longer computation times. Baran (2014) considered ensembles with nonexchangeable members, and multiple groups containing exchangeable members within each group, but of course this model can be implemented for exchangeable ensemble members by constraining all the w_k, a_k, and b_k to be the same for each ensemble member.

A forecast precipitation distribution is usually more difficult to model, as it often consists of a discrete probability for exactly zero, and a continuous probability density for the nonzero amounts. Sloughter, Raftery, Gneiting, and Fraley (2007) describe BMA for such precipitation distributions, specifying the probability of exactly zero precipitation with the logistic regression (Section 3.2.4)

$$p_{t,k} = \frac{\exp\left[a_{0,k} + a_{1,k}x_{t,k}^{1/3} + a_{2,k}I(x_{t,k} = 0)\right]}{1 + \exp\left[a_{0,k} + a_{1,k}x_{t,k}^{1/3} + a_{2,k}I(x_{t,k} = 0)\right]} \tag{3.43}$$

and a gamma distribution for the nonzero amounts, yielding the mixed discrete-continuous component distributions

$$f_k(y_t) = p_{t,k}I(y_t = 0) + (1 - p_{t,k})I(y_{t,k} > 0)\frac{(y_t/\beta_{t,k})^{\alpha_{i,k}-1}\exp(-y_t/\beta_{t,k})}{\beta_{t,k}\Gamma(\alpha_{t,k})} \tag{3.44}$$

Schmeits and Kok (2010) modified this approach slightly, specifying equal probabilities of zero precipitation for each ensemble member's component dressing distribution using the logistic regression

$$p_{t,k} = \frac{\exp\left[a_0 + a_1\sum_{k=1}^{m}x_{t,k}^{1/3}\right]}{1 + \exp\left[a_0 + a_1\sum_{k=1}^{m}x_{t,k}^{1/3}\right]} \tag{3.45}$$

rather than Eq. (3.43). Eqs. (3.43)–(3.45) indicate that Sloughter et al. (2007) and Schmeits and Kok (2010) found best results when working with cube-root transformed precipitation. Sloughter et al. (2007) specified the links between the gamma distribution parameters and the ensemble statistics using

$$\mu_{t,k} = b_{0,k} + b_{1,k}x_{t,k}^{1/3} \tag{3.46a}$$

and

$$\sigma_{t,k}^2 = c + dx_{t,k} \tag{3.46b}$$

For other data sets, specifying Eq. (3.46b) in terms of the ensemble variance, and/or allowing distinct variance regression parameters for each ensemble member, might be preferable. The parameters in Eqs. (3.43), (3.46a) and (3.46b) were estimated separately by Sloughter et al. (2007) using regression, after which the BMA weights and variance parameters c and d were estimated using maximum likelihood through the EM algorithm. Eqs. (3.46a) and (3.46b) relate to the gamma shape and scale parameters through $\mu_{t,k} = \alpha_{t,k}\beta_{t,k}$ and $\sigma_{t,k}^2 = \alpha_{t,k}\beta_{t,k}^2$. Sloughter et al. (2007) consider a nonexchangeable ensemble, but as before their model could be adapted to the case of exchangeable ensemble members by forcing all parameters to be the same for each member.

Fig. 3.6 illustrates the construction of a BMA predictive distribution for precipitation using these mixed discrete and continuous component distributions. The heavy vertical bar at zero indicates that the weighted sum of the probabilities specified by $w_k p_{t,k}$ is approximately 0.37. The weighted component gamma distributions are shown as the light curves, each having area equal to $w_k(1 - p_{t,k})$, and their sum is the nonzero part of the predictive distribution indicated by the heavy curve. Note that this mixed discrete-continuous distribution form is different from both the truncated and censored distributions exemplified by Figs. 3.1 and 3.2, respectively, in that no part of the continuous distribution is defined for $y_{t,k} \leq 0$, but is similar in spirit to the regression mixture distribution (Eq. 3.28) proposed by Bentzien and Friederichs (2012). Each $f_k(y_t)$ in Eq. (3.44) will integrate to unit probability because the second term contains the scaling factor $(1 - p_{t,k})$. Probability calculations for this model are therefore

$$\Pr\{y_t \leq q\} = \sum_{k=1}^{m}w_k\left[p_{t,k} + (1 - p_{t,k})F_{\gamma(\alpha_{t,k})}(q/\beta_{t,k})\right], \quad y_t \geq 0 \tag{3.47}$$

since $F_{\gamma(\alpha)}(y_t) = 0$ for $y_t = 0$, where $F_{\gamma(\alpha)}(y_t)$ is the CDF for the standard gamma distribution.

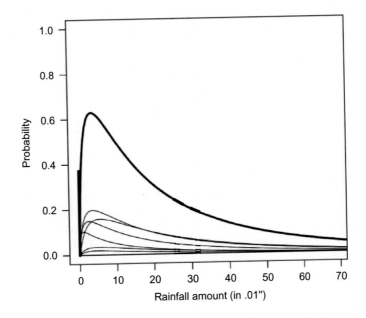

FIG. 3.6

Example BMA predictive distribution *(heavy vertical bar and curve)* composed of component distributions that are mixtures of a discrete component at zero and gamma distributions for nonzero precipitation amounts *(light curves)*.

Modified from Sloughter, J. M., Raftery, A. E., Gneiting, T., & Fraley, C. (2007). Probabilistic quantitative precipitation forecasting using Bayesian model averaging. Monthly Weather Review, 135, *3209–3220.*

3.3.2 OTHER ENSEMBLE DRESSING METHODS

The dressing approach to ensemble postprocessing is closely allied to kernel density smoothing (e.g., Silverman, 1986; Wilks, 2011), in which a probability distribution is estimated from a finite set of data by centering a characteristic shape (the kernel) at each data point (here, each corrected ensemble member), and summing or averaging these. Ensemble dressing as a statistical postprocessing idea was introduced by Roulston and Smith (2003), as a nonparametric method that will be described later, in Section 3.5.3. They proposed that the dressing kernels should represent the distribution of forecast errors around each ensemble member, assuming that member provided the "best" of the individual members' forecasts on that occasion, in order that the dressing kernels should represent only uncertainty not already reflected by the ensemble dispersion.

Wang and Bishop (2005) extended the best-member dressing idea of Roulston and Smith (2003) as a parametric method by proposing use of continuous Gaussian kernels centered on each corrected ensemble member, all of which having the same variance

$$\sigma_D^2 = s_{\mu_t - y_t}^2 - \left(1 + \frac{1}{m}\right)\bar{s}_t^2 \qquad (3.48)$$

where the first term on the right-hand side indicates the error variance for the corrected (Eq. 3.6, although Wang and Bishop assume in effect $b = 1$) ensemble-mean forecasts, and the second term approaches the average ensemble variance over the training period as the ensemble size m increases. Accordingly the second-moment constraint in Eq. (3.48) can be viewed as reflecting a partition of the total error variance $s^2_{\mu_t - y_t}$ for the corrected ensemble-mean forecasts into uncertainty due to the average ensemble spread, $\bar{s}^2_t$, plus uncertainty σ^2_D around each ensemble member. Probability forecasts are then computed using

$$\Pr\{y_t \leq q\} = \frac{1}{m}\sum_{k=1}^{m} \Phi\left(\frac{q - \mu_{t,k}}{\sigma_D}\right) \tag{3.49}$$

where $\mu_{t,k}$ denotes the kth corrected ensemble member (Eq. 3.34). Eq. (3.49) amounts to a simplification relative to the calculation for BMA predictive probabilities (Eq. 3.37), with all weights $w_k = 1/m$, and the common dressing variance estimated using Eq. (3.48) rather than Eq. (3.36). This Gaussian ensemble dressing approach is effective for the usual case of underdispersed ensembles, but it cannot reduce the predictive variances of overdispersed ensembles, and will fail by specifying negative dressing variance if the ensembles are sufficiently overdispersed. If the difference of the two terms in Eq. (3.48) is positive but small, the mixture distribution will by noisy and unrealistic (Bishop & Shanley, 2008).

Fortin, Favre, and Said (2006) note that different predictive mixture-distribution weights for each ensemble member should be appropriate in the best-member setting, depending on each member's rank within the ensemble, even if the raw ensemble members are exchangeable. The basic idea is that more extreme ensemble members are more likely to be the best member when the dynamical forecast model is underdispersive, whereas the more central ensemble members are more likely to be best when the raw ensemble is overdispersive. Fortin et al. (2006) model the mixture probabilities using beta distributions, and show that the resulting postprocessed forecasts can correct both over- and underdispersion of the raw ensemble. Furthermore, for overdispersed ensembles, the dressing kernels should be centered between their corrected ensemble members and the ensemble mean, at least for the more extreme ensemble members. The method also allows different component distributional forms to be associated with the corrected ensemble members depending on their ranks, and so could be expressed in the form of Eq. (3.33).

Bröcker and Smith (2008) derived an ensemble dressing extension that they call affine kernel dressing (AKD). They proposed centering component Gaussian dressing distributions at the corrected values

$$\mu_{t,k} = a + b_1 x_{t,k} + b_2 \bar{x}_t \tag{3.50}$$

and setting the variances for these dressing distributions as

$$\sigma^2_t = c + b_1^2 d s_t^2 \tag{3.51}$$

where the parameters a, b_1, b_2, c, and d must be estimated from the training data. Eq. (3.51) reduces to Eq. (3.3), for $b_1 = 1$. The linkage of Eqs. (3.50) and (3.51) through the parameter b_1 allows AKD to correct both over- and underdispersion in the raw ensembles because the variance of the resulting predictive mixture distribution is

$$\sigma^2_{y_t} = c + (1 + d)b_1^2 s_t^2 \tag{3.52}$$

which can be smaller than the ensemble variance.

Bröcker and Smith (2008) also propose adding an additional "ensemble member," consisting of the climatological distribution for y, to the dressing procedure in order to make it more robust to the occasional particularly bad forecast ensemble. Including this climatological Gaussian distribution, with mean μ_C and standard deviation σ_C, AKD predictive probabilities are computed using

$$\Pr\{y_t \leq q\} = \frac{1 - w_C}{m} \sum_{k=1}^{m} \Phi\left(\frac{q - \mu_{t,k}}{\sigma_t}\right) + w_C \Phi\left(\frac{q - \mu_C}{\sigma_C}\right) \tag{3.53}$$

Thus each of the actual ensemble members is given equal weight and equal dressing variance, although this variance changes from forecast to forecast depending on the raw ensemble variance (Eq. 3.51). All six of the parameters a, b_1, b_2, c, d, and w_C were estimated simultaneously using maximum likelihood. Bröcker and Smith (2008) found values for the weighting parameter for the climatological distribution w_C ranging from approximately 0.02 to 0.06 in their example, with larger values chosen for longer lead times.

Unger, van den Dool, O'Lenic, and Collins (2009) calculate regression-based ensemble-dressing probabilities for ensembles with exchangeable members using Eq. (3.49), but compute corrections to the individual ensemble members using the same correction parameters for each member,

$$\mu_{t,k} = a + b x_{t,k} \tag{3.54}$$

where the parameters a and b are fit through regression between the ensemble mean and the observations. Using results from regression, they set the standard deviation for the component Gaussian distributions to be

$$\sigma_D = \left[\frac{n}{n-2} s_y^2 \left(1 - \frac{r_M^2}{r_x}\right)\right]^{1/2} \tag{3.55}$$

where s_y^2 is the (climatological) sample variance of the predictand, r_M is the correlation between the ensemble means and the predictand, and r_x is the correlation between the individual ensemble members and the predictand, over the training period. Glahn et al. (2009) and Veenhuis (2013) present a similar approach, using empirically-based formulations for σ_D.

3.4 FULLY BAYESIAN ENSEMBLE POSTPROCESSING APPROACHES

Although BMA (Section 3.3.1) has a grounding in Bayesian ideas, it is not a fully Bayesian procedure (Di Narzo & Cocchi, 2010). Bayesian analyses are based on the relationship

$$f(y \mid x) = \frac{f(x \mid y) f(y)}{\int\limits_y f(x \mid y) f(y)} \tag{3.56}$$

which is known both as Bayes' Theorem and Bayes' Rule. Here y denotes the target of inference (in the present context the quantity being forecast), and x represents the available relevant training data. The left-hand side of Eq. (3.56) is called the posterior distribution for the target given the data; the first term in the numerator is called the likelihood, characterizing the process generating the training data given a particular value for y; and the second term in the numerator is the prior distribution, characterizing knowledge about y before the data have been observed. The integral in the denominator of Eq. (3.56) serves only as a scaling constant ensuring that the posterior distribution integrates to unity and so is a proper probability density function.

Most implementations of Bayesian forecast postprocessing have assumed Gaussian distributions for the prior and likelihood distributions. This assumption is convenient because, if the variance associated with the Gaussian likelihood can be specified externally to the Bayesian analysis as a single value, the posterior distribution is Gaussian also and its parameters can be specified analytically. Krzysztofowicz (1983) was apparently the first to use this framework for postprocessing weather forecasts, using Gaussian distributions to forecast a temperature variable y by postprocessing a single (i.e., nonprobabilistic) dynamical forecast x, although x in Eq. (3.56) can as easily be regarded as the ensemble mean in the context of ensemble forecasting. Krzysztofowicz and Evans (2008) extend this postprocessing framework to a much broader range of possible distribution forms, by transformation to Gaussian distributions.

When the variable y to be postprocessed is an observed weather quantity, a natural choice for the prior distribution $f(y)$ is its climatological distribution. Long climatological records for y are generally available, and one strength of the Bayesian approach in the context of forecast postprocessing is that these long records can be brought naturally into the analysis even if the training data relating x and y are much more limited. In this context, the likelihood encodes probabilistic information about past forecast errors within the training sample, characterizing the degree to which a forecast x reduces uncertainty about the predictand y. Accordingly, Eq. (3.56) expresses a modification or updating of the prior information $f(y)$ in light of the past observed performance of the forecasts. However, the climatological distribution need not necessarily provide the prior information: for example Coelho, Pezzulli, Balmaseda, Doblas-Reyes, and Stephenson (2004) use the forecast distribution from a statistical model as the prior, together with a likelihood encoding performance of a dynamical model, to combine the two predictive information sources through Bayes' theorem.

Because the likelihood in Eq. (3.56) denotes conditional distributions for the (often, ensemble mean) forecast, given particular values of the observed variable y, it characterizes the discrimination (e.g., Wilks, 2011; Jolliffe & Stephenson, 2012) of the forecasts in the training data set. Coelho et al. (2004) and Luo, Wood, and Pan (2007) estimate Gaussian likelihoods using linear regressions where the predictand is the ensemble mean, and the predictor is the observation y, so that the mean function for the Gaussian likelihood is $\mu_L = a + by$, and the variance σ_L^2 is the regression prediction variance. Fig. 3.7, from Coelho et al. (2004), illustrates the procedure for an example where the colder forecasts are nearly unbiased but the warmer forecasts exhibit a marked warm bias. Because the regression prediction variance σ_L^2 is specified as a point value external to Eq. (3.56), the resulting predictive distributions are also Gaussian with mean $\mu_{P,t}$ and standard deviation $\sigma_{P,t}$ for the tth forecast so that probabilities are calculated using

$$\Pr\{y_t \leq q\} = \Phi\left(\frac{q - \mu_{P,t}}{\sigma_{P,t}}\right) \tag{3.57}$$

Luo et al. (2007) formulate the Gaussian predictive parameters as

$$\mu_{P,t} = \sigma_P^2 \left[\frac{\mu_C}{\sigma_C^2} + \frac{b(\bar{x}_t - a)}{\sigma_L^2 + \sigma_e^2}\right] \tag{3.58a}$$

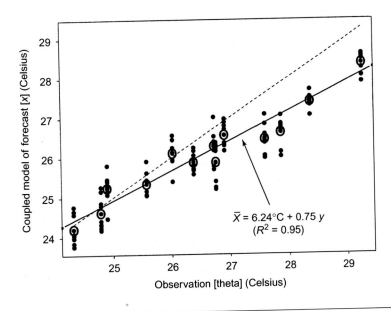

FIG. 3.7

Individual ensemble members *(small dots)* and ensemble means *(circles)* as functions of observed December Niño 3.4 temperatures. Solid line is the weighted regression for the ensemble means defining the Bayesian likelihood. The dashed 1:1 line would correspond to perfect forecasts.

Modified from Coelho, C. A. S., Pezzulli, S., Balmaseda, M., Doblas-Reyes, F. J., & Stephenson, D. B. (2004). Forecast calibration and combination: a simple Bayesian approach for ENSO. Journal of Climate, 17, 1504–1516.

and

$$\sigma_P = \left[\frac{\left(\sigma_L^2 + \overline{\sigma}_e^2\right)\sigma_C^2}{\sigma_L^2 + \overline{\sigma}_e^2 + b^2 \sigma_C^2} \right]^{1/2} \tag{3.58b}$$

where μ_C and σ_C^2 are the (prior) climatological mean and variance, and $\overline{\sigma}_e^2$ is the average ensemble variance.

In a method they call Forecast Assimilation, Stephenson, Coelho, Doblas-Reyes, and Balmaseda (2005) extend Gaussian-based Bayesian calibration using a multivariate normal likelihood distribution for individual nonexchangeable ensemble members, rather than the ensemble mean. Reggiani, Renner, Weerts, and van Gelder (2009) implement Gaussian-based Bayesian recalibration (after transformation of the positively skewed hydrological variables) individually for each ensemble member, and then construct the predictive distribution as the average of the m individual predictive density functions, analogously to Eq. (3.33) with $w_k = 1/m$. Hodyss, Satterfield, McLay, Hamill, and Scheuerer (2016) accommodate nonexchangeability of individual ensemble members by representing the likelihood as the product of conditional distributions for each member, but at the cost of requiring the estimation of a large number of regression parameters to define the likelihood. Siegert et al. (2016b) describe a more elaborate Bayesian framework for ensemble-mean recalibration that does not have an analytic result for the posterior distribution, and so requires resampling from the final predictive distribution.

Friederichs and Thorarinsdottir (2012) describe a Bayesian approach to postprocessing peak wind speed forecasts, using the GEV (Eq. 3.15) as the form of the likelihood distribution, where the location (μ_t) and scale (σ_t) parameters are specified as linear functions of covariates, in forms similar to Eq. (3.2), while the shape parameter ξ is assumed to be the same for all forecasts. They use noninformative distributions (independent Gaussian distributions with very large variances) for the prior, and a computationally intensive parameter estimation procedure.

A different approach to Bayesian ensemble calibration was proposed by Satterfield and Bishop (2014). Their target of inference is the forecast error variance as predicted by the current ensemble variance, so that the procedure specifically seeks to represent the spread-skill relationship of the ensemble forecasts. Inverse-gamma prior distributions and shifted-gamma likelihoods are assumed, so that again the predictive distribution must be evaluated by resampling.

The Bayesian predictions summarized so far in this section are analogous to the regression methods reviewed in Section 3.2, in that the output is a single predictive distribution that in most cases is of a known parametric form. In contrast, Bishop and Shanley (2008), Di Narzo and Cocchi (2010), and Marty, Fortin, Kuswanto, Favre, and Parent (2015) formulate full Bayesian analyses for the ensemble-dressing setting, which is an alternative approach to the methods described in Section 3.3. Bishop and Shanley (2008) propose using the BMA mixture-distribution formulation as the Bayesian likelihood rather than the predictive distribution. Di Narzo and Cocchi (2010) employ a hierarchical Bayesian model in which a latent variable represents the "best member." Marty et al. (2015) combine the Krzysztofowicz and Evans (2008) Bayesian approach with conventional BMA. These approaches incorporate ensemble-variance information into the predictive distribution by construction, and so allow features such as bimodality in the raw ensemble to carry through to the postprocessed predictive distribution.

3.5 NONPARAMETRIC ENSEMBLE POSTPROCESSING METHODS

3.5.1 RANK HISTOGRAM RECALIBRATION

Nonparametric ensemble postprocessing methods are wholly or mostly data-based, in contrast to the methods described in previous sections that rely on parametric probability distributions (i.e., prespecified mathematical forms). Hamill and Colucci (1997) proposed the earliest of these nonparametric methods, which estimates postprocessed ensemble probabilities on the basis of the verification rank histogram. The verification rank histogram tabulates relative frequencies p_j in the training data that the observed value y_t is larger than j of its forecast ensemble members $x_{t,j}$, plus 1 (e.g., Hamill & Colucci, 1997; Thorarinsdottir and Schuhen, 2018, Chapter 6 of this book). For example, p_1 is the proportion of training-sample forecasts for which the observation was smaller than all ensemble members, and p_{m+1} is the proportion of forecasts where the observation is larger than all ensemble members.

The Hamill and Colucci (1997) recalibration procedure operates on the rank histogram for the unconditionally debiased ensemble members

$$\widetilde{x}_{t,k} = x_{t,k} - \sum_{t=1}^{n} (\bar{x}_t - y_t) \tag{3.59}$$

and aims to achieve flatness of the recalibrated rank histogram. When a quantile of interest is not outside the range of the ensemble, probabilities are estimated by linear interpolation:

$$\Pr\{y_t \le q\} = \sum_{j=1}^{k} p_j + p_{j+1} \frac{q - \widetilde{x}_{t,(k)}}{\widetilde{x}_{t,(k+1)} - \widetilde{x}_{t,(k)}}, \quad \widetilde{x}_{t,(k)} \le q \le \widetilde{x}_{t,(k+1)} \tag{3.60}$$

The parenthetical subscripts denote that the ensemble members have been sorted in ascending order. When a quantile of interest is outside the range of the ensemble, the probabilities represented in p_1 and p_{m+1} must be extrapolated in some way. Because their forecasts were for nonnegative precipitation amounts, Hamill and Colucci (1997) extrapolated the left tail (when all ensemble members were nonzero) by linear interpolation:

$$\Pr\{0 \le y_t \le q\} = \frac{p_1 q}{\widetilde{x}_{t,(1)}}, \quad 0 \le q \le \widetilde{x}_{t,(1)} \tag{3.61}$$

and extrapolated the right tail using fitted Gumbel distributions. Alternatively, Gaussian tails can be extrapolated using (Wilks, 2006)

$$\Pr\{y_t \le q\} = p_1 \frac{\Phi(z_q)}{\Phi(z_{\widetilde{x}_{(1)}})}, \quad q \le \widetilde{x}_{(1)} \tag{3.62a}$$

and

$$\Pr\{y_t \le q\} = (1 - p_{m+1}) + p_{m+1} \frac{\Phi(z_q) - \Phi(z_{\widetilde{x}_{(m)}})}{1 - \Phi(z_{\widetilde{x}_{(m)}})}, \quad q > \widetilde{x}_{(m)} \tag{3.62b}$$

where z indicates standardization by subtraction of the (de-biased) ensemble mean, and dividing by the ensemble standard deviation. When, as is usual, the ensembles are underdispersed, substantial portions of both the left and right tails must be extrapolated.

A similar method, based on correcting the ensemble miscalibration exhibited in the reliability diagram (Sanders, 1963; Wilks, 2011) rather than the rank histogram, is described by Flowerdew (2014).

3.5.2 QUANTILE REGRESSION

Bremnes (2004) introduced the use of quantile regression (Koenker & Bassett, 1978) for ensemble postprocessing of continuous predictands. Quantile regression can be seen as a nonparametric counterpart to methods such as NGR, where the predictive probability distribution for y_t is represented by a finite collection of its quantiles. A quantile q_p is the magnitude of the random variable exceeding $p \times 100\%$ of the probability in its distribution. For example, $q_{.50}$ denotes the median, or 50th percentile.

The predictand considered by Bremnes (2004) was the precipitation amount, which has a mixed discrete (finite probability mass at zero) and continuous (nonzero precipitation amounts) distribution. He separately forecast the probability of exactly zero precipitation using probit regression (which is very similar to logistic regression, Section 3.2.5), forecast selected nonzero precipitation quantiles with quantile regression, and combined the two according to the mathematical definition of conditional probability, yielding

$$\Pr\{y_t \le q_p\} = 1 - \frac{1 - p_{y>0}}{1 - \Pr\{y_t = 0\}} \tag{3.63}$$

where $p_{y>0}$ is an appropriate cumulative probability derived from the quantile regression describing the nonzero amounts. For example, in order to calculate the median ($q_{.50}$) of the predictive distribution when the probability of zero precipitation is 0.2, the precipitation amount of interest would be the quantile of the predictive distribution for nonzero amounts corresponding to $p_{y>0} = 0.6$. When $p_{y>0} \leq$ $\Pr\{y_t = 0\}$ then $q_p = 0$ is implied. The method can of course be used, and indeed is more easily implemented, for strictly positive predictands where $\Pr\{y_t = 0\} = 0$.

Each selected quantile of the predictive distribution for the nonzero y_t's is forecast using the linear prediction

$$q_p(\mathbf{x}_t) = a_q + \sum_{i=1}^{I} b_{q,i} x_{t,i} \tag{3.64}$$

where $\mathbf{x}_t$ denotes the vector of I predictor variables $x_{t,1}, x_{t,2}, ..., x_{t,I}$. The predictors $x_{t,i}$ can be different for forecast equations pertaining to different quantiles. The regression coefficients a_q and $b_{q,i}$ are estimated separately for each quantile by numerically minimizing

$$\sum_{t=1}^{n^*} \rho_p \left[y_t - q_p(\mathbf{x}_t) \right] \tag{3.65}$$

where n^* is the number of nonzero y_t's in the training data, and

$$\rho_p(u) = \begin{cases} up, & u \geq 0 \\ u(p-1), & u < 0 \end{cases} \tag{3.66}$$

is called the check function.

For each forecast predictand quantile q_p an appropriate list of ensemble predictors $\mathbf{x}_t$ must be selected, as is also the case for ordinary least-squares regression (e.g., Wilks, 2011), although Ben Bouallègue (2016) suggests use of a lasso-penalized (Tibshirani, 1996) version of Eq. (3.65) to choose predictor variables. Bremnes (2004) considered the five predictand quantiles $q_{.05}$, $q_{.25}$, $q_{.50}$, $q_{.75}$, and $q_{.95}$, and for each of the five quantile regressions he used the same $I = 2$ predictors, being the two quartiles of the ensemble. Fig. 3.8 illustrates the resulting forecasts as functions of the 75th percentiles of the ensembles, for three levels of the ensemble 25th percentiles. These show greater forecast uncertainty (larger distances between the forecast quantiles) for increasing values of both but especially for $q_{.25}$.

Bentzien and Friederichs (2012) proposed a modified quantile regression procedure, where Eq. (3.65) is replaced by minimization of

$$\sum_{t=1}^{n^*} \rho_p \left[y_t - \max\left\{ 0, q_p(\mathbf{x}_t) \right\} \right] \tag{3.67}$$

They call this implementation censored quantile regression, because in effect the regression function in Eq. (3.64) is constrained to be nonnegative. Bentzien and Friederichs (2012) operated on cube-root transformed precipitation amounts, modeled the probability of zero precipitation using logistic regression, and used as the $I = 2$ predictors the ensemble mean and the empirical ensemble quantile corresponding to the forecast quantile.

One potential problem with quantile regressions arises as a consequence of the prediction equations for each target quantile being derived independently of one another. Especially in settings where training data are limited the procedure may yield probabilistically incoherent results, so that the forecast

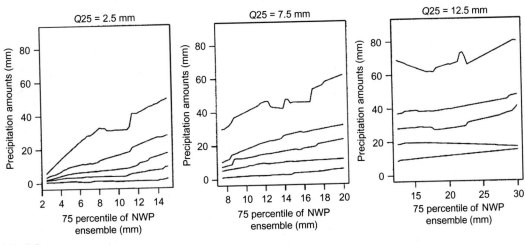

FIG. 3.8

Quantile regression forecasts for (from top to bottom) the $q_{.95}$, $q_{.75}$, $q_{.50}$, $q_{.25}$, and $q_{.05}$ quantiles of the predictive distribution of nonzero precipitation, for three levels of the ensemble lower quartile, as functions of the ensemble upper quartile.

From Bremnes, J. B. (2004). Probabilistic forecasts of precipitation in terms of quantiles using NWPmodel output.
Monthly Weather Review, 132, *338–347.*

cumulative probability for a larger quantile may be smaller than the probability for a lower quantile. Such cases would be reflected in plots such as those in Fig. 3.8 exhibiting functions that cross. Recently Noufaily and Jones (2013) have proposed a parametric version of quantile regression that, by construction, cannot exhibit this problem.

3.5.3 ENSEMBLE DRESSING

Although ensemble dressing is usually implemented using parametric kernels (Section 3.3.2), it was originally proposed by Roulston and Smith (2003) as a nonparametric ensemble postprocessing approach. They "dressed" each corrected ensemble member using a random sample from the collection of errors that were defined relative to the ensemble member closest to the observation on each occasion (the "best member"), in an archive of past forecasts. Fig. 3.9 shows example dressing distributions in the form of time trajectories for forecasts of the Lorenz '96 (Lorenz, 2006) system. Here there are 32 ensemble members (medium gray curves), each of which has been dressed using a sample of 16 best-member error trajectories (light curves), so that the predictive distribution is represented by a collection of 512 trajectories. Fig. 3.9a shows a low-predictability case, Fig. 3.9b shows a high-predictability case, and in both panels the heavy curves indicate the true trajectories. The same catalog of past best-member errors has been sampled in both examples, but the spread of the predictive distributions for the low-predictability case is larger because the underlying ensemble spread is larger. Messner and Mayr (2011) propose a similar resampling approach, where the discrete dressing kernels are derived from close analogs in the training data to the current ensemble members, following the method of Hamill and Whitaker (2006).

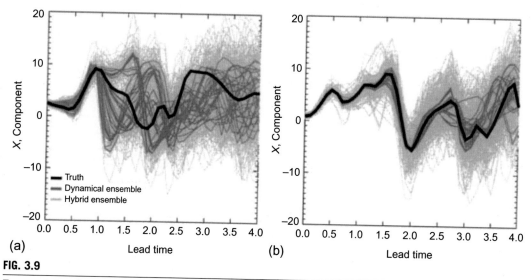

FIG. 3.9

True evolution *(heavy curves)*, 32 raw ensemble members *(medium curves)*, and the dressed predictive distribution represented by 16 best-member trajectories around each raw ensemble member *(light curves)*, for two forecast cases of the Lorenz '96 system. Panel (a) shows a low-predictability case and panel (b) shows a high-predictability case.

From Roulston, M. S., & Smith, L. A. (2003). Combining dynamical and statistical ensembles. Tellus A, 55A, *16–30.*

3.5.4 INDIVIDUAL ENSEMBLE-MEMBER ADJUSTMENTS

Another nonparametric approach to ensemble postprocessing involves transforming each of the ensemble members individually, leading to a corrected ensemble of finite size m. This has been termed the member-by-member postprocessing (MBMP) approach (Van Schaeybroeck & Vannitsem, 2015).

Eckel, Allen, and Sittel (2012) proposed the "shift-and-stretch" transformation

$$y_{t,k} = (\bar{x}_t - a) + c(x_{t,k} - \bar{x}_t) \tag{3.68}$$

where the double subscript on the left-hand side emphasizes that each raw ensemble member $x_{t,k}$ maps to a different corrected ensemble member $y_{t,k}$. Here it is assumed that the ensemble members are exchangeable. The postprocessed ensemble members are thus composed of the sum of a bias-corrected ensemble mean, plus a scaled deviation of each raw ensemble member from the raw ensemble mean. Eckel et al. (2012) computed the bias (shift) and scaling (stretch) parameters from the training data using, respectively,

$$a = \frac{1}{n}\sum_{t=1}^{n}(\bar{x}_t - y_t) \tag{3.69a}$$

which is the average error of the ensemble-mean forecast, and

$$c = \left[\frac{\dfrac{m}{(m+1)n}\sum_{t=1}^{n}(\bar{x}_t - y_t)^2}{\dfrac{1}{(m-1)n}\sum_{t=1}^{n}\sum_{k=1}^{m}(x_{t,k}-\bar{x}_t)^2} \right]^{1/2} \tag{3.69b}$$

which is the square root of the ratio of the mean-squared error for the ensemble-mean forecasts to the average ensemble variance. Johnson and Swinbank (2009) present a similar method appropriate to multimodel (nonexchangeable) ensembles. Eq. (3.68) is an extension of the "inflation" method sometimes used in downscaling climate projections (e.g., Von Storch, 1999), where $a = 0$. Taylor, McSharry, and Buizza (2009) also employed Eq. (3.68), jointly with a kernel-density (effectively, an ensemble dressing) smoothing, where the parameters a, c, and the width of the dressing kernels were optimized jointly by minimizing the Ignorance score.

More generally, MBMP adjustments can be expressed as

$$y_{t,k} = (a + b\bar{x}_t) + \gamma_t(x_{t,k} - \bar{x}_t) \tag{3.70}$$

where again each raw ensemble member maps to a distinct corrected counterpart, $y_{t,k}$. The parameters a and b define unconditional and conditional bias corrections, and the parameter γ_t controls a forecast-dependent expansion or contraction relative to the ensemble mean. Various definitions for these parameters have been proposed (and Eq. 3.35 can be seen as a special case).

Eade et al. (2014) estimate the parameters in Eq. (3.70) using

$$a = \bar{y} - b\bar{\bar{x}} \tag{3.71a}$$

$$b = \frac{s_{y_t}}{s_{\bar{x}_t}} r_{y_t,\bar{x}_t} \tag{3.71b}$$

and

$$\gamma_t = \frac{s_{y_t}}{s_t}\sqrt{1 - r^2_{y_t,\bar{x}_t}} \tag{3.71c}$$

where s_{y_t} is the climatological predictand standard deviation, the correlations in Eqs. (3.71b and 3.71c) relate the predictand and the ensemble means, and

$$\bar{\bar{x}} = \frac{1}{n}\sum_{t=1}^{n}\bar{x}_t \tag{3.72}$$

is the average of the ensemble means over the training data. Eqs. (3.71a) and (3.71b) are the ordinary least-squares regression coefficients relating the ensemble mean to the predictand y, and Eq. (3.71c) varies from forecast to forecast according to the ensemble standard deviation s_t in the denominator. Doblas-Reyes, Hagedorn, and Palmer (2005) and Johnson and Bowler (2009) have proposed an equivalent model assuming that both the forecast ensembles and the observations have been centered, so that $a = 0$ in Eq. (3.71a).

Van Schaeybroeck and Vannitsem (2015) allow the "stretch" coefficient γ_t to depend on the ensemble spread according to

$$\gamma_t = c + \frac{d}{\delta_t} \tag{3.73a}$$

where

$$\delta_t = \frac{1}{m(m-1)}\sum_{j=1}^{m}\sum_{k=1}^{m}|x_{t,j} - x_{t,k}| \tag{3.74}$$

is the average absolute difference between the uncorrected ensemble members, and both c and d are constrained to be nonnegative. Thus Eq. (3.70) can be viewed as a nonparametric counterpart to the nonhomogeneous regressions described in Sections 3.2.1–3.2.3 (Schefzik, 2017). Van Schaeybroeck and Vannitsem (2015) describe estimation of the parameters a, b, c, and d in two ways. The first is using the method maximum likelihood, assuming the errors of the corrected ensemble-mean forecasts are exponentially distributed, with mean δ_t, and so maximizing

$$L_{\exp} = -\sum_{t=1}^{n} \left[\frac{\bar{y}_t - y_t}{\delta_t} + \ln(\delta_t) \right] \tag{3.75}$$

with respect to a, b, c, and d. Here

$$\bar{y}_t = \frac{1}{m} \sum_{k=1}^{m} y_{t,k} \tag{3.76}$$

is the average of the individually-corrected ensemble members for forecast case t.

   Alternatively, the parameters can be estimated by minimizing the ensemble CRPS, which is based on the relationship (Gneiting & Raftery, 2007)

$$\text{CRPS}(F, x) = E_F |X - x| - \frac{1}{2} E_F |X - X'| \tag{3.77}$$

Here x is the observed quantity being verified, and X and X' are random draws from the predictive distribution F with respect to which the expectations are evaluated. Regarding the discrete ensemble as a random sample from F and substituting sample averages for the statistical expectations in Eq. (3.77) leads to the expression for the ensemble CRPS

$$\text{CRPS}_e = \frac{1}{n} \sum_{t=1}^{n} \left[\frac{1}{m} \sum_{k=1}^{m} (|y_{t,k} - y_t|) - \frac{\delta_t}{2} \right] \tag{3.78}$$

where δ_t characterizes the dispersion of the ensemble according to Eq. (3.74). Minimization of Eq. (3.78) is fully nonparametric, but carries more computational overhead than the likelihood maximization in Eq. (3.75).

   Williams (2016) uses the nonparametric adjustment procedure in Eq. (3.70), but defines the "stretch" coefficient as

$$\gamma_t = \frac{\sqrt{d + c s_t^2}}{s_t} = \sqrt{c + d / s_t^2} \tag{3.79}$$

With this formulation, method-of-moments estimators can be derived for the required parameters, the computation of which will be relatively fast. The two bias-correction parameters a and b are again the least-squares regression parameters defined in Eqs. (3.71a) and (3.71b). The method-of-moments estimates for the "stretch" parameters are

$$c = \frac{\text{cov}\left(s_t^2, y_t^2\right) - 2ab\,\text{cov}\left(s_t^2, \bar{x}_t\right) - b^2\,\text{cov}\left(s_t^2, \bar{x}_t\right)}{\text{Var}\left(s_t^2\right)} \tag{3.80a}$$

and

$$d = s_{y_t}^2 - c\bar{s}^2 - b^2 s_{\bar{x}_t}^2 \tag{3.80b}$$

where

$$\bar{s}^2 = \frac{1}{n}\sum_{t=1}^{n} s_t^2 \qquad (3.81)$$

is the average ensemble variance over the training period, and

$$s_{\bar{x}_t}^2 = \frac{1}{n-1}\sum_{t=1}^{n}(\bar{x}_t - \bar{\bar{x}})^2 \qquad (3.82)$$

is the variance of the ensemble means over the training period. Again both c and d are constrained to be nonnegative. Eqs. (3.73a) and (3.79) were found to perform similarly in Wilks (2018).

Fig. 3.10 illustrates the MBMP adjustment method. Fig. 3.10a (solid line) shows the debiasing function defined by the parameters a and b in Eqs. (3.71a) and (3.71b), indicating that ensembles with moderate and large means were positively biased in the training data, and that ensembles having small

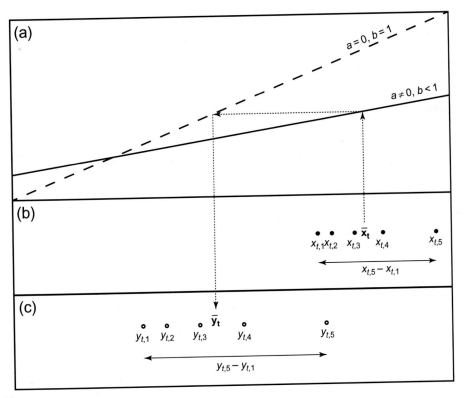

FIG. 3.10

Illustration of the MBMP adjustment method of Van Schaeybroeck and Vannitsem (2015) and Williams (2016), for (b) a hypothetical $m=5$ member underdispersive and positively biased raw ensemble *(filled circles)*, which has been transformed to (c) the corrected ensemble *(open circles)* using the debiasing function *(solid line)* in (a) and the "stretch" coefficient $\gamma_t = 1.5$.

ensemble means were negatively biased. Fig. 3.10b shows a hypothetical $m = 5$ member ensemble (filled circles) to be corrected. The dotted arrows originating at the raw ensemble mean in Fig. 3.10b locate the corrected ensemble mean in Fig. 3.10c. The corrected ensemble members (open circles) in Fig. 3.10c retain their distributional shape, but have been expanded away from the corrected ensemble mean, by the factor $\gamma_t = 1.5$ relative to the raw ensemble in Fig. 3.10b, so that the indicated corrected ensemble range in Fig. 3.10c is larger than the ensemble range of the example underdispersive raw ensemble in Fig. 3.10b by the factor 1.5.

Because the identities of the individual ensemble members are preserved in these adjustments, different forecast variables that have been subjected to independent postprocessing will continue to exhibit the rank correlation structures inherited from the raw ensemble. Thus correlations in the raw ensemble among different variables at a single location, and spatio-temporal correlations for a given variable, will all be carried forward to the postprocessed ensemble. Therefore MBMP can also serve as the basis for a multivariate postprocessing algorithm, as described in Chapter 4. On the other hand, postprocessing predictands such as wind speed or precipitation may be problematic without constraints requiring all postprocessed ensemble members to be nonnegative.

3.5.5 "STATISTICAL LEARNING" METHODS FOR ENSEMBLE POSTPROCESSING

Statistical learning methods (Hastie, Tibshirani, & Friedman, 2009) are computationally intensive algorithms that have only recently begun to be applied to ensemble postprocessing problems. They require very large training data sets, but are extremely flexible and data-adaptive, potentially including many types of nonlinearities in their structure.

Roebber (2013, 2015) has proposed use of genetic algorithms, which are meant to mimic the natural selection of genes maximizing reproductive fitness in a biological population. In the statistical learning setting each individual, i, in the population is an algorithm or prediction equation, which aggregates the effects of a finite number of "genes," of which any number and mathematical form appropriate to the problem at hand may be defined. Roebber (2013, 2015) defines 10 genes, indexed by the subscript j, for each individual i, in the form

$$\zeta_{t,i,j} = \begin{cases} = \left(c_{1,i,j}v_{1,i,j}\right)O_{1,i,j} \left(c_{2,i,j}v_{2,i,j}\right)O_{2,i,j} \left(c_{3,i,j}v_{3,i,j}\right), & \text{if } v_{4,i,j}O_{R,i,j}v_{5,i,j} \\ = 0, & \text{otherwise} \end{cases} \tag{3.83}$$

Here $v_{1,i,j}$, $v_{2,i,j}$, $v_{3,i,j}$, $v_{4,i,j}$, and $v_{5,i,j}$, have been selected from among the available predictors for the training data indexed by the subscript t, which in the ensemble postprocessing context will often be members of the forecast ensemble, ensemble statistics such as the ensemble mean or variance, or other meteorologically relevant covariates, normalized according to their range in the training data to the interval $[-1, 1]$. The variables $c_{1,i,j}$, $c_{2,i,j}$, and $c_{3,i,j}$, are multiplicative constants in the range -1 to $+1$, $O_{1,i,j}$, and $O_{2,i,j}$ are either addition or multiplication operators, and $O_{R,i,j}$ is one of the relational operators "$\leq$" or "$>$". (The underlined groups in Eq. 3.83 will be explained at the end of the next paragraph.) The overall reproductive fitness of an individual i is defined by its prediction MSE over the training data,

$$\text{MSE}_i = \frac{1}{n}\sum_{t=1}^{n} \left(\widehat{y}_t - \sum_{j=1}^{10} \zeta_{t,i,j}\right)^2 \tag{3.84}$$

where the arc accent denotes transformation of the predictand y_t to the interval $[-1, 1]$ according to its range in the training data.

The genetic algorithm is initialized by selecting each of the 11 elements for each gene in Eq. (3.83), separately for each of the i individuals, randomly and with uniform probabilities from among their possible values. The result is that most of the initial individuals' forecasts will be quite poor (large MSE_i), although within a large population some individuals will represent plausible but rough forecast models by chance. Individuals for which Eq. (3.84) is above a minimum MSE threshold, or for which the climatological distribution over the training data are sufficiently unlike that of the y_t, "die" and cannot reproduce. Those that survive might reproduce, either by (with specified probabilities) copying themselves; by undergoing "mutations," meaning random replacement of one of the 11 elements of one of the genes in Eq. (3.83) by another of its possibilities before copying; or by "genetic crossover," whereby one of the four underlined elements for one of the genes in Eq. (3.83) replaces its counterpart in one of the other 10 genes for that individual before it is copied.

The genetic algorithm is iterated over successive "generations," with the minimum MSE criterion gradually tightened as the "gene pool" is progressively modified, until a stopping criterion is reached. Depending on the specific logic of $v_{4,i,j} OR_{i,j} v_{5,i,j}$ in Eq. (3.83), the jth gene for a given individual i may be always active, sometimes active, or never active. At the end of the process, the predictive distribution is composed of the surviving population members' predictions for y_t, when current values for $v_{1,i,j}$, $v_{2,i,j}$, $v_{3,i,j}$, $v_{4,i,j}$, and $v_{5,i,j}$ have been substituted into Eq. (3.83).

A second statistical learning approach for ensemble postprocessing was proposed by Taillardat et al. (2016), who describe use of quantile regression forests. Despite the similarity in name, the method is quite different from the quantile regressions of Section 3.5.2, in that the entire predictive distribution is forecast, rather than a limited number of preselected predictand quantiles. A quantile regression forest is composed of random-forest "trees." An individual tree is built through the sequence of binary splittings of the training data on the basis of the predictor variables, that progressively decreases the sum of variances of the training-sample predictand y_t across the groups defined by the binary splits. In the context of ensemble postprocessing, the predictor variables will be such quantities as the ensemble mean, the ensemble variance, selected quantiles of the raw ensemble, etc. At the first step of the algorithm (the "base of the tree trunk"), all possible definitions for two groups g_1 and g_2 of the y_t, defined by binary splits of each of the predictors at each of its possible values, are examined to find the minimum of the combined variance

$$V = \frac{1}{n_1 - 1} \sum_{y_t \in g_1} \left(y_t - \bar{y}_{g_1} \right)^2 + \frac{1}{n_2 - 1} \sum_{y_t \in g_2} \left(y_t - \bar{y}_{g_2} \right)^2 \tag{3.85}$$

The search for minimum-variance binary splits of one of the previously defined groups based on one of the predictors continues for the second and subsequent steps of the algorithm, yielding a sequence of branches for the tree that ends with a stopping criterion being satisfied, at which point $s + 1$ groups have been defined on the basis of s algorithm steps.

A random forest is a collection of trees, which differ from each other in that each is computed based on a different bootstrap sample of the training data, which is called bootstrap aggregation, or bagging (Hastie et al., 2009). Differences among trees in the random forest are further increased (and the required computations also reduced) by allowing only a random subset of the predictors to be considered as candidates for defining the binary splitting at each new branch. Forecasts are derived from the

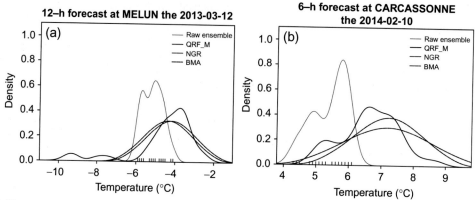

FIG. 3.11

Example quantile-random-forest predictive distributions (*solid*), compared to the raw ensemble (*tick-marks on horizontal axes*) and smoothed predictive distributions based on them (*dotted*). Also shown are the corresponding NGR (*dashed*) and BMA (*dot-dashed*) predictive distributions.

Modified from Taillardat, M., Mestre, O., Zamo, M., & Naveau, P. (2016). Calibrated ensemble forecasts using quantile regression forests and ensemble model output statistics. Monthly Weather Review, 144, *2375–2393.*

random forest by following the branches for each tree that are directed by the predictor variables pertaining to the case in question, to the terminal branches ("leaves"). The subsets of training-sample predictands at each of these leaves define an empirical predictive distribution for that tree. The quantile-regression-forest predictive distribution is then formed as the average of the empirical distributions at the terminal leaves, over all trees in the random forest. Because the resulting predictive distribution is the average of the empirical distributions for the terminal leaves, incoherent probability forecasts cannot be produced. Also, nonzero probability cannot be assigned to impossible outcomes, such as negative wind speed or precipitation amount. On the other hand, neither can nonzero probability be assigned to predictand values outside their range in the training data, so that extreme-value forecasts may be problematic.

The predictive distributions produced by a random quantile forest can take on a variety of forms, and may be quite different from forecast to forecast, as illustrated by Fig. 3.11, modified from Taillardat et al. (2016). The two quantile-random-forest predictive distributions shown in Fig. 3.11 (solid) are very different from each other, and from their raw ensembles (tick marks on the horizontal axis, and dotted smoothed version). In contrast, the NGR (dashed) and BMA (dot-dashed) predictive distributions are quite similar in shape, although, as expected, all three postprocessing methods yield predictive distributions more similar to each other than to the raw ensemble.

3.6 COMPARISONS AMONG METHODS

With so many ensemble postprocessing methods having been proposed, it may be difficult to choose among them. Many of these methods have been described only recently and so relatively little experience with them has yet accumulated. Most of the papers reviewed in this chapter compare the methods proposed therein to a selected benchmark approach, but few papers have appeared in which comparison among methods is a primary goal. Table 3.1 summarizes the results of these papers, where '+' indicates generally good performance of the method, '±' indicates moderately good performance, and '−' denotes comparatively poor performance. Of course not all of these papers investigated each of the methods.

Table 3.1 Summary of Available Comparisons Among Ensemble Postprocessing Methods

	Data Type	NR	LR	XLR	BMA	OED	FA	RH	QR	MBM
Wilks (2006)	synthetic	+	±		±	+	±	−		
Williams et al. (2014)	synthetic	+	−		+	+				
Boucher, Perreault, Anctil, and Favre (2015)	synthetic	+				+				
Van Schaeybroeck and Vannitsem (2015)	synthetic & temp.	+								+
Prokosch (2013)	temp.	+			+					
Vannitsem and Hagedorn (2011)	temp.	+								+
Wilks and Hamill (2007)	temp. & ppt.	+	±			±				
Schmeits and Kok (2010)	ppt.			+	+					
Bentzien and Friederichs (2012)	ppt.	±	+						+	
Ruiz and Saulo (2012)	ppt.		±	+	+			−		
Scheuerer (2014)	ppt.	+		+	±					
Mendoza, Rajagopalan, Clark, Ideda, and Rasmussen (2015)	ppt.		±		+				+	

'+' = good performance, '±' = moderately good performance, '−' = comparatively poor performance, NR = nonhomogeneous regression (Gaussian and non-Gaussian), LR = logistic regression, XLR = extended logistic regression, BMA = Bayesian model averaging, OED = other ensemble dressing methods, FA = forecast assimilation, RH = rank histogram recalibration, QR = quantile regression, and MBM = member-by-member. All methods outperformed the raw ensembles upon which they operated. The data types temp. = temperature, and ppt. = precipitation.

The worst-performing method among those reviewed is clearly rank histogram recalibration (RH). Although RH is a very early and somewhat crude approach, it (and all other methods investigated) outperformed the corresponding raw ensemble forecasts and so is a successful method in that sense. Conventional logistic regressions (LR) exhibit comparatively weak performance, although several of the studies indicating only moderately good performance for LR involved relatively short training samples. The remaining methods summarized in Table 3.1 yielded generally good results.

The most frequently compared methods are BMA and the nonhomogeneous regressions (NR), particularly considering that XLR can be interpreted as a nonhomogeneous regression with logistic-distributed errors (Messner et al., 2014a). The nonhomogeneous regression methods have the advantage that both bias and dispersion errors can be represented and corrected while requiring estimation of relatively few parameters, particularly if the ensemble mean is the only predictor variable in Eq. (3.2). On the other hand, these methods constrain the predictive distributions to have a particular, predefined and usually unimodal, form. In contrast, BMA and allied methods are much more flexible in the shapes of predictive distributions that can be represented, including multimodal distributions if suggested by a particular underlying forecast ensemble. However, these methods may have difficulty correcting ensemble overdispersion.

The three grades of performance have been assigned somewhat subjectively in Table 3.1, because even within a given study a method may be strong in terms of particular performance metrics or for a subset of considered lead times, but weaker for others. This variation in performance among the methods according to context also confounds any attempt to find a uniquely best method: each of the more competitive methods has its own strengths and weaknesses, and none dominates all others. Many of the described methods have been proposed only recently, and no consensus has yet emerged regarding the best ensemble postprocessing approaches, either overall or for particular predictands.

REFERENCES

Baran, S. (2014). Probabilistic wind speed forecasting using Bayesian model averaging with truncated normal components. *Computational Statistics and Data Analysis, 75*, 227–238.

Baran, S., & Lerch, S. (2015). Log-normal distribution based ensemble model output statistics models for probabilistic wind-speed forecasting. *Quarterly Journal of the Royal Meteorological Society, 141*, 2289–2299.

Baran, S., & Lerch, S. (2016a). Mixture EMOS model for calibrating ensemble forecasts of wind speed. *Environmetrics, 27*, 116–130.

Baran, S., & Lerch, S. (2016b). *Combining predictive distributions for statistical post-processing of ensemble forecasts.* (31 pp.), arXiv:1607.08096v2.

Baran, S., & Nemoda, D. (2016). Censored and shifted gamma distribution based EMOS model for probabilistic quantitative precipitation forecasting. *Environmetrics, 27*, 280–292.

Ben Bouallègue, Z. (2016). Statistical postprocessing of ensemble global radiation forecasts with penalized quantile regression. *Meteorologische Zeitschrift, 26*, 253–264.

Bentzien, S., & Friederichs, P. (2012). Generating and calibrating probabilistic quantitative precipitation forecasts from the high-resolution NWP model COSMO-DE. *Weather and Forecasting, 27*, 988–1002.

Bishop, C. H., & Shanley, K. T. (2008). Bayesian model averaging's problematic treatment of extreme weather and a paradigm shift that fixes it. *Monthly Weather Review, 136*, 4641–4652.

Boucher, M.-A., Perreault, L., Anctil, F., & Favre, A.-C. (2015). Exploratory analysis of statistical post-processing methods for hydrological ensemble forecasts. *Hydrological Processes, 29*, 1141–1155.

Box, G., & Cox, D. (1964). An analysis of transformations. *Journal of the Royal Statistical Society B*, *26*, 211–252.

Bremnes, J. B. (2004). Probabilistic forecasts of precipitation in terms of quantiles using NWP model output. *Monthly Weather Review*, *132*, 338–347.

Brier, G. W. (1950). Verification of forecasts expressed in terms of probabilities. *Monthly Weather Review*, *78*, 1–3.

Bröcker, J., & Smith, L. A. (2008). From ensemble forecasts to predictive distribution functions. *Tellus A*, *60A*, 663–678.

Buizza, R. (2018). Ensemble forecasting and the need for calibration. In S. Vannitsem, D. S. Wilks, & J. W. Messner (Eds.), *Statistical Postprocessing of Ensemble Forecasts*. Amsterdam: Elsevier.

Coelho, C. A. S., Pezzulli, S., Balmaseda, M., Doblas-Reyes, F. J., & Stephenson, D. B. (2004). Forecast calibration and combination: a simple Bayesian approach for ENSO. *Journal of Climate*, *17*, 1504–1516.

Coles, S. (2001). *An Introduction to Statistical Modeling of Extreme Values* (208 pp.). Springer.

Delle Monache, L., Eckel, F. A., Rife, D. L., Nagarajan, B., & Searight, K. (2013). Probabilistic weather prediction with an analog ensemble. *Monthly Weather Review*, *141*, 3498–3516.

Di Narzo, A. F., & Cocchi, D. (2010). A Bayesian hierarchical approach to ensemble weather forecasting. *Journal of the Royal Statistical Society C*, *59*, 405–422.

Doblas-Reyes, F. J., Hagedorn, R., & Palmer, T. N. (2005). The rationale behind the success of multi-model ensembles in seasonal forecasting. II. Calibration and combination. *Tellus A*, *57*, 234–252.

Duan, Q., Ajami, N. K., Gao, X., & Sorooshian, S. (2007). Multi-model ensemble hydrologic prediction using Bayesian model averaging. *Advances in Water Resources*, *30*, 1371–1386.

Eade, R., Smith, D., Scaife, A., Wallace, E., Dunstone, N., Hermanson, L., et al. (2014). Do seasonal-to-decadal climate predictions underestimate the predictability of the real world? *Geophysical Research Letters*, *41*, 5620–5628.

Eckel, F. A., Allen, M. S., & Sittel, M. C. (2012). Estimation of ambiguity in ensemble forecasts. *Weather and Forecasting*, *27*, 50–69.

Ferro, C. A. T., Richardson, D. S., & Weigel, A. P. (2008). On the effect of ensemble size on the discrete and continuous ranked probability scores. *Meteorological Applications*, *15*, 19–24.

Flowerdew, J. (2014). Calibrating ensemble reliability whilst preserving spatial structure. *Tellus A*, *66*, 22662.

Fortin, V., Favre, A.-C., & Said, M. (2006). Probabilistic forecasting from ensemble prediction systems: Improving upon the best-member method by using a different weight and dressing kernel for each member. *Quarterly Journal of the Royal Meteorological Society*, *132*, 1349–1369.

Fraley, C., Raftery, A. E., & Gneiting, T. (2010). Calibrating multimodel forecast ensembles with exchangeable and missing members using Bayesian model averaging. *Monthly Weather Review*, *138*, 190–202.

Friederichs, P., & Thorarinsdottir, T. L. (2012). Forecast verification for extreme value distributions with an application to probabilistic peak wind prediction. *Environmetrics*, *23*, 579–594.

Gebetsberger, M., Messner, J. W., Mayr, G. J., & Zeileis, A. (2017a). Fine-tuning nonhomogeneous regression for probabilistic precipitation forecasts: unanimous predictions, heavy tails, and link functions. *Monthly Weather Review*, *145*, 4693–4708.

Gebetsberger, M., Messner, J. W., Mayr, G. J., & Zeileis, A. (2017b). *Estimation methods for non- homogeneous regression—minimum CRPS vs. maximum likelihood*. University of Innsbruck *working paper in economics and statistics 23*. https://EconPapers.repec.org/RePEc:inn:wpaper:2017-23.

Glahn, H. R., & Lowry, D. A. (1972). The use of model output statistics (MOS) in objective weather forecasting. *Journal of Applied Meteorology*, *19*, 769–775.

Glahn, H. R., Peroutka, M., Wiedenfeld, J., Wagner, J., Zylstra, G., & Schuknecht, B. (2009). MOS uncertainty estimates in an ensemble framework. *Monthly Weather Review*, *137*, 246–268.

Gneiting, T., Balabdaoui, F., & Raftery, A. E. (2007). Probabilistic forecasts, calibration and sharpness. *Journal of the Royal Statistical Society B*, *69*, 243–268.

Gneiting, T., & Raftery, A. E. (2007). Strictly proper scoring rules, prediction, and estimation. *Journal of the American Statistical Association, 102,* 359–378.

Gneiting, T., Raftery, A. E., Westveld, A. H., & Goldman, T. (2005). Calibrated probabilistic forecasting using ensemble model output statistics and minimum CRPS estimation. *Monthly Weather Review, 133,* 1098–1118.

Good, I. J. (1952). Rational decisions. *Journal of the Royal Statistical Society B, 14,* 107–114.

Hamill, T. M. (2007). Comments on "Calibrated surface temperature forecasts from the Canadian ensemble prediction system using Bayesian model averaging" *Monthly Weather Review, 135,* 4226–4230.

Hamill, T. M., & Colucci, S. J. (1997). Verification of Eta-RSM short-range ensemble forecasts. *Monthly Weather Review, 125,* 1312–1327.

Hamill, T. M., & Whitaker, J. S. (2006). Probabilistic quantitative precipitation forecasts based on reforecast analogs: Theory and application. *Monthly Weather Review, 134,* 3209–3229.

Hamill, T. M., Whitaker, J. S., & Wei, X. (2004). Ensemble reforecasting: Improving medium-range forecast skill using retrospective forecasts. *Monthly Weather Review, 132,* 1434–1447.

Harris, I. R. (1989). Predictive fit for natural exponential families. *Biometrika, 76,* 675–684.

Hastie, T., Tibshirani, R., & Friedman, J. (2009). *The Elements of Statistical Learning* (2nd ed., 745 pp.). New York: Springer.

Hemri, S., Haiden, T., & Pappenberger, F. (2016). Discrete postprocessing of total cloud cover ensemble forecasts. *Monthly Weather Review, 144,* 2565–2577.

Hemri, S., Lisniak, D., & Klein, B. (2015). Multivariate postprocessing techniques for probabilistic hydrological forecasting. *Water Resources Research, 51,* 7436–7451.

Hemri, S., Scheuerer, M., Pappenberger, F., Bogner, K., & Haiden, T. (2014). Trends in the predictive performance of raw ensemble weather forecasts. *Geophysical Research Letters, 41,* 9197–9205.

Hodyss, D., Satterfield, E., McLay, J., Hamill, T. M., & Scheuerer, M. (2016). Inaccuracies with multi-model postprocessing methods involving weighted, regression-corrected forecasts. *Monthly Weather Review, 144,* 1649–1668.

Jewson, S., Brix, A., & Ziehmann, C. (2004). A new parametric model for the assessment and calibration of medium-range ensemble temperature forecasts. *Atmospheric Science Letters, 5,* 96–102.

Johnson, C., & Bowler, N. (2009). On the reliability and calibration of ensemble forecasts. *Monthly Weather Review, 137,* 1717–1720.

Johnson, C., & Swinbank, R. (2009). Medium-range multimodel ensemble combination and calibration. *Quarterly Journal of the Royal Meteorological Society, 135,* 777–794.

Johnson, N. L., Kotz, S., & Balakrishnan, N. (1994). *Continuous Univariate Distributions* (Vol. 1, 756 pp.). New York: Wiley.

Jolliffe, I. T., & Stephenson, D. B. (2012). *Forecast Verification, a Practitioner's Guide in Atmospheric Science* (2nd ed., 274 pp.). Chichester: Wiley-Blackwell.

Junk, C., Delle Monache, L., & Alessandrini, S. (2015). Analog-based ensemble model output statistics. *Monthly Weather Review, 143,* 2909–2917.

Koenker, R., & Bassett, B. (1978). Regression quantiles. *Econometrica, 46,* 33–49.

Krzysztofowicz, R. (1983). Why should a forecaster and decision maker use Bayes theorem. *Water Resources Research, 19,* 327–336.

Krzysztofowicz, R., & Evans, W. B. (2008). Probabilistic forecasts from the National Digital Forecast database. *Weather and Forecasting, 23,* 270–289.

Lerch, S., & Thorarinsdottir, T. L. (2013). Comparison of non-homogeneous regression models for probabilistic wind speed forecasting. *Tellus A, 65,* 21206.

Leith, C. E. (1974). Theoretical skill of Monte-Carlo forecasts. *Monthly Weather Review, 102,* 409–418.

Lorenz, E. N. (2006). Predictability—A problem partly solved. In T. Palmer & R. Hagedorn (Eds.), *Predictability of weather and climate* (pp. 40–58). Cambridge, UK: Cambridge University Press.

Luo, L., Wood, E. F., & Pan, M. (2007). Bayesian merging of multiple climate model forecasts for seasonal hydrological predictions. *Journal of Geophysical Research, D112*, D10102.

Marty, R., Fortin, V., Kuswanto, H., Favre, A.-C., & Parent, E. (2015). Combining the Bayesian processor of output with Bayesian model averaging for reliable ensemble forecasting. *Applied Statistics, 64*, 75–92.

Matheson, J. E., & Winkler, R. L. (1976). Scoring rules for continuous probability distributions. *Management Science, 22*, 1087–1096.

McCullagh, P. (1980). Regression models for ordinal data. *Journal of the Royal Statistical Society B, 42*, 109–142.

McLachlan, G. J., & Krishnan, T. (1997). *The EM Algorithm and Extensions* (274 pp.). Hoboken, NJ: Wiley.

Mendoza, P. A., Rajagopalan, B., Clark, M. P., Ideda, K., & Rasmussen, R. M. (2015). Statistical postprocessing of high-resolution regional climate model output. *Monthly Weather Review, 143*, 1533–1553.

Messner, J. W., & Mayr, G. J. (2011). Probabilistic forecasts using analogs in the idealized Lorenz '96 setting. *Monthly Weather Review, 139*, 1960–1971.

Messner, J. W., Mayr, G. J., Wilks, D. S., & Zeileis, A. (2014a). Extending extended logistic regression: extended versus separate versus ordered versus censored. *Monthly Weather Review, 142*, 3003–3014.

Messner, J. W., Mayr, G. J., & Zeileis, A. (2017). Nonhomogeneous boosting for predictor selection in ensemble postprocessing. *Monthly Weather Review, 145*, 137–147.

Messner, J. W., Mayr, G. J., Zeileis, A., & Wilks, D. S. (2014b). Heteroscedastic extended logistic regression for postprocessing of ensemble guidance. *Monthly Weather Review, 142*, 448–456.

Möller, A., & Groß, J. (2016). Probabilistic temperature forecasting based on an ensemble autoregressive modification. *Quarterly Journal of the Royal Meteorological Society, 142*, 1385–1394.

Murphy, A. H., & Winkler, R. L. (1987). A general framework for forecast verification. *Monthly Weather Review, 115*, 1330–1338.

Nelder, J., & Wedderburn, R. (1972). Generalized linear models. *Journal of the Royal Statistical Society A, 135*, 370–384.

Noufaily, A. N., & Jones, M. C. (2013). Parametric quantile regression based on the generalized gamma distribution. *Journal of the Royal Statistical Society C, 62*, 723–740.

Prokosch, J. (2013). *Bivariate Bayesian model averaging and ensemble model output statistics* (M.S. thesis) (85 pp.). Norwegian University of Science and Technology. http://www.diva-portal.org/smash/get/diva2:656466/FULLTEXT01.pdf

Raftery, A. E., Gneiting, T., Balabdaoui, F., & Polakowski, M. (2005). Using Bayesian model averaging to calibrate forecast ensembles. *Monthly Weather Review, 133*, 1155–1174.

Reggiani, P., Renner, M., Weerts, A. H., & van Gelder, P. A. H. J. M. (2009). Uncertainty assessment via Bayesian revision of ensemble streamflow predictions in the operational river Rhine forecasting system. *Water Resources Research, 45*, W02428.

Roebber, P. J. (2013). Using evolutionary programming to generate skillful extreme value probabilistic forecasts. *Monthly Weather Review, 141*, 3170–3185.

Roebber, P. J. (2015). Evolving ensembles. *Monthly Weather Review, 143*, 471–490.

Roulston, M. S., & Smith, L. A. (2002). Evaluating probabilistic forecasts using information theory. *Monthly Weather Review, 130*, 1653–1660.

Roulston, M. S., & Smith, L. A. (2003). Combining dynamical and statistical ensembles. *Tellus A, 55A*, 16–30.

Ruiz, J. J., & Saulo, C. (2012). How sensitive are probabilistic precipitation forecasts to the choice of calibration algorithms and the ensemble generation method? Part I. Sensitivity to calibration methods. *Meteorological Applications, 19*, 302–313.

Sanders, F. (1963). On subjective probability forecasting. *Journal of Applied Meteorology, 2*, 191–201.

Sansom, P. G., Ferro, C. A. T., Stephenson, D. B., Goddard, L., & Mason, S. J. (2016). Best practices for postprocessing ensemble climate forecasts. Part I. Selecting appropriate calibration methods. *Journal of Climate, 29*, 7247–7264.

Satterfield, E. A., & Bishop, C. H. (2014). Heteroscedastic ensemble postprocessing. *Monthly Weather Review*, *142*, 3484–3502.

Schefzik, R. (2017). Ensemble calibration with preserved correlations: Unifying and comparing ensemble copula coupling and member-by-member postprocessing. *Quarterly Journal of the Royal Meteorological Society*, *143*, 999–1008.

Schefzik, R., & Möller, A. (2018). Multivariate ensemble postprocessing. In S. Vannitsem, D. S. Wilks, & J. W. Messner (Eds.), *Statistical Postprocessing of Ensemble Forecasts*. Amsterdam: Elsevier.

Scheuerer, M. (2014). Probabilistic quantitative precipitation forecasting using ensemble model output statistics. *Quarterly Journal of the Royal Meteorological Society*, *140*, 1086–1096.

Scheuerer, M., & Hamill, T. M. (2015). Statistical postprocessing of ensemble precipitation forecasts by fitting censored, shifted gamma distributions. *Monthly Weather Review*, *143*, 4578–4596.

Scheuerer, M., & Möller, D. (2015). Probabilistic wind speed forecasting on a grid based on ensemble model output statistics. *Annals of Applied Statistics*, *9*, 1328–1349.

Schmeits, M. J., & Kok, K. J. (2010). A comparison between raw ensemble output, (modified) Bayesian model averaging, and extended logistic regression using ECMWF ensemble precipitation forecasts. *Monthly Weather Review*, *138*, 4199–4211.

Siegert, S., Sansom, P. G., & Williams, R. M. (2016a). Parameter uncertainty in forecast recalibration. *Quarterly Journal of the Royal Meteorological Society*, *142*, 1213–1221.

Siegert, S., Stephenson, D. B., Sansom, P. G., Scaife, A. A., Eade, R., & Arribas, A. (2016b). A Bayesian framework for verification and recalibration of ensemble forecasts: How uncertain is NAO predictability? *Journal of Climate*, *29*, 995–1012.

Silverman, B. W. (1986). *Density Estimation for Statistics and Data Analysis* (pp. 175). Chapman and Hall.

Sloughter, J. M., Gneiting, T., & Raftery, A. E. (2010). Probabilistic wind speed forecasting using ensembles and Bayesian model averaging. *Journal of the American Statistical Association*, *105*, 25–35.

Sloughter, J. M., Raftery, A. E., Gneiting, T., & Fraley, C. (2007). Probabilistic quantitative precipitation forecasting using Bayesian model averaging. *Monthly Weather Review*, *135*, 3209–3220.

Stauffer, R., Mayr, G. J., Messner, J. W., Umlauf, N., & Zeileis, A. (2017). Ensemble post-processing of daily precipitation sums over complex terrain using censored high-resolution standardized anomalies. *Monthly Weather Review*, *145*, 955–969.

Stephenson, D. B., Coelho, C. A. S., Doblas-Reyes, F. J., & Balmaseda, M. (2005). Forecast assimilation: A unified framework for the combination of multi-model weather and climate predictions. *Tellus A*, *57*, 253–264.

Taillardat, M., Mestre, O., Zamo, M., & Naveau, P. (2016). Calibrated ensemble forecasts using quantile regression forests and ensemble model output statistics. *Monthly Weather Review*, *144*, 2375–2393.

Taylor, J. W., McSharry, P. E., & Buizza, R. (2009). Wind power density forecasting using ensemble predictions and time series models. *IEEE Transactions on Energy Conversion*, *24*, 775–782.

Thorarinsdottir, T. L., & Gneiting, T. (2010). Probabilistic forecasts of wind speed: Ensemble model output statistics by using heteroscedastic censored regression. *Journal of the Royal Statistical Society A*, *173*, 371–388.

Thorarinsdottir, T. L., & Schuhen, N. (2018). Verification: assessment of calibration and accuracy. In S. Vannitsem, D. S. Wilks, & J. W. Messner (Eds.), *Statistical Postprocessing of Ensemble Forecasts*. Amsterdam: Elsevier.

Tibshirani, R. (1996). Regression shrinkage and selection via the lasso. *Journal of the Royal Statistical Society B*, *58*, 267–288.

Tobin, J. (1958). Estimation of relationships for limited dependent data. *Econometrica*, *26*, 24–36.

Unger, D. A., van den Dool, H., O'Lenic, E., & Collins, D. (2009). Ensemble regression. *Monthly Weather Review*, *137*, 2365–2379.

Vannitsem, S., & Hagedorn, R. (2011). Ensemble forecast postprocessing over Belgium: Comparison of deterministic-like and ensemble regression methods. *Meteorological Applications*, *18*, 94–104.

Van Schaeybroeck, B., & Vannitsem, S. (2015). Ensemble post-processing using member-by-member approaches: Theoretical aspects. *Quarterly Journal of the Royal Meteorological Society, 141,* 807–818.

Veenhuis, B. A. (2013). Spread calibration of ensemble MOS forecasts. *Monthly Weather Review, 141,* 2467–2482.

Von Storch, H. (1999). On the use of "inflation" in statistical downscaling. *Journal of Climate, 12,* 3505–3506.

Wang, X., & Bishop, C. H. (2005). Improvement of ensemble reliability with a new dressing kernel. *Quarterly Journal of the Royal Meteorological Society, 131,* 965–986.

Wilks, D. S. (2006). Comparison of ensemble-MOS methods in the Lorenz '96 setting. *Meteorological Applications, 13,* 243–256.

Wilks, D. S. (2009). Extending logistic regression to provide full-probability-distribution MOS forecasts. *Meteorological Applications, 16,* 361–368.

Wilks, D. S. (2011). *Statistical Methods in the Atmospheric Sciences* (3rd ed., 676 pp.). Amsterdam: Academic Press.

Wilks, D. S. (2018). Enforcing calibration in ensemble postprocessing. *Quarterly Journal of the Royal Meteorological Society, 144,* 76–84. https://doi.org/10.1002/qj.3185.

Wilks, D. S., & Hamill, T. M. (2007). Comparison of ensemble-MOS methods using GFS reforecasts. *Monthly Weather Review, 135,* 2379–2390.

Wilks, D. S., & Livezey, R. E. (2013). Performance of alternative "normals" for tracking climate changes, using homogenized and nonhomogenized seasonal U.S. surface temperatures. *Journal of Climate and Applied Meteorology, 52,* 1677–1687.

Williams, R. M. (2016). *Statistical methods for post-processing ensemble weather forecasts* (Ph.D. dissertation) (197 pp.). University of Exeter. https://ore.exeter.ac.uk/repository/bitstream/handle/10871/21693/WilliamsR.pdf.

Williams, R. M., Ferro, C. A. T., & Kwasniok, F. (2014). A comparison of ensemble post-processing methods for extreme events. *Quarterly Journal of the Royal Meteorological Society, 140,* 1112–1120.

Wilson, L. J., Beauregard, S., Raftery, A. E., & Verret, R. (2007). Reply. *Monthly Weather Review, 135,* 4231–4236.

Yeo, L.-K., & Johnson, R. A. (2000). A new family of power transformations to improve normality or symmetry. *Biometrika, 87,* 954–959.

ENSEMBLE POSTPROCESSING METHODS INCORPORATING DEPENDENCE STRUCTURES

4

Roman Schefzik*, Annette Möller[†]

German Cancer Research Center (DKFZ), Heidelberg, Germany* Clausthal University of Technology, Clausthal-Zellerfeld, Germany[†]

CHAPTER OUTLINE

4.1 INTRODUCTION

Despite addressing the major sources of uncertainty, ensembles are often subject to biases and/or dispersion errors (Buizza, 2018, Chapter 2 of this book). Thus, ensemble forecasts need to be statistically postprocessed. Wilks (2018, Chapter 3 of this book) has reviewed state-of-the-art ensemble

Statistical Postprocessing of Ensemble Forecasts. https://doi.org/10.1016/B978-0-12-812372-0.00004-2

postprocessing techniques that yield univariate predictive distributions for individual scalar prognostic variables, including the Bayesian model averaging (BMA) method (Raftery, Gneiting, Balabdaoui, & Polakowski, 2005) and the nonhomogeneous regression (NR) approach (Gneiting, Raftery, Westveld, & Goldman, 2005), also sometimes known as ensemble model output statistics (EMOS). In this chapter, we will discuss ensemble postprocessing methods that are able to account for multivariate dependencies. The focus will be on the key example of weather forecasting, but some of the presented methods and principles may also apply in broader contexts.

Univariate postprocessing methods such as BMA and NR can substantially improve the predictive performance of dynamical weather forecast ensemble output (Hagedorn, Buizza, Hamill, Leutbecher, & Palmer, 2012; Hemri, Scheuerer, Pappenberger, Bogner, & Haiden, 2014; Wilks & Hamill, 2007). However, they typically apply to a single weather variable at a single location and a single lead time, and therefore may fail to properly incorporate intervariable, spatial, and temporal dependence structures. In many applications, it is nevertheless crucial to account for dependence patterns. For instance, probabilistic forecasts of wind fields are required for air traffic control (Chaloulos & Lygeros, 2007), and spatiotemporal weather trajectories are important for the management of renewable energy resources (Pinson, 2013; Pinson, Madsen, Nielsen, Papaefthymiou, & Klöckl, 2009; Pinson & Messner, 2018, Chapter 9 of this book).

To capture intervariable, spatial, and/or temporal dependence structures, which are ignored if statistical postprocessing proceeds individually for each weather variable, location, and lead time, a variety of different approaches has been developed. One strategy is to design methods that yield truly multivariate predictive distributions. Such multivariate ensemble postprocessing techniques can be either parametric, fitting a specific multivariate distribution, or nonparametric, essentially based on reordering notions. In low-dimensional settings, or if there is a specific intervariable, spatial, or temporal structure, parametric approaches for the modeling of multivariate dependence patterns in the forecast errors are adequate (Berrocal, Raftery, & Gneiting, 2007; Schuhen, Thorarinsdottir, & Gneiting, 2012; Sloughter, Gneiting, & Raftery, 2013, among others). However, applications, particularly in weather forecasting, may involve much higher dimensions than can be handled by a parametric model. In such cases, nonparametric methods such as ensemble copula coupling (ECC) (Schefzik, Thorarinsdottir, & Gneiting, 2013) and the Schaake shuffle (Clark, Gangopadhyay, Hay, Rajagopalan, & Wilby, 2004) appear to be a more appropriate choice. In these approaches, univariate samples from individually postprocessed predictive distributions are arranged with respect to the rank order structure of a specific multivariate template, which is referred to as a "dependence template" hereafter (Schefzik, 2016b; Wilks, 2015). In particular, the postprocessed samples adopt the pairwise rank correlation structure from the dependence template.

From a mathematical point of view, the majority of the postprocessing methods yielding truly multivariate distributions can be interpreted either explicitly or implicitly in terms of so-called copula functions (Nelsen, 2006), which are a well-established tool in stochastic dependence modeling (Joe, 2014). In particular, parametric approaches may rely on the use of parametric, especially Gaussian, copula families, while nonparametric methods typically employ empirical copulas (Schefzik, 2015).

An alternative strategy comprises approaches that indeed yield univariate parametric distributions, but account for spatial or temporal dependencies by means of the design of the estimation procedure for the model parameters. For example, to ensure spatial consistency, constraints can be put on BMA or NR coefficients, so that they vary smoothly across locations (Kleiber, Raftery, Baars, et al., 2011; Scheuerer & Büermann, 2014, for instance). Univariate approaches incorporating spatial dependencies may also be fully Bayesian in nature (Möller, Thorarinsdottir, Lenkoski, & Gneiting, 2016).

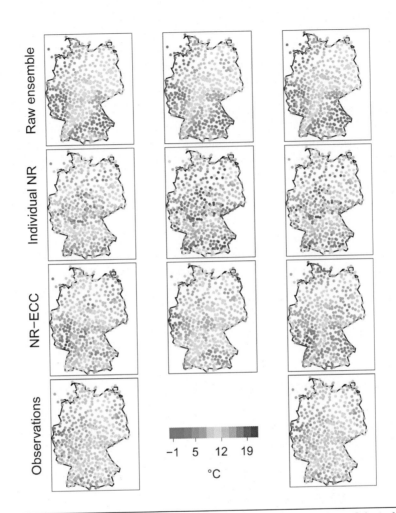

FIG. 4.1

Illustration of the effect of a multivariate postprocessing approach. The *top row* shows 3 (out of 50) randomly selected ECMWF raw ensemble member forecasts for surface temperature (in °C) at 508 stations in Germany, valid October 3, 2010, 0000 UTC. The *second row* shows 3 (out of 50) randomly selected individual NR samples, while the *third row* presents the respective NR-ECC samples, with the rank structure adopted from the ensemble members shown in the *top row*. The *bottom row* shows the respective observations, for reasons of symmetry one copy in the *left* and one in the *right panel*.

An illustration of the need for and the effect of incorporating dependence structures in postprocessing is given in Fig. 4.1, considering surface temperature forecasts (in °C) over different stations in Germany, valid October 3, 2010, 0000 UTC. The top row shows 3 (out of 50) randomly selected ensemble member forecasts from the European Center for Medium-Range Weather Forecasts (ECMWF) ensemble (ECMWF Directorate, 2012). The temperature maps issued by the three raw ECMWF ensemble members in each case exhibit pronounced spatial correlations between the forecasts at neighboring stations, which reasonably can be expected. However, compared with the actual observations shown twice in the bottom row, the raw forecasts exhibit some local biases. For instance, predictions in the western part of Germany are too warm, and predictions in the southeastern part too cool. These biases can be corrected by applying the standard NR method, which performs an individual site-by-site postprocessing of the raw temperature forecasts. In the second row of Fig. 4.1, 3 (out of 50) randomly selected individual NR postprocessed samples are shown. These are able to remove the biases to some extent, but lack the realistic spatial structure present in the raw ensemble, instead producing noisy and incoherent temperature fields. This shortcoming can be addressed by combining the individual NR postprocessed forecasts with the ECC procedure, where the respective resulting NR-ECC postprocessed ensemble members are shown in the third row of Fig. 4.1. The NR-ECC postprocessed ensemble inherits the bias-corrected margins from the individual NR postprocessed forecast and simultaneously retains the spatial correlation structure of the raw ensemble forecasts. Specifically, each NR-ECC ensemble member in the third row of Fig. 4.1 adopts the rank order structure from the respective raw ensemble member in the top row.

The remainder of this chapter is organized as follows. Section 4.2 provides the theoretical background about copulas. In Section 4.3, parametric multivariate ensemble postprocessing approaches that model intervariable, spatial, or temporal dependencies are discussed. Section 4.4 deals with nonparametric multivariate ensemble postprocessing methods based on empirical copulas, including ECC and the Schaake shuffle. Univariate ensemble postprocessing approaches that model spatial or temporal dependencies are presented in Section 4.5. Finally, Section 4.6 concludes with a discussion including a comparison of the methods reviewed in this chapter.

4.2 DEPENDENCE MODELING VIA COPULAS

To model dependencies adequately and obtain truly multivariate postprocessed distributions, we make use of the concept of copulas and the famous and fundamental Sklar's theorem. We review parametric, in particular Gaussian, copula families, as well as empirical copulas, which permit the adoption of a rank order structure from a dependence template.

4.2.1 COPULAS AND SKLAR'S THEOREM

Introduced by Sklar (1959), copulas are valuable and established tools with respect to the modeling of stochastic dependence (Joe, 2014; Nelsen, 2006). A copula $C: [0,1]^L \to [0,1]$ is an L-variate cumulative distribution function (CDF) with standard uniform univariate marginal CDFs, where $L \in \mathbb{N}, L \geq 2$. Copulas are not only of theoretical interest (Nelsen, 2006; Sempi, 2011), but have also been employed successfully in a vast range of application areas, such as in climatology, hydrology, risk management, and finance. In particular, they are appropriate to handle dependencies in ensemble postprocessing, especially in the meteorological context focused on here. The practical relevance and importance of copulas originates from the following famous theorem of Sklar (1959).

Theorem (Sklar, 1959) *For any L-variate CDF F with marginal CDFs F_1, ..., F_L, there exists a copula C such that*

$$F(y_1,...,y_L) = C(F_1(y_1),...,F_L(y_L)) \tag{4.1}$$

for $y_1,...,y_L \in \mathbb{R}$. Moreover, C is unique on the range of the margins.

Conversely, given a copula C and marginal CDFs $F_1, \ldots, F_L$, the function F as defined in Eq. (4.1) is an L-variate CDF.

In the context of multivariate ensemble postprocessing, assume that we have a postprocessed predictive CDF F_ℓ for each univariate weather quantity Y_ℓ, where $\ell \in \{1, \ldots, L\}$. Here, ℓ can be interpreted as a multiindex $\ell := (i, j, k)$ that refers to weather variable $i \in \{1, \ldots, I\}$, location $j \in \{1, \ldots, J\}$ and lead time or time point $k \in \{1, \ldots, K\}$, with $L := I \times J \times K$ (see the methods in Section 4.4). In purely intervariable (i.e., $J = 1$ and $K = 1$), purely spatial (i.e., $I = 1$ and $K = 1$), or purely temporal (i.e., $I = 1$ and $J = 1$) settings, ℓ may specifically refer only to weather variables or locations or lead times/time points, respectively (see the methods in Section 4.3).

We strive for a physically coherent multivariate predictive CDF F with margins $F_1, \ldots, F_L$. Due to Sklar's theorem and Eq. (4.1), such a multivariate CDF F can be obtained by specifying both the univariate marginal CDFs $F_1, \ldots, F_L$ and a copula C modeling the dependence structure. In our setting, the CDFs $F_1, \ldots, F_L$ can be derived via univariate postprocessing methods as discussed by Wilks (2018, Chapter 3 of this book), for instance via BMA or NR, applied to each location, weather variable, and prediction horizon separately. The remaining challenge that has to be addressed is: How to choose the copula C? Based on the specification of the copula, the resulting truly multivariate ensemble postprocessing methods can be divided into two categories: first, parametric approaches that are based on parametric families of copulas, and second, nonparametric approaches relying on empirical copulas.

4.2.2 PARAMETRIC, IN PARTICULAR GAUSSIAN, COPULAS

If the dimension L in a specific scenario is small, or if one can make use of a particular (e.g., intervariable, spatial, or temporal) structure, parametric or semiparametric families of copulas are adequate choices.

Many parametric approaches are based on a Gaussian copula framework, under which the multivariate CDF F has the form

$$F(y_1, \ldots, y_L \mid \mathbf{\Gamma}) = \Phi_L\big(\Phi^{-1}(F_1(y_1)), \ldots, \Phi^{-1}(F_L(y_L)) \mid \mathbf{\Gamma}\big) \tag{4.2}$$

for $y_1, \ldots, y_L \in \mathbb{R}$. In Eq. (4.2), $\Phi_L(\cdot \mid \mathbf{\Gamma})$ denotes the CDF of an L-variate normal distribution $\mathcal{N}_L(\mathbf{0}, \mathbf{\Gamma})$ with mean vector $\mathbf{0}$ and correlation matrix $\mathbf{\Gamma}$, and Φ^{-1} is the quantile function corresponding to the CDF Φ of the univariate standard normal distribution $\mathcal{N}(0, 1)$ with mean 0 and variance 1. The Gaussian copula itself is defined as

$$C_{\text{Gauss}}(u_1, \ldots, u_L) := \Phi_L\big(\Phi^{-1}(u_1), \ldots, \Phi^{-1}(u_L) \mid \mathbf{\Gamma}\big)$$

for $u_1, \ldots, u_L \in [0, 1]$. Because only the correlation matrix $\mathbf{\Gamma}$ needs to be modeled, Gaussian copulas are particularly convenient. According to Eq. (4.2), the Gaussian copula model leads to an L-variate normal distribution in the special case when the margins $F_1, \ldots, F_L$ are normal.

An illustration of two bivariate Gaussian copulas with different strengths of correlation is given in Fig. 4.2.

Gaussian copulas have been employed in many application areas, including climatology (Schoelzel & Friederichs, 2008) and hydrology (Genest & Favre, 2007; Hemri, Lisniak, & Klein, 2015). In our key setting of weather forecasting, the postprocessing approaches of Berrocal et al. (2007), Berrocal, Raftery, and Gneiting (2008), Gel, Raftery, and Gneiting (2004), Möller, Lenkoski, and Thorarinsdottir (2013), Pinson et al. (2009), and Schuhen et al. (2012) invoke Gaussian

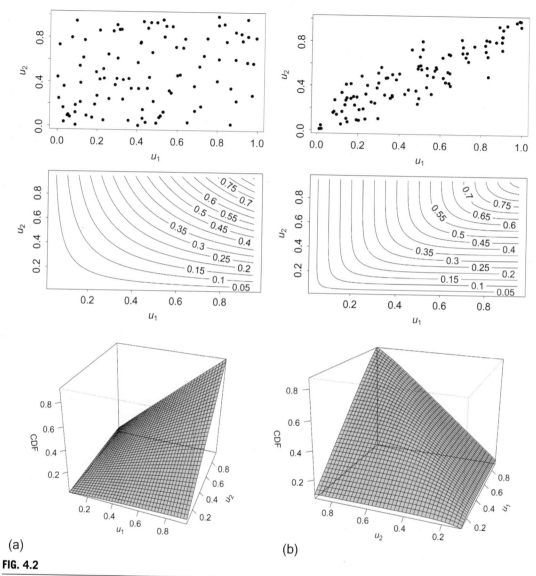

(a)   (b)

FIG. 4.2

Illustration of two different Gaussian copulas. The *left and right columns* correspond to a bivariate Gaussian copula with correlation parameters (a) 0.2 (low correlation between the two variables) and (b) 0.9 (strong correlation between the two variables), respectively. The *top row* shows bivariate scatter plots of random samples of size 100, the *middle row* contour plots, and the *bottom row* the respective surface plots of the CDFs.

copulas, whether explicitly or implicitly. There are also approaches, such as that of Sloughter et al. (2013), which use mixtures of multivariate normal distributions, where each mixture component can be linked to a Gaussian copula.

In geostatistics, the employment of Gaussian copula techniques is long standing and referred to as anamorphosis (Chilès & Delfiner, 2012). In the spatial setting, the correlation matrix Γ in Eq. (4.2) is assumed to be highly structured, fulfilling conditions such as spatial stationarity and/or isotropy. Examples include the approaches for temperature and precipitation field prediction of Gel et al. (2004) and Berrocal et al. (2007, 2008). In a similar manner, Gaussian copulas have been used to incorporate dependence patterns over consecutive look-ahead times in postprocessed predictive CDFs (Pinson et al., 2009; Schoelzel & Hense, 2011). In the case of normal margins $F_1, ..., F_L$, the underlying stochastic model is that of a Gaussian process or Gaussian random field. Then, selecting a parameterization for the correlation matrix Γ corresponds to the choice of a parametric correlation model in spatial statistics (Cressie & Wikle, 2011).

In many cases, Gaussian copulas yield convenient stochastic models, and the aforementioned Gaussian copula approaches will be discussed in more detail in Section 4.3. However, there are also parametric or semiparametric alternatives, such as elliptical copulas (Demarta & McNeil, 2005), Archimedean copulas (McNeil & Nešlehová, 2009), pair copulas (Aas, Czado, Frigessi, & Bakken, 2009), or vine copulas (Kurowicka & Joe, 2011).

4.2.3 EMPIRICAL COPULAS

In case the dimension L is huge and one cannot take advantage of a specific structure, parametric methods may become inappropriate, or at least cumbersome. Then, one typically is drawn to nonparametric approaches relying on the use of empirical copulas. Referring to Sklar's theorem and Eq. (4.1), we consider the case in which $F_1, ..., F_L$ are the empirical CDFs defined by samples from univariate postprocessed predictive CDFs and C is a so-called empirical copula (Rüschendorf, 2009), also known as "empirical dependence function" (Deheuvels, 1979).

An empirical copula is deduced from a specific discrete data set, which in the context of ensemble postprocessing takes the role of a multivariate dependence template (Wilks, 2015) as described in Section 4.1 (see also the methods in Section 4.4). For a formal description, let

$$I_N := \left\{0, \frac{1}{N}, \frac{2}{N}, ..., \frac{N-1}{N}, 1\right\}$$

and $I_N^L := \underbrace{I_N \times \cdots \times I_N}_{L \text{ times}}$, where $N \in \mathbb{N}$. Moreover, let

$$\mathbf{z} := \{(z_1^1, ..., z_N^1), ..., (z_1^L, ..., z_N^L)\}$$

denote a data set comprising L tuples of size N with real-valued entries. Furthermore, let $\text{rank}(z_n^\ell)$ be the rank of z_n^ℓ in $\{z_1^\ell, ..., z_N^\ell\}$, with $n \in \{1, ..., N\}$ and $\ell \in \{1, ..., L\}$, where we assume for convenience that there are no ties, that is, $z_n^1 \neq z_\nu^1, ..., z_n^L \neq z_\nu^L$ for $n, \nu \in \{1, ..., N\}$, $n \neq \nu$.

The empirical copula $E_N : I_N^L \to I_N$ induced by the data set $\mathbf{z}$ is then defined by

$$E_N\left(\frac{i_1}{N}, ..., \frac{i_L}{N}\right) := \frac{1}{N}\sum_{n=1}^{N} \mathbb{1}_{\left\{\text{rank}(z_n^1) \leq i_1, ..., \text{rank}(z_n^L) \leq i_L\right\}} = \frac{1}{N}\sum_{n=1}^{N}\prod_{\ell=1}^{L} \mathbb{1}_{\left\{\text{rank}(z_n^\ell) \leq i_\ell\right\}}, \tag{4.3}$$

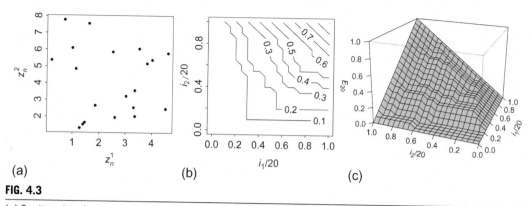

FIG. 4.3

(a) Scatter plot of a synthetic data set consisting of $N = 20$ bivariate points ($L = 2$), (b) (stabilized) contour plot of the corresponding empirical copula, and (c) perspective plot of the corresponding empirical copula.

where $0 \leq i_1, \ldots, i_L \leq N$ are integers, and $\mathbb{1}_A$ denotes the indicator function, whose value is 1 if the event A materializes, and 0 otherwise. According to the definition in Eq. (4.3), the empirical copula can be interpreted as the empirical distribution of the rank-transformed data given by $\mathbf{z}$.

Any empirical copula is a discrete copula (Schefzik, 2015), which again is a special type of copula. The properties of discrete copulas have been studied by Kolesárová, Mesiar, Mordelová, and Sempi (2006), Mayor, Suñer, and Torrens (2005, 2007), and Mesiar (2005) in the bivariate case ($L = 2$), while Schefzik (2015) has extended both the notion of discrete copulas and some results to the general multivariate case. In particular, a multivariate discrete version of the celebrated Sklar's theorem tailored to situations with empirical copulas for data without ties has been proven. Moreover, the equivalence of empirical copulas, stochastic arrays, and Latin hypercubes can be shown, see Schefzik (2015) for mathematical details.

An illustration of a bivariate (i.e., $L = 2$) empirical copula is given in Fig. 4.3, where we consider a synthetic data set $\mathbf{z}$ consisting of $N = 20$ points (z_n^1, z_n^2), $n \in \{1, \ldots, 20\}$, with z_n^1 and z_n^2 drawn form a uniform distribution on [0, 5] and [1, 8], respectively. In Fig. 4.3a, the scatter plot corresponding to $\mathbf{z}$ is exhibited. Fig. 4.3b and c shows the corresponding empirical copula induced by $\mathbf{z}$ in a (stabilized) contour plot and a perspective plot, respectively, where we note that the empirical copula is only defined on the set $\{0, 1/20, \ldots, 19/20, 1\} \times \{0, 1/20, \ldots, 19/20, 1\}$ and not on the whole interval [0, 1].

As we will see in Section 4.4, empirical copulas provide the theoretical frame for nonparametric multivariate ensemble postprocessing methods, which are essentially based on reordering notions. Empirical copulas enable the adoption of a multivariate rank order structure from a suitably chosen dependence template $\mathbf{z}$.

4.3 PARAMETRIC MULTIVARIATE APPROACHES

Developing genuinely multivariate parametric ensemble postprocessing models that account for specific types of dependence structure has gained increased interest and has become a quite active area of research. Of particular interest in applications are methods that jointly model multiple weather quantities (such as temperature, wind speed, precipitation, pressure, or components of wind vectors, as there are well-known physical relationships between such quantities), or models that incorporate dependencies between spatial locations. Such models are able to produce spatially coherent forecast fields over a region. While models accounting for temporal dependencies are quite common in climate and environmental research, there also exist some ensemble postprocessing approaches that model temporal correlations explicitly.

In this section, an overview on some prominent parametric postprocessing approaches that yield truly multivariate predictive distributions is given.

4.3.1 INTERVARIABLE DEPENDENCIES

The very first approaches to jointly model multiple weather quantities have dealt with wind vectors, describing direction and velocity of wind, encoded in the real-valued u (west/east) and v (south/north) wind vector components. These approaches are based on a bivariate normal distribution that is utilized in different ways; thus, the dimension in these examples is $L = 2$.

Pinson (2012) proposes an approach that outputs a corrected forecast ensemble instead of fitting a full predictive probability distribution based on the ensemble forecast. This allows us to retain the original multivariate structure inherent in a dynamical ensemble, while improving calibration at the same time. The approach of Pinson (2012) is an example for member-by-member postprocessing (MBMP) discussed in Section 4.4.2.

The random vector $\mathbf{Y} := (U, V)^\top$ representing the wind vector is assumed to have a bivariate normal distribution with mean vector $\boldsymbol{\mu} := (\mu_u, \mu_v)^\top$ and covariance matrix $\boldsymbol{\Sigma}$:

$$\mathbf{Y} \sim \mathcal{N}_2(\boldsymbol{\mu}, \boldsymbol{\Sigma}).$$

The individual forecast $\mathbf{x}_m := (u_m, v_m)^\top$ of ensemble member $m \in \{1, ..., M\}$ and the forecast error of the empirical ensemble mean vector $\overline{\mathbf{x}} := (\overline{u}, \overline{v})^\top$ are assumed to follow bivariate normal distributions as well. Pinson (2012) introduces models for the mean and variance of the bivariate normal distribution and obtains so-called translation (corresponds to bias correction) factors $\boldsymbol{\tau} := (\tau_u, \tau_v)^\top$ and dilation (corresponds to variance correction) factors $\boldsymbol{\xi} := (\xi_u, \xi_v)^\top$. A corrected ensemble member $\hat{\mathbf{x}}_m := (\hat{u}_m, \hat{v}_m)^\top$, $m \in \{1, ..., M\}$, is then given by

$$\hat{\mathbf{x}}_m = \mathbf{x}^* + \begin{pmatrix} \xi_u & 0 \\ 0 & \xi_v \end{pmatrix} (\mathbf{x}_m - \overline{\mathbf{x}}),$$

where $\mathbf{x}^* := \overline{\mathbf{x}} + \boldsymbol{\tau}$.

The translation and dilation factors are obtained by fitting the following linear models to the u and v components, where $\hat{\mu}_u$ and $\hat{\mu}_v$ are the corrected means, and $\hat{\sigma}_u$ and $\hat{\sigma}_v$ are the corrected standard deviations of the generating process:

$$\hat{\mu}_u = \boldsymbol{\theta}_u^\top \mathbf{q} \quad \text{and} \quad \hat{\mu}_v = \boldsymbol{\theta}_v^\top \mathbf{q}$$
$$\hat{\sigma}_u = \exp(\boldsymbol{\gamma}_u)^\top \mathbf{z}_u \quad \text{and} \quad \hat{\sigma}_v = \exp(\boldsymbol{\gamma}_v)^\top \mathbf{z}_v.$$

Here, $\mathbf{q} := (1, \overline{u}, \overline{v})^\top$, $\mathbf{z}_u := (1, s_u)^\top$, and $\mathbf{z}_v := (1, s_v)^\top$, with s_u and s_v denoting the respective empirical ensemble standard deviations, and the parameter vectors $\boldsymbol{\theta}_u$, $\boldsymbol{\theta}_v$, $\boldsymbol{\gamma}_u$, and $\boldsymbol{\gamma}_v$ are estimated using a recursive maximum-likelihood approach with exponential forgetting of past observations.

Schuhen et al. (2012) modify and extend the ensemble member correction approach of Pinson (2012) by developing a bivariate NR model for wind vectors, yielding a full bivariate normal predictive probability distribution. A further extension compared with the method of Pinson (2012) is the explicit modeling of the correlation ρ_{uv} between the u and v components. Schuhen et al. (2012) fit models for the mean and the variance of the u and v components, respectively, as well. However, they do not employ these models to obtain correction factors applied to the ensemble members themselves, but use the estimated mean vector and covariance matrix as plug-in estimates for the parameters of the bivariate normal distribution. As in the univariate NR approach, the means μ_u and μ_v are modeled as linear functions of the empirical ensemble mean values $\bar{u}$ and $\bar{v}$, respectively:

$$\mu_u = a_u + b_u\bar{u}$$
$$\mu_v = a_v + b_v\bar{v},$$

with bias correction parameters a_u, a_v, b_u, and b_v. Equivalently, the variances σ_u^2 and σ_v^2 are modeled as linear functions of the empirical ensemble variances s_u^2 and s_v^2, respectively:

$$\sigma_u^2 = c_u + d_u s_u^2$$
$$\sigma_v^2 = c_v + d_v s_v^2,$$

where the variance parameters c_u, c_v, d_u, and d_v are constrained to be nonnegative. In addition, the correlation ρ_{uv} is explicitly modeled using a trigonometric function of the ensemble mean wind direction θ:

$$\rho_{uv} = r\cos\left(\frac{2\pi}{360}(k\theta + \phi)\right) + q,$$

where r, k, ϕ, and q are additional parameters that need to be estimated.

While Schuhen et al. (2012) develop an NR model based on the bivariate normal distribution to postprocess wind vector forecasts, Sloughter et al. (2013) propose a quite similar approach, but based on a bivariate BMA model. Analogous to the case of univariate BMA, the mean $\boldsymbol{\mu}_m$ of the bivariate normal distribution associated with an individual ensemble member $m \in \{1, \ldots, M\}$ is assumed to be a linear function of the ensemble member forecast vector:

$$\boldsymbol{\mu}_m = \mathbf{a}_m + \mathbf{B}_m\mathbf{x}_m,$$

where $\mathbf{a}_m \in \mathbb{R}^2$ and $\mathbf{B}_m \in \mathbb{R}^{2\times 2}$, which is a (2 × 2)-matrix with real-valued entries, are model coefficients to be estimated, and $\mathbf{x}_m$ is the wind vector forecast of ensemble member m. That is, the proposed model assumes

$$\mathbf{Y}|\mathbf{x}_m \sim \mathcal{N}_2(\boldsymbol{\mu}_m, \boldsymbol{\Sigma}),$$

with some covariance matrix $\boldsymbol{\Sigma}$, which is assumed to be the same for all ensemble members.

Equivalently, for the forecast error $\mathbf{e}_m(\mathbf{y}) := \mathbf{y} - \boldsymbol{\mu}_m$ of each ensemble member m, with $\mathbf{y}$ denoting the corresponding observation vector, it holds that

$$\mathbf{e}_m(\mathbf{y}) \sim \mathcal{N}_2(\mathbf{0}, \boldsymbol{\Sigma}). \tag{4.4}$$

To account for slightly heavier tails in the forecast errors $\mathbf{e}_m$ in their specific data, which cannot be captured by a bivariate normal distribution, Sloughter et al. (2013) consider a power-transformed forecast error vector using polar coordinates (θ_m, r_m) and transforming the r_m:

$$\mathbf{e}_m^*(\mathbf{y}) = \left(r_m^{4/5}\cos(\theta_m), r_m^{4/5}\sin(\theta_m)\right)^\top.$$

They then assume the bivariate normal distribution for the transformed error vector:

$$\mathbf{e}_m^*(\mathbf{y}) \sim \mathcal{N}_2(\mathbf{0}, \mathbf{\Sigma}). \tag{4.5}$$

Sloughter et al. (2013) state that different transformations of the forecast error vector, if any, might be more appropriate for other data.

The final BMA predictive probability density function (PDF)

$$p(\mathbf{y}|\mathbf{x}_1, \dots, \mathbf{x}_M) := \sum_{m=1}^{M} w_m\, g_m(\mathbf{y}|\mathbf{x}_m)$$

with weights w_m is then given analogously to the univariate BMA predictive PDF, but with the kernel densities $g_m(\mathbf{y}|\mathbf{x}_m)$ of each individual member m defined by the bivariate densities of $\mathbf{y}$ implied by (4.4) or (4.5), respectively.

When multiple weather quantities of arbitrary type with marginal distributions potentially different from being normal are to be modeled jointly, the postprocessing model needs to be more flexible than in the case where all margins are normally distributed and the joint distribution can thus be assumed to be multivariate normal. In this regard, Möller et al. (2013) propose a flexible approach that is also applicable in slightly higher-dimensional settings. As the univariate distributions being suitable to describe the individual weather quantities can be of quite different nature, as for instance in the case of temperature and precipitation, joint parametric modeling can be a difficult task. Möller et al. (2013) employ a Gaussian copula to fit the joint distribution of temperature, pressure, wind speed, and precipitation. This approach is highly flexible, as the margins associated with the individual weather quantities are allowed to be of any desired type. Further, the Gaussian copula approach has the advantage that it only requires the modeling of the marginal distributions and the joint correlation structure, which can simply be fitted in separate steps, by any method of choice.

Möller et al. (2013) propose using appropriate univariate ensemble postprocessing models to fit the individual margins. In their case study, they apply BMA models to each of the considered quantities. However, the approach is not limited to the use of BMA; any postprocessing model of choice or any other type of model can be used to obtain estimates of the margins. For instance, Baran and Möller (2017) use NR models to fit the individual margins in the Gaussian copula approach.

For L random variables $Y_1, \dots, Y_L$, representing the weather quantities of interest and having marginal CDFs $F_1, \dots, F_L$, the joint predictive distribution F obtained by assuming a Gaussian copula takes the form displayed in Eq. (4.2). The CDFs $F_1, \dots, F_L$ can be fitted individually with any postprocessing model of choice, as for example BMA or NR.

At this point, the advantage of this approach should be highlighted. The Gaussian copula models the joint correlation structure *after* postprocessing the margins, thus capturing any residual correlation remaining after the individual postprocessing.

The analytic structure of the joint distribution may not be simple and not be available in closed form. However, the joint distribution F can be described by obtaining a large number of samples $\hat{\mathbf{Y}}$ from F. To obtain such samples, a link between a sample $\hat{\mathbf{Y}}$ from the joint distribution F and a latent L-dimensional Gaussian variable

$$\mathbf{Z} := (Z_1, \dots, Z_L)^\top \sim \mathcal{N}_L(\mathbf{0}, \mathbf{\Gamma})$$

with correlation matrix $\boldsymbol{\Gamma}$ can be constructed. When defining

$$Y_\ell := F_\ell^{-1}(\Phi(Z_\ell))$$

for each $\ell \in \{1, ..., L\}$, with F_ℓ^{-1} denoting the quantile function of the respective marginal CDF F_ℓ, then

$$\mathbf{Y} := (Y_1, ..., Y_L)^\top \sim F.$$

This relationship also highlights that each individual random variable Y_ℓ is marginally distributed according to F_ℓ. For fully continuous marginal distributions F_ℓ, it follows directly that $Z_\ell = \Phi^{-1}(F_\ell(Y_\ell))$. Thus, a latent observation z_ℓ can be obtained in a straightforward way from an observation y_ℓ and the respective CDF F_ℓ. For marginal distributions F_ℓ that are not fully continuous, the definition of the latent variable Z_ℓ needs to be refined according to the needs of the underlying distribution. For details in case of precipitation, see Möller et al. (2013). The construction described herein allows a sample $\hat{\mathbf{Y}} := (\hat{Y}_1, ..., \hat{Y}_L)^\top$ to be obtained from F by first sampling

$$\mathbf{Z} \sim \mathcal{N}_L(\mathbf{0}, \hat{\boldsymbol{\Gamma}})$$

and then setting

$$\hat{Y}_\ell := \hat{F}_\ell^{-1}(\Phi(Z_\ell)),$$

where $\hat{\boldsymbol{\Gamma}}$ denotes the estimated correlation matrix, and $\hat{F}_\ell$ is the estimated marginal distribution of random variable Y_ℓ, which can be obtained by fitting any appropriate univariate postprocessing model of choice to the respective data. Möller et al. (2013) assume a correlation matrix that is constant over time. This allows estimation of $\boldsymbol{\Gamma}$ (e.g., simply as the sample correlation matrix) from latent Gaussian variables $\mathbf{Z}$ based on past observations of the random variables $Y_1, ..., Y_L$. The assumption of a constant correlation matrix can be relaxed, but at the cost of additional computational burden.

As a special case of joint modeling of multiple weather quantities, Baran and Möller (2015) and Baran and Möller (2017) consider the joint postprocessing of wind speed and temperature (i.e., again an $L = 2$-dimensional setting). Let the random vector $\mathbf{Y} := (Y_W, Y_T)^\top$ contain the random variables representing wind speed (W) and temperature (T), respectively. In Baran and Möller (2015) and Baran and Möller (2017), wind speed is modeled using a normal distribution truncated from below at zero, while for temperature the classical normal distribution is employed. A bivariate truncated normal distribution, denoted by $\mathcal{N}_2^0(\boldsymbol{\kappa}, \boldsymbol{\Theta})$, is introduced, which has the PDF

$$g(\mathbf{y}|\boldsymbol{\kappa}, \boldsymbol{\Theta}) := \frac{(\det(\boldsymbol{\Theta}))^{-1/2}}{2\pi\Phi(\kappa_W/\vartheta_W)} \exp\left(-\frac{1}{2}(\mathbf{y}-\boldsymbol{\kappa})^\top \boldsymbol{\Theta}^{-1}(\mathbf{y}-\boldsymbol{\kappa})\right) \mathbb{1}_{\{y_W \geq 0\}} \tag{4.6}$$

where $\mathbf{y} := (y_W, y_T)^\top$, $\boldsymbol{\kappa} := (\kappa_W, \kappa_T)^\top$ is a location vector, and

$$\boldsymbol{\Theta} := \begin{pmatrix} \vartheta_W^2 & \vartheta_{WT} \\ \vartheta_{WT} & \vartheta_T^2 \end{pmatrix}$$

a scale matrix. For the relationship of $\boldsymbol{\kappa}$ and $\boldsymbol{\Theta}$ to the mean vector $\boldsymbol{\mu}$ and the covariance matrix $\boldsymbol{\Sigma}$ of the distribution, see Baran and Möller (2015). While Baran and Möller (2015) propose a bivariate BMA model based on the truncated normal distribution, Baran and Möller (2017) develop a respective bivariate NR version.

In line with the corresponding univariate BMA models for temperature and wind speed, Baran and Möller (2015) assume that the location vector $\boldsymbol{\kappa}_m := \left(\kappa_m^W, \kappa_m^T\right)^\top$ of the BMA mixture component $m \in \{1, \ldots, M\}$ is a linear function of the respective ensemble member forecast vector $\mathbf{x}_m := \left(x_m^W, x_m^T\right)^\top$, and that the scale matrix $\boldsymbol{\Theta}$ is equal for all mixture components m. The assumption of a common scale parameter was already used in the univariate BMA variants and in the bivariate BMA approach of Sloughter et al. (2013), as it reduces the number of parameters. Then, the bivariate predictive BMA PDF is given by

$$p(\mathbf{y} \,|\, \mathbf{x}_1, \ldots, \mathbf{x}_M) := \sum_{m=1}^{M} w_m g_m(\mathbf{y} \,|\, \mathbf{a}_m + \mathbf{B}_m \mathbf{x}_m, \boldsymbol{\Theta}),$$

where w_m are weights, g_m is the PDF defined by Eq. (4.6) associated with member m, and $\mathbf{a}_m \in \mathbb{R}^2$ and $\mathbf{B}_m \in \mathbb{R}^{2 \times 2}$ are model coefficients to be estimated.

An even more parsimonious model is given by additionally assuming the same model coefficients $\mathbf{a}_m$ and $\mathbf{B}_m$, respectively, for each component m in the BMA mixture, that is, $\mathbf{a}_m = \mathbf{a}$ and $\mathbf{B}_m = \mathbf{B}$ for all m. The respective bivariate predictive BMA PDF then simplifies to

$$q(\mathbf{y} \,|\, \mathbf{x}_1, \ldots, \mathbf{x}_M) := \sum_{m=1}^{M} w_m g_m(\mathbf{y} \,|\, \mathbf{a} + \mathbf{B} \mathbf{x}_m, \boldsymbol{\Theta}).$$

In contrast, the bivariate NR version for wind speed and temperature developed by Baran and Möller (2017) assumes the bivariate predictive distribution

$$\mathbf{Y} | \mathbf{x}_1, \ldots, \mathbf{x}_M \sim \mathcal{N}_2^0 \left(\mathbf{a} + \mathbf{B}_1 \mathbf{x}_1 + \cdots + \mathbf{B}_M \mathbf{x}_M, \mathbf{C} + \mathbf{D} \mathbf{S} \mathbf{D}^\top \right)$$

with the ensemble covariance matrix

$$\mathbf{S} := \frac{1}{M-1} \sum_{m=1}^{M} (\mathbf{x}_m - \bar{\mathbf{x}})(\mathbf{x}_m - \bar{\mathbf{x}})^\top.$$

Here, $\bar{\mathbf{x}} := \left(\bar{x}_W, \bar{x}_T\right)^\top$ denotes the empirical ensemble mean vector, and $\mathbf{a} \in \mathbb{R}^2$ and $\mathbf{B}_1, \ldots, \mathbf{B}_M, \mathbf{C}, \mathbf{D} \in \mathbb{R}^{2 \times 2}$ are model coefficients to be estimated, where $\mathbf{C}$ is assumed to be symmetric and nonnegative definite.

4.3.2 SPATIAL DEPENDENCIES

Many applications call for spatially coherent forecast fields over a region of interest, instead of considering locations individually and ignoring possible spatial correlations. Thus, it has become of increased interest to develop postprocessing models that estimate the model parameters in a spatially adaptive way, incorporating dependencies between (neighboring) locations by using, for example, spatial or geostatistical models.

Some of these approaches formally yield univariate models, which do not set up explicit multivariate predictive distributions (as in the intervariable cases described in Section 4.3.1), but fit the parameters in a specific way. Such methods will be discussed in Section 4.5.1. Other approaches are able to define and fit a truly multivariate predictive distribution and are discussed in this section.

In the following, let $\{1, \ldots, L\}$ be a finite set of indices, where each index $\ell \in \{1, \ldots, L\}$ refers to a distinct model grid point or observation site $\mathbf{s} \in D \subset \mathbb{R}^2$ in a spatial region D.

The first approach that aims at obtaining spatial forecast fields in a computationally feasible way was introduced by Gel et al. (2004). This approach was named the geostatistical perturbation output (GOP) method, as it applies a geostatistical model to forecast errors. The motivation was given by the fact that smaller weather centers not having the computational capacities to run full ensemble models still wish to obtain probabilistic forecasts from their deterministic model forecast. Gel et al. (2004) propose to obtain an ensemble afterwards by adding perturbations to the deterministic model output. These perturbations incorporate spatial dependencies through a geostatistical model, thus leading to an ensemble of spatially consistent forecast fields. However, in contrast to ensembles obtained from a dynamical forecast model, the GOP ensemble is obtained by perturbing the output of a dynamical model, not the input, which is the more typical approach to generate an ensemble.

The goal is to predict a spatial field $\mathbf{Y} := \{Y(\mathbf{s}): \mathbf{s} \in D\}$ at all locations $\mathbf{s} \in D \subset \mathbb{R}^2$ simultaneously for a fixed verification time point and forecast horizon by using a single deterministic forecast field $\mathbf{x} := \{x(\mathbf{s}): \mathbf{s} \in D\}$. The random variable $Y(\mathbf{s})$ describes the weather quantity of interest, for instance temperature, while $x(\mathbf{s})$ is the respective deterministic forecast for $Y(\mathbf{s})$ at location $\mathbf{s} \in D$, obtained for example from a dynamical model.

A parsimonious version of the general model introduced by Gel et al. (2004) uses simple additive and multiplicative bias correction parameters a and b, resulting in the model

$$Y(\mathbf{s}) = a + b\,x(\mathbf{s}) + e(\mathbf{s}), \tag{4.7}$$

where $\mathbf{s} \in D$, and $e(\mathbf{s})$ is a mean-zero stationary Gaussian process.

To model spatial correlation, an exponential variogram model (Cressie, 1993) is assumed:

$$\frac{1}{2}\,\mathrm{Var}(e(\mathbf{s}_1) - e(\mathbf{s}_2)) = \rho^2 + \tau^2(1 - \exp(-\|\mathbf{s}_1 - \mathbf{s}_2\|/r)), \tag{4.8}$$

where $\mathbf{s}_1, \mathbf{s}_2 \in D$, $\mathbf{s}_1 \neq \mathbf{s}_2$, and $\|\cdot\|$ denotes the Euclidean norm. The parameter ρ^2 can be interpreted as the variance of the measurement error of the observations, and $\rho^2 + \tau^2$ as the marginal variance of the process $e(\mathbf{s})$. The parameter r is a range parameter controlling the rate at which the spatial correlation of the error process decays.

The model specified by Eqs. (4.7) and (4.8) defines a fully multivariate predictive distribution for $\mathbf{Y}$, for example a multivariate normal distribution in the case of temperature.

After having estimated the model parameters, an ensemble of forecast fields can be generated by simulating realizations of the spatial process defined by the models (4.7) and (4.8), given the current deterministic forecast $x(\mathbf{s})$.

While the GOP method incorporates spatial dependencies via a geostatistical model, it utilizes only a single deterministic forecast, instead of a full forecast ensemble. Thus, GOP ignores the flow-dependent information hidden in the forecast ensemble. On the contrary, univariate postprocessing methods such as NR and BMA utilize the complete forecast ensemble. However, these approaches estimate location-specific postprocessed predictive distributions and do not incorporate spatial correlation. To combine the advantages of both approaches, Berrocal et al. (2007) introduce a spatial BMA model for normally distributed quantities such as temperature and extend it to precipitation forecasts in Berrocal et al. (2008).

Similar to the GOP model, let $\mathbf{Y} := \{Y(\mathbf{s}): \mathbf{s} \in D\}$ denote the spatial field of the weather quantity of interest at all locations $\mathbf{s} \in D$. In spatial BMA, not only a single deterministic forecast field

$\mathbf{x} := \{x(\mathbf{s}): \mathbf{s} \in D\}$ is considered, but an ensemble of M forecast fields $\mathbf{x}_1 := \{x_1(\mathbf{s}): \mathbf{s} \in D\}$, ..., $\mathbf{x}_M := \{x_M(\mathbf{s}): \mathbf{s} \in D\}$. Thus, the GOP method considers the special case of $M = 1$.

The predictive distribution of the field $\mathbf{Y}$ is modeled by the multivariate BMA PDF

$$p(\mathbf{y}|\mathbf{x}_1, ..., \mathbf{x}_M) := \sum_{m=1}^{M} w_m \, g_m(\mathbf{y}|\mathbf{x}_m),$$

with weights w_m. Dealing with temperature here, multivariate normal kernel densities g_m are assumed, where each g_m, $m \in \{1, ..., M\}$, is centered at $a_m\mathbf{1} + b_m \mathbf{x}_m$, with a_m and b_m being parameters and $\mathbf{1}$ denoting the vector of ones with length L. The corresponding predictive multivariate distribution of $\mathbf{Y}$ given $\mathbf{x}_m$ then reads

$$\mathbf{Y}|\mathbf{x}_m \sim \mathcal{N}_L(a_m\mathbf{1} + b_m \mathbf{x}_m, \mathbf{\Sigma}_m^*), \tag{4.9}$$

where $\mathbf{\Sigma}_m^*$ is a member-specific spatially structured covariance matrix given by

$$\mathbf{\Sigma}_m^* := \frac{\sigma^2}{\rho_m^2 + \tau_m^2} \mathbf{\Sigma}_m$$

for each ensemble member m. Here, σ^2 is the variance of the univariate BMA model, and $\mathbf{\Sigma}_m$ is the covariance matrix obtained when applying GOP to member m. The deflation factor $\sigma^2/(\rho_m^2 + \tau_m^2)$ describes the ratio of the BMA variance to the GOP variance of member m for the errors.

Parameter estimation is conducted in two steps. In the first step, the univariate BMA model is fitted. Given the BMA estimates, the GOP parameters are estimated by fitting a GOP model to each member individually.

A spatial BMA ensemble of arbitrary size can then be generated by randomly sampling from (4.9).

Feldmann, Scheuerer, and Thorarinsdottir (2015) introduce a spatial NR approach for normally distributed quantities, where they combine NR models with the GOP method in a similar way as for spatial BMA. Thus, in case of using the original NR model, the multivariate predictive distribution of $\mathbf{Y}$ is defined conditionally on all M ensemble members as

$$\mathbf{Y}|\mathbf{x}_1, ..., \mathbf{x}_M \sim \mathcal{N}_L(a\mathbf{1} + b_1 \mathbf{x}_1 + \cdots, b_M \mathbf{x}_M, \mathbf{\Sigma}^*).$$

In the spatial NR model, $a, b_1, ..., b_M$ are parameters to be estimated, and $\mathbf{\Sigma}^*$ is defined as $\mathbf{\Sigma}^* := \mathbf{D\Gamma D}$, where $\mathbf{D}$ is a diagonal matrix with the univariate NR predictive standard deviations on the diagonal, and $\mathbf{\Gamma}$ is the correlation matrix corresponding to the GOP covariance matrix $\mathbf{\Sigma}$.

Both the GOP method (Gel et al., 2004) and the spatial NR technique (Feldmann et al., 2015) are technically Gaussian copula approaches, where additionally a parametric correlation model is assumed.

4.3.3 TEMPORAL DEPENDENCIES

While modeling intervariable and spatial dependencies by means of truly multivariate ensemble postprocessing methods has become quite prominent, the consideration of temporal dependencies between different lead times has gained some interest as well.

Pinson et al. (2009) fit a multivariate normal distribution to forecast errors of wind power forecasts at multiple look-ahead times, which corresponds to a Gaussian copula model. However, dependence structures besides multivariate normal are possible and can be accounted for by using other copula models.

Long-term variations that the dependence structure of the prediction errors may exhibit are accounted for by recursive estimation of the covariance matrix of the multivariate distribution.

Schoelzel and Hense (2011) introduce a multivariate Gaussian kernel dressing approach to fit a bivariate predictive PDF for temperature changes in South Germany. The bivariate predictive PDF jointly models temporal averages and trends of an ensemble of regional climate simulations.

Hemri et al. (2015) combine univariate NR models fitted to runoff ensemble forecasts at each lead time individually within a Gaussian copula model to obtain a fully multivariate predictive distribution that incorporates dependencies between the lead limes, thus ensuring temporally coherent predictions.

4.4 NONPARAMETRIC MULTIVARIATE APPROACHES

Parametric multivariate ensemble postprocessing methods to model dependencies as discussed in Section 4.3 are mainly suitable in low-dimensional settings and typically address either intervariable or spatial or temporal correlations.

In high-dimensional or more general, nonspecific settings it is not clear how to set up and estimate a reasonable parametric model, thus it can be more appropriate to make use of nonparametric multivariate postprocessing methods. These are based on the use of empirical copulas and can deal with intervariable, spatial, and temporal dependencies simultaneously.

In this section, we first introduce the general frame of nonparametric, empirical copula-based multivariate ensemble postprocessing. Then, we explicitly review two representatives, namely ensemble copula coupling (ECC) (Schefzik et al., 2013) and approaches based on the Schaake shuffle (Clark et al., 2004; Schefzik, 2016b).

4.4.1 EMPIRICAL COPULA-BASED ENSEMBLE POSTPROCESSING

In the following, let $\ell := (i, j, k)$ be a multiindex summarizing a weather variable $i \in \{1, ..., I\}$, a location $j \in \{1, ..., J\}$ and a lead time or time point $k \in \{1, ..., K\}$, and let $L := I \times J \times K$. Further, let M denote the number of raw ensemble members, and N the desired number of members the postprocessed ensemble shall comprise.

Then, multivariate empirical copula-based ensemble methods to postprocess a raw ensemble forecast

$$\mathbf{x} := \{(x_1^1, ..., x_M^1), ..., (x_1^L, ..., x_M^L)\}$$

proceed according to the following steps (Schefzik, 2016b).

1. **Specification of the dependence template** To model dependence structures, derive an empirical copula E_N via Eq. (4.3) from a suitably chosen data set

$$\mathbf{z} := \{(z_1^1, ..., z_N^1), ..., (z_1^L, ..., z_N^L)\}$$

serving as a dependence template.

Equivalently, compute the univariate order statistics $z_{(1)}^\ell \leq \cdots \leq z_{(N)}^\ell$ for $\ell \in \{1, ..., L\}$, which induce the permutation $\pi_\ell(n) := \mathrm{rank}(z_n^\ell)$ for $n \in \{1, ..., N\}$, with ties resolved at random. Apart from randomization, which is a natural approach in the case of ties, other allocation schemes are possible and do not pose technical challenges. Regardless of the allocation method, Eq. (4.3) continues to apply.

While in principle any template **z** could be used to determine a dependence structure, **z** has actually to be chosen carefully, because an improvident specification might lead to physically incoherent forecasts in the following procedure. To ensure physical consistency, an adequate dependence template **z** should represent actual intervariable, spatial, and temporal correlations as accurately as possible. As will be discussed later, plausible dependence templates may, for instance, rely on the raw ensemble forecast, as in the ECC approach (Schefzik et al., 2013), or on historical verifying observations, as in Schaake shuffle-based approaches (Clark et al., 2004; Schefzik, 2016b).

2. **Univariate postprocessing** For each margin ℓ, apply univariate postprocessing to the raw ensemble forecast $x_1^\ell, \ldots, x_M^\ell$ and obtain a corresponding postprocessed predictive CDF F_ℓ. In principle, any univariate ensemble postprocessing can be employed, with BMA and NR being prominent choices.

3. **Quantization/Sampling** Draw a sample $\tilde{x}_1^\ell, \ldots, \tilde{x}_N^\ell$ of size N from each marginal CDF F_ℓ. Conveniently, the equally spaced quantiles

$$\tilde{x}_1^\ell := F_\ell^{-1}\left(\frac{1}{N+1}\right), \ldots, \tilde{x}_N^\ell := F_\ell^{-1}\left(\frac{N}{N+1}\right) \tag{4.10}$$

of F_ℓ can be used as a sample. An alternative option is to take a random sample of the form

$$\tilde{x}_1^\ell := F_\ell^{-1}(u_1), \ldots, \tilde{x}_N^\ell := F_\ell^{-1}(u_N), \tag{4.11}$$

with $u_1, \ldots, u_N$ being independent standard uniform random variates.

The quantized values may be ordered, as in the case of (4.10), or may not be ordered, as for (4.11).

The use of equidistant quantiles as samples appears to be somewhat natural, with Bröcker (2012) providing theoretical support in favor of the particular choice of (4.10), which maintains the calibration of the univariate ensemble forecasts. Another choice would be, for instance, to take

$$\tilde{x}_1^\ell := F_\ell^{-1}\left(\frac{\frac{1}{2}}{N}\right), \tilde{x}_2^\ell := F_\ell^{-1}\left(\frac{\frac{3}{2}}{N}\right), \ldots, \tilde{x}_N^\ell := F_\ell^{-1}\left(\frac{N-\frac{1}{2}}{N}\right)$$

which fails to maintain calibration to some extent, but is optimal in terms of some specific criteria (Bröcker, 2012; Graf & Luschgy, 2000).

Alternatively, Hu et al. (2016) propose the use of a stratified sampling approach, in which the interval $(0, 1]$ is first partitioned into N disjoint intervals $(0, \frac{1}{N}], (\frac{1}{N}, \frac{2}{N}], \ldots, (\frac{N-1}{N}, 1]$. Then, a sample $\tilde{x}_1^\ell, \ldots, \tilde{x}_N^\ell$ from F_ℓ is obtained by setting

$$\tilde{x}_1^\ell := F_\ell^{-1}(v_1), \ldots, \tilde{x}_N^\ell := F_\ell^{-1}(v_N), \tag{4.12}$$

where v_n is a random number drawn from a uniform distribution on the interval $(\frac{n-1}{N}, \frac{n}{N}]$ for $n \in \{1, \ldots, N\}$.

4. **Reordering according to the dependence template** Apply the empirical copula E_N determined in the first step to the samples from the quantization step.

Equivalently, arrange these samples in terms of the rank-order structure of the dependence template **z** as specified in the first step. Hence, the final empirical copula-based postprocessed ensemble $\hat{x}_1^\ell, \ldots, \hat{x}_N^\ell$ for each margin ℓ is given by

$$\hat{x}_1^\ell := \tilde{x}_{(\pi_\ell(1))}^\ell, \ldots, \hat{x}_N^\ell := \tilde{x}_{(\pi_\ell(N))}^\ell. \tag{4.13}$$

While the permutation π_ℓ is determined via the dependence template $\mathbf{z}$, the procedure in (4.13) applies this permutation to the postprocessed and quantized forecasts from the third step.

As the critical reordering step (4.13) is computationally cheap, empirical copula-based postprocessing methods virtually come for free, once the univariate postprocessing has been performed. Empirical copula-based postprocessing is conceptually simple and intuitive and does not require complex modeling or sophisticated parameter fitting, hence offering a natural benchmark. The concept is general and may be applied (with adaptations if necessary) to broader settings apart from weather prediction.

As pointed out, the reordering step (4.13) is crucial, because it transfers the intervariable, spatial, and temporal rank dependence pattern of the template $\mathbf{z}$ to the postprocessed ensemble. If we stopped after the quantization/sampling stage, we would get a postprocessed ensemble that does not necessarily retain dependence structures. In the following, an ensemble of that type will be called an individually postprocessed ensemble.

For a more technical description of the preceding scheme, let $Z_1, \ldots, Z_L$ be discrete random variables that take values in $\{z_1^1,\ldots,z_N^1\},\ldots,\{z_1^L,\ldots,z_N^L\}$, respectively, according to the dependence template $\mathbf{z}$. Assuming for simplicity that there are no ties among the corresponding margins, the marginal empirical CDFs $H_1, \ldots, H_L$ of the multivariate random vector $\mathbf{Z} := (Z_1, \ldots, Z_L)^\top$ take values in $I_N = \{0, \frac{1}{N},\ldots, \frac{N-1}{N},1\}$. The multivariate empirical CDF $H : \mathbb{R}^L \to I_N$ of $\mathbf{Z}$ also maps into I_N. According to the multivariate discrete variant of Sklar's theorem tailored to such a framework (Schefzik, 2015), there exists a uniquely determined empirical copula $E_N : I_N^L \to I_N$ such that

$$H(y_1,\ldots,y_L) = E_N(H_1(y_1),\ldots,H_L(y_L)) \tag{4.14}$$

for $y_1,\ldots,y_L \in \mathbb{R}$. Conversely, if E_N is taken to be the empirical copula induced by the dependence template $\mathbf{z} = \{(z_1^1,\ldots,z_N^1),\ldots,(z_1^L,\ldots,z_N^L)\}$, and $H_1, \ldots, H_L$ are the univariate empirical CDFs of the corresponding margins, then H as defined in Eq. (4.14) is a multivariate empirical CDF.

The preceding considerations apply analogously to an individually postprocessed ensemble

$$\tilde{\mathbf{x}} := \{(\tilde{x}_1^1,\ldots,\tilde{x}_N^1),\ldots,(\tilde{x}_1^L,\ldots,\tilde{x}_N^L)\}$$

and the empirical copula-based postprocessed ensemble

$$\hat{\mathbf{x}} := \{(\hat{x}_1^1,\ldots,\hat{x}_N^1),\ldots,(\hat{x}_1^L,\ldots,\hat{x}_N^L)\}$$

according to (4.13), respectively. Let $\tilde{F}$ and $\hat{F}$ denote the multivariate empirical CDFs induced by $\tilde{\mathbf{x}}$ and $\hat{\mathbf{x}}$, respectively, and $\tilde{F}_1,\ldots,\tilde{F}_L$ the marginal empirical CDFs of the individually postprocessed ensemble. Moreover, let $\tilde{E}_N$ denote the empirical copula induced by the individually postprocessed ensemble $\tilde{\mathbf{x}}$, in contrast to the empirical copula E_N induced by the dependence template $\mathbf{z}$. Similarly as before,

$$\tilde{F}(y_1,\ldots,y_L) = \tilde{E}_N(\tilde{F}_1(y_1),\ldots,\tilde{F}_L(y_L)) \tag{4.15}$$

and

$$\hat{F}(y_1,\ldots,y_L) = E_N(\tilde{F}_1(y_1),\ldots,\tilde{F}_L(y_L)) \tag{4.16}$$

for $y_1,\ldots,y_L \in \mathbb{R}$.

As revealed by Eqs. (4.14)–(4.16), the individually postprocessed ensemble and the empirical copula-based postprocessed ensemble have the same marginal distributions, while the dependence template and the empirical copula-based postprocessed ensemble share the empirical copula. The empirical copula-based postprocessed ensemble conserves the multivariate rank dependence pattern as well as the pairwise Spearman rank correlation coefficients in the dependence template. In particular, two-dimensional scatter plots of the dependence template and the respective empirical copula-based postprocessed ensemble, see for instance Fig. 4.4, both exhibit the same Spearman correlation.

In a nutshell, empirical copula-based ensemble postprocessing methods come up with an empirical L-dimensional distribution, according to Eq. (4.16) constructed from univariate empirical CDFs $\tilde{F}_1, ..., \tilde{F}_L$ and an empirical copula E_N. While the CDFs $\tilde{F}_1, ..., \tilde{F}_L$ are determined by the samples drawn from the respective predictive CDFs $F_1, ..., F_L$ obtained by univariate postprocessing, E_N is induced by the suitably chosen dependence template $\mathbf{z}$.

An extension of the nonparametric, reordering-based postprocessing scheme presented before is given by the approach of Schefzik (2016a). In this method, a high-dimensional ensemble postprocessing problem is partitioned into multiple low-dimensional building blocks. Each of these blocks is postprocessed via a suitable parametric multivariate approach, with potentially useful techniques having been discussed in Section 4.3. From each postprocessed low-dimensional distribution, a (multivariate) sample is drawn. Subsequently, each sample is rearranged according to the multidimensional rank structure of an adequately chosen dependence template, where an appropriate ranking concept in

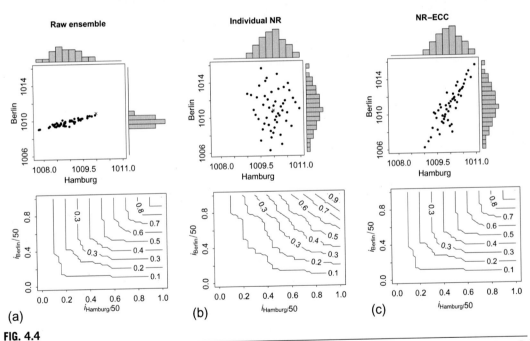

FIG. 4.4

Illustration of ECC using 24 hour-ahead pressure forecasts (in hectopascals) at Berlin and Hamburg, valid June 8, 2010, 0000 UTC. Using (a) the ECMWF raw ensemble as a dependence template and (b) NR for univariate postprocessing of the raw ensemble at each station individually leads to (c) the NR-ECC postprocessed ensemble. *First row*: scatter plots, where ensemble forecasts are indicated by the dots, and the materializing observation by the cross; *second row*: (stabilized) contour plots of the empirical copulas induced by the corresponding ensembles.

multivariate settings (Gneiting, Stanberry, Grimit, Held, & Johnson, 2008; Thorarinsdottir, Scheuerer, & Heinz, 2016) is required. Lastly, all such reordered samples are aggregated to obtain the final postprocessed ensemble. Note that in the approach of Schefzik (2016a), the samples that get reordered may originate from multivariate, and not necessarily univariate, distributions. This is the main difference to the empirical copula-based postprocessing discussed before, to which the procedure of Schefzik (2016a) reduces in the special case in which all chosen parametric postprocessing methods are univariate.

In what follows, we will stick to empirical copula-based ensemble postprocessing techniques as extensively introduced before and will explicitly discuss ECC and Schaake shuffle-based approaches as reference examples.

4.4.2 ENSEMBLE COPULA COUPLING (ECC)

The ECC approach (Schefzik et al., 2013), originally hinted at by Bremnes (2007) and Krzysztofowicz and Toth (2008), is a particularly attractive example of an empirical copula-based ensemble postprocessing method. It makes use of the rank order information given by the raw ensemble forecast and relies on the implicit assumption that the ensemble members are exchangeable, that is, statistically indistinguishable. Moreover, it is assumed that the raw ensemble is able to represent observed intervariable, spatial, and temporal dependence structures reasonably well. This may be expected, because dynamical forecast models discretize the equations governing the physics of the atmosphere. However, it is recommended to check empirically whether the dependence patterns in an ensemble forecast are consistent with historical observations.

The main feature of the ECC postprocessed ensemble is the conservation of the multivariate rank dependence structure within the raw ensemble. Referring to the scheme in Section 4.4.1, ECC uses the raw ensemble forecast as a dependence template (i.e., $\mathbf{z} = \mathbf{x}$). Hence, the ECC postprocessed ensemble is restricted to have the same size as the raw ensemble (i.e., $N = M$). As ECC adopts the empirical copula of the (raw) ensemble, thereby restoring its rank dependence structure, the term "ensemble copula coupling" is indeed justified.

Fig. 4.4 illustrates ECC in a real-data example, in which we consider forecasts of the $M = 50$-member ECMWF ensemble. We focus on a bivariate setting (i.e., $L = 2$) dealing with 24 hour-ahead forecasts for pressure at Berlin and Hamburg, valid 0000 UTC on June 8, 2010. Scatter plots of (a) the raw ensemble forecast, (b) an individually postprocessed ensemble, where the univariate postprocessing for each location separately has been performed via NR (Gneiting et al., 2005), and (c) the postprocessed NR ensemble with ECC on top (NR-ECC) are shown in the first row of Fig. 4.4. In this context, each dot represents an ensemble member forecast, and the verifying observation is indicated by the cross. In the second row of Fig. 4.4, the corresponding (stabilized) contour plots of the empirical copulas linked to the three different ensembles are shown. The raw ensemble members in (a) reveal a pronounced positive spatial correlation, which appears to be reasonable. However, the verifying observations lie outside of the ensemble ranges. While the individually site-by-site NR postprocessed ensemble in (b) corrects biases and dispersion errors, it exhibits no correlation structure, in that the realistic bivariate rank dependence pattern of the unprocessed forecast from (a) is lost. The postprocessed NR-ECC ensemble in (c) also employs the ensemble forecasts obtained by the individual site-by-site NR postprocessing, but in addition retains the rank dependence structure of the raw ensemble. Hence, even though the individually NR postprocessed and the NR-ECC ensemble share the marginal distributions, as indicated by the histograms on the margins in the top row, they differ strongly in their bivariate rank dependence patterns. As pointed out before, the raw and the NR-ECC ensemble are related to the same empirical copula. In contrast, the individually NR postprocessed ensemble is associated with a different empirical copula, whose (stabilized) contour plot meaningfully looks a lot like that of the so-called independence copula (Nelsen, 2006).

The ECC approach combines analytical, statistical, and numerical modeling. It is conceptually simple and easy to implement, with essentially no computational costs beyond the univariate postprocessing, thus providing a natural benchmark. Approaches of the ECC type have gained prominence at several weather centers (Flowerdew, 2012; Roulin & Vannitsem, 2012), and the ECC concept has been used as a reference technique in several papers, for instance in the context of meteorology (Ben Bouallègue, Heppelmann, Theis, & Pinson, 2016; Wilks, 2015) and hydrology (Hemri et al., 2015) or when designing verification methods (Scheuerer & Hamill, 2015; Thorarinsdottir et al., 2016).

While the focus is on the key example of weather forecasting here, the general frame of ECC can likely be used in other application areas, too. Basically, ECC can be employed whenever a raw ensemble of simulation runs is at hand, the ensemble is able to appropriately represent multivariate correlation patterns, and there is a sufficient amount of training data to statistically adjust the univariate margins.

The limitations of ECC are basically set by its defining property, i.e., the adoption of the rank dependence structure of the raw ensemble. The size of the ECC postprocessed ensemble is restricted to equal that of the raw ensemble. Therefore, ECC is especially attractive in case the raw ensemble is reasonably large. However, the raw ensemble size may be rather small in practice. An option to address this drawback is to employ a recycling procedure that is based on a repeated implementation using the quantization scheme (4.11) from Section 4.4.1 that can generate postprocessed ensembles in an ECC manner whose size is an integer multiple of the raw ensemble size (Wilks, 2015). Another possibility to circumvent this restriction is opened up by employing Bayesian postprocessing models, where repeated draws from the posterior distribution also allow to form an ECC-like ensemble with the size being an integer multiple of the raw ensemble size. Such an approach is described in Section 4.5.1.

ECC relies on the perfect model assumption that the ensemble prediction system correctly describes the multivariate dependence structures across weather quantities, locations, and lead times. This assumption appears to be reasonably adequate for modern dynamical weather forecast models. However, it is quite strong and cannot be expected to be valid each and every day. In particular, the assumption may not hold for spatial dependence in case the coarse resolution of a global forecast model is unable to resolve small-scale spatial variability. In general, numerical models might exhibit errors in dependence structures, and ECC cannot remove any inconsistencies between the postprocessed marginal distributions themselves.

According to Ben Bouallègue et al. (2016), ECC can create unrealistic scenarios when the postprocessing largely increases the ensemble spread in an indiscriminate manner. Nonrepresentative dependence patterns in the unprocessed ensemble are amplified after calibration, yielding unrealistic forecast variability. Hence, applying ECC to ensembles with poor reliability can worsen the ensemble information content. To overcome this shortcoming, Ben Bouallègue et al. (2016) propose a modification of ECC, called dual-ECC. Their approach focuses on temporal dependencies and aims at combining the structure of the raw ensemble with a component accounting for the autocorrelation of the forecast error over consecutive lead times, exploiting the assumption of stationarity of the forecast errors.

Parametric multivariate approaches as discussed in Section 4.3 may be able to correct any biases in the ensemble's representation of conditional dependence patterns and likely outperform ECC in low-dimensional settings. However, they typically require sophisticated statistical models to be fitted and usually either model intervariable or spatial or temporal correlations. On the contrary, the computational resources needed by ECC are almost negligible, and ECC can deal with cross-variable, spatial, and temporal dependencies at the same time, being applicable to model output of virtually any dimensionality.

As pointed out by Schefzik et al. (2013), the ECC approach can be seen as a unifying framework for various seemingly unrelated methods scattered in the literature. Examples include, but are not limited to, the works of Flowerdew (2012, 2014), Pinson (2012), Roulin and Vannitsem (2012), or Van Schaeybroeck and Vannitsem (2015). The common thread of these approaches lies in the adoption of the empirical copula of the raw ensemble, thereby conserving its rank dependence pattern.

As discussed by Schefzik (2017), a striking connection exists between ECC with standard NR being used for univariate postprocessing and the so-called member-by-member postprocessing (MBMP) concept, which will be elucidated in what follows.

We specifically consider the NR approach of Gneiting et al. (2005), which fits and adjusts a Gaussian distribution to univariately postprocess the raw ensemble forecast $x_1^\ell, \ldots, x_M^\ell$ for a fixed weather quantity, location, and lead time, with $\ell \in \{1, \ldots, L\}$ referring to the respective multiindex introduced in Section 4.4.1. Further considering the ensemble members to be exchangeable, the NR postprocessing model for quantity Y_ℓ reads

$$Y_\ell \mid x_1^\ell, \ldots, x_M^\ell \sim \mathcal{N}(a + b\bar{x}_\ell, c + ds_\ell^2), \tag{4.17}$$

where the corresponding NR CDF shall be denoted by F_ℓ. In (4.17), $\bar{x}_\ell$ and s_ℓ^2 denote the empirical raw ensemble mean and variance, respectively, and a, b, c, and d are the NR model parameters to be estimated, where c and d are constrained to be nonnegative. To obtain a postprocessed sample $\tilde{x}_1^\ell, \ldots, \tilde{x}_M^\ell$ from F_ℓ, we do not employ the quantization schemes (4.10), (4.11), or (4.12) discussed in Section 4.4.1, but follow an alternative procedure in which a parametric, continuous CDF R_ℓ to the raw ensemble forecast $x_1^\ell, \ldots, x_M^\ell$ is fitted first. Subsequently, the transformation scheme

$$\tilde{x}_1^\ell := F_\ell^{-1}(R_\ell(x_1^\ell)), \quad \ldots, \quad \tilde{x}_M^\ell := F_\ell^{-1}(R_\ell(x_M^\ell)) \tag{4.18}$$

is applied (Schefzik et al., 2013). In case F_ℓ and R_ℓ belong to the same location-scale family, meaning that $R_\ell(x) = G((x - \mu)/\sigma)$ and $F_\ell(x) = G((x - \mu')/\sigma')$ for some continuous CDF G with $\mu, \mu' \in \mathbb{R}$ and $\sigma, \sigma' > 0$, the transformation from the raw forecast x_m^ℓ to the postprocessed forecast $\tilde{x}_m^\ell$ for ensemble member $m \in \{1, \ldots, M\}$ is linear (Schefzik, 2017), where

$$\tilde{x}_m^\ell = F_\ell^{-1}(R_\ell(x_m^\ell)) = \mu' + \frac{\sigma'}{\sigma}(x_m^\ell - \mu). \tag{4.19}$$

Due to the monotonicity of the respective mapping when using sampling procedure (4.18) and the specific setting (4.19), respectively, the rank dependence structure of the raw ensemble is conserved by construction, such that the final reordering step in the ECC approach becomes superfluous in this case, that is, $\hat{\mathbf{x}} = \tilde{\mathbf{x}}$ in the notation of Section 4.4.1.

Even though derived in a different way, the result in Eq. (4.19) coincides with a finding in the context of the so-called error-in-variables EMOS method with one predictor proposed by Vannitsem (2009). This shows a strong relationship of the transformation (4.18) to the linear fitting in the error-in-variables EMOS approach, which has been designed to account for errors of both the predictor and the predictant.

For weather variables with continuous probability distributions, physically consistent postpro-cessed ensembles can alternatively be obtained using MBMP approaches (Doblas-Reyes, Hagedorn, & Palmer, 2005; Johnson & Bowler, 2009; Pinson, 2012; Van Schaeybroeck & Vannitsem, 2011, 2015; Wood & Schaake, 2008). MBMP methods have arisen and enjoyed popularity before and independently from the evolution of ECC. In an MBMP method, the raw ensemble forecast $x_1^\ell, \ldots, x_M^\ell$ is directly converted to a postprocessed ensemble forecast $\hat{x}_1^\ell, \ldots, \hat{x}_M^\ell$ using the linear transformation

$$\hat{x}_m^\ell = \zeta + \eta \bar{x}_\ell + \kappa \left(x_m^\ell - \bar{x}_\ell \right) \tag{4.20}$$

for each ensemble member $m \in \{1, \ldots, M\}$ separately, where the MBMP techniques differ in terms of the design and estimation of the parameters ζ, η, and κ. MBMP approaches automatically retain the rank dependence structure of the raw ensemble and do not change the ensemble nature of the forecast, while NR postprocessing inherently produces a full predictive distribution, from which then a sample has to be extracted in order to obtain an ensemble forecast.

According to Schefzik (2017), an MBMP method (4.20) can be regarded as a special NR-ECC var-iant based on (4.17), (4.18), and (4.19), respectively, by letting $a := \zeta, b := \eta, c := 0$, and $d := \kappa^2$, such that F_ℓ is the NR CDF of a $\mathcal{N}(\zeta + \eta \bar{x}_\ell, \kappa^2 s_\ell^2)$-distribution, and letting R_ℓ be the CDF of a $\mathcal{N}(\bar{x}_\ell, s_\ell^2)$-distribution.

4.4.3 SCHAAKE SHUFFLE-BASED APPROACHES

To reconstruct spatiotemporal structure in forecasted temperature and precipitation fields, Clark et al. (2004) propose the so-called Schaake shuffle, which, as ECC, works within the frame of empirical copula-based ensemble postprocessing.

Unlike ECC, the dependence template in the Schaake shuffle is not specified by the raw ensemble forecasts, but by historical observations

$$\mathbf{y} := \{(y_1^1, \ldots, y_N^1), \ldots, (y_1^L, \ldots, y_N^L)\}$$

taken from N different dates of a past data record, where the multiindex $\ell \in \{1, \ldots, L\}$ may comprise combinations of weather quantities, locations, lead times, and time points. That is, for a fixed forecast instance, observations from the same N dates are used for all weather variables and locations. Thus, $\mathbf{z} := \mathbf{y}$ in the scheme from Section 4.4.1, and particularly, N is not restricted to be equal to the raw ensemble size M.

In the Schaake shuffle's original implementation (Clark et al., 2004), the N dates are picked from all years in the historical record, except for the year of the current forecast to be postprocessed. Moreover, they lie within 7 days before and after the verification date, regardless of the year. Alternatively, a more general version may employ observations from arbitrary dates in the whole past record (Schefzik, 2016b). Once the observation-based dependence template $\mathbf{z} = \mathbf{y}$ has been determined, the Schaake shuffle follows the general scheme from Section 4.4.1, including the interpretation as an empirical copula-based postprocessing technique.

The Schaake shuffle has proven to be beneficial in various applications, in which it regains observed intervariable and spatial dependence patterns, and also temporal persistence (Clark et al., 2004; Schaake et al., 2007; Voisin, Pappenberger, Lettenmaier, Buizza, & Schaake, 2011;

Vrac & Friederichs, 2015; Wilks, 2015). However, the original Schaake shuffle does not condition the dependence structure on present or predicted atmospheric states.

Based on a suggestion by Clark et al. (2004), Schefzik (2016b) implements a specific variant of the Schaake shuffle, called the SimSchaake approach, to resolve this drawback. In this method, the observations $\mathbf{y}$ that specify the dependence template $\mathbf{z}$ are taken from historical dates at which the ensemble forecast was similar to the current ensemble prediction in terms of a specifically chosen similarity criterion. Thus, the dependence template $\mathbf{z}$ in the SimSchaake approach is built by using an analog ensemble concept (Delle Monache, Nipen, Liu, Roux, & Stull, 2011; Hamill & Whitaker, 2006). Having obtained $\mathbf{z}$ in such a manner, the SimSchaake approach proceeds further according to the scheme in Section 4.4.1. An adequate selection of the similarity criterion in the SimSchaake approach is essential and may strongly depend on the application one is interested in (Schefzik, 2016b).

Another specific implementation of the Schaake shuffle is proposed by Scheuerer, Hamill, Whitin, He, and Henkel (2017). In their method, the historical dates used to build the dependence template $\mathbf{z}$ are selected in such a manner that the marginal distributions of the sampled observation trajectories are similar to those of the postprocessed marginal forecast distributions, where similarity is quantified in terms of the divergence of two distributions. According to Scheuerer et al. (2017), this approach weakens the standard Schaake shuffle's implicit assumption of state-independent spatiotemporal rank correlations, as the values of $\mathbf{z}$ are closer to the forecast values to which they are mapped during the reordering process.

Both ECC and the Schaake shuffle-based approaches have shown good performances in several case studies (Clark et al., 2004; Schefzik, 2016b; Schefzik et al., 2013; Vrac & Friederichs, 2015) and can be employed as a benchmark. However, the size N of the standard ECC ensemble is constrained to equal the raw ensemble size M (unless for instance a Bayesian approach is utilized, see Section 4.5.1). Consequently, the raw ensemble should be reasonably large to provide reliable information about the correlation structure when applying ECC. The Schaake shuffle-based approaches have a somewhat broader applicability, in that they can generate an ensemble of arbitrary size N, but provided that sufficiently many historical observations are available. However, the original Schaake shuffle does not account for atmospheric flow, in contrast to ECC. The answer to the question whether to prefer ECC or Schaake shuffle-based techniques may depend on the specific setting at hand and different factors, such as ensemble size or data availability, and Wilks (2015) provides an initial study in this context.

4.5 UNIVARIATE APPROACHES ACCOUNTING FOR DEPENDENCIES

In Sections 4.3 and 4.4, we have reviewed parametric and nonparametric ensemble postprocessing approaches, respectively, which incorporate dependence structures by providing truly multivariate predictive distributions. However, there are also several postprocessing approaches that formally yield univariate predictive distributions, but account for spatial or temporal correlations by specific estimation procedures for the model parameters. Such techniques are discussed in this section.

4.5.1 SPATIAL DEPENDENCIES

There are several approaches accounting for spatial correlations that lead to univariate predictive distributions, while the model parameters are estimated location-specifically. Such so-called spatially adaptive techniques include NR or BMA models with parameters that vary smoothly across space, thus yielding locally calibrated univariate predictive distributions. At the same time, such models allow interpolation of the postprocessed forecasts to locations not present in the training data, or even to the full dynamical model grid.

Kleiber, Raftery, Baars, et al. (2011) introduce geostatistical model averaging (GMA) for normally distributed temperature forecasts, extending the original BMA method to have spatially adaptive, location-specific model parameters. By employing a geostatistical model, the parameters, and thus the postprocessed forecasts, can be interpolated to locations not present in the (training) data. Kleiber, Raftery, and Gneiting (2011) modify the original GMA model such that it becomes suitable for precipitation forecasts.

Let Y_ℓ denote the random variable representing the weather quantity of interest, and y_ℓ the corresponding realization at location $\ell \in \{1, ..., L\}$. Analogously to the original BMA model, the univariate GMA predictive PDF at location ℓ, conditionally on the M ensemble member forecasts $x_1^\ell, ..., x_M^\ell$, is given by

$$p(y_\ell | x_1^\ell, ..., x_M^\ell) := \sum_{m=1}^{M} w_m \, g_m(y_\ell | x_m^\ell), \tag{4.21}$$

with weights w_m. If Y_ℓ is assumed to be normally distributed, the predictive distribution of Y_ℓ given the mth ensemble forecast at location ℓ, associated with the kernel density g_m, is given by

$$Y_\ell | x_m^\ell \sim \mathcal{N}(x_m^\ell - a_m^\ell, c\sigma_\ell^2),$$

where σ_ℓ^2 is the nonmember-specific variance, c is a positive variance deflation factor, and a_m^ℓ is a member-specific additive bias correction parameter. In contrast to the original BMA model, Kleiber, Raftery, Baars, et al. (2011) employ only an additive bias correction parameter and no multiplicative one. Both a_m^ℓ and σ_ℓ^2 are estimated from training data using the usual sample versions.

In order to interpolate a_m^ℓ and σ_ℓ^2 to unknown locations, geostatistical models are considered for both parameters. Specifically, the estimates $\{\hat{a}_m^\ell\}_{\ell=1}^{L}$ and $\{\hat{v}_\ell\}_{\ell=1}^{L}$, where $\hat{v}_\ell := \log(\hat{\sigma}_\ell^2)$, are viewed as samples from stationary Gaussian random fields (GRFs) with an exponential covariance function.

For any known location $\ell \in \{1, ..., L\}$ that is part of the training data, the predictive PDF is explicitly given by Eq. (4.21). For other locations ℓ_0 not being part of the training data, the estimates $a_m^{\ell_0}$ and $\sigma_{\ell_0}^2$ are unknown. However, they can be obtained by applying a geostatistical interpolation technique called kriging (Cressie, 1993; Stein, 1999). Plugging in these interpolated estimates into Eq. (4.21) for the location ℓ_0 yields the predictive PDF for Y_{ℓ_0}.

Scheuerer and Büermann (2014) and Scheuerer and König (2014) introduce a spatially adaptive NR model for temperature. In contrast to the original NR model, they employ local observation and forecast anomalies as response and covariates in the regression model. Here, anomalies are defined as differences between observations or forecasts and the respective averages over a historic time period (climatological means). This implicitly leads to location-specific bias correction, although the NR parameters themselves are not location-specific.

The resulting NR model is given by

$$y_\ell = \bar{y}_\ell + \sum_{m=1}^{M} b_m(x_m^\ell - \bar{x}_m^\ell) + \varepsilon_\ell, \tag{4.22}$$

where b_m is a bias correction parameter, $\bar{y}_\ell$ is the empirical mean over all observations in the training period at location ℓ, and $\bar{x}_m^\ell$ the empirical mean over all forecasts of member m in the training period at location ℓ. The most important component of this model is $\bar{y}_\ell$, which allows incorporation of spatial small-scale variability that is not resolved by the dynamical forecast model.

For the error term ε_ℓ, it is assumed that

$$\varepsilon_\ell \sim \mathcal{N}(0, \sigma_\ell^2),$$

with $\sigma_\ell^2 = c_1 \xi_\ell^2 + c_2 s_\ell^2$, where c_1 and c_2 are nonnegative variance parameters, s_ℓ^2 is the empirical ensemble variance at location ℓ, and ξ_ℓ^2 reflects the local uncertainty at location ℓ over the training period.

The model defined in Eq. (4.22) yields a predictive distribution for locations ℓ that are present in the training data. However, as for GMA, at unknown locations ℓ_0, the model parameters $\bar{y}_{\ell_0}$ and $\xi_{\ell_0}^2$ are unknown. Similar to the GMA method, geostatistical models are fitted for the model parameters. In the approaches of Scheuerer and Büermann (2014) and Scheuerer and König (2014), intrinsic GRFs are employed.

Dabernig, Mayr, Messner, and Zeileis (2017b) present a further modification of the model in Scheuerer and Büermann (2014), employing standardized observation and forecast anomalies. That is, the anomalies are further divided by their climatological standard deviations. This approach removes seasonal and location-specific characteristics, thus yielding a model that is implicitly valid at any arbitrary location.

The approaches described herein utilize anomalies, which can be seen as forecasts and observations corrected for location-specific characteristics. Many postprocessing models implicitly assume spatial stationarity, and the preceding methods highlight that this assumption is not always fulfilled and alternative methods that can deal with this fact become more and more important.

Recently, Möller et al. (2016) proposed a fully Bayesian NR version for temperature that accounts for spatial structures in the observations and ensemble forecasts. The proposed model utilizes the so-called stochastic partial differential equation (SPDE) approach (Lindgren, Rue, & Lindström, 2011) for efficient spatial modeling, and model fitting is performed in a Bayesian framework by using the integrated nested Laplace approximation (INLA) methodology (Rue, Martino, & Chopin, 2009) that allows for fast and accurate Bayesian estimation.

While a local version of the NR method can model location-specific parameters as well, the estimates are solely based on the data of the respective station, and data from other (neighboring) stations are not taken into account. Thus, possible dependencies between stations are not incorporated. Möller et al. (2016) extend the classical local NR approach by interpreting the (location-specific) bias correction parameters as realizations from GRFs. This allows the parameters to be adapted to local conditions, while using the data from all locations in the estimation process. For efficient estimation, the SPDE approach (Lindgren et al., 2011) is utilized, which allows a Gaussian Markov random field (GMRF) to be obtained, that is a finite multivariate representation for a GRF. The basic methodology of Lindgren et al. (2011) utilized for the Bayesian postprocessing model in Möller et al. (2016) uses a stationary SPDE, thus implicitly assuming stationarity in space. However, the SPDE method is not limited to this setting. There are also nonstationary and anisotropic versions available, but some of the implementations for these more advanced methods are still under construction.

The spatially adaptive Bayesian version of NR proposed by Möller et al. (2016) is called Markovian NR (MNR), due to the Markovian properties implied by the model for the parameters. Model fitting is performed in a Bayesian fashion within the SPDE-INLA framework, implemented in the R-INLA package (Lindgren & Rue, 2015).

The MNR predictive distribution at a spatial location **s** is given by

$$Y(\mathbf{s})|\bar{x}(\mathbf{s}), a(\mathbf{s}), b(\mathbf{s}), \sigma \sim \mathcal{N}(a(\mathbf{s}) + b(\mathbf{s})\bar{x}(\mathbf{s}), \sigma^2), \tag{4.23}$$

where $a(\mathbf{s})$ and $b(\mathbf{s})$ are GRFs assumed for the bias-correction parameters of the NR model, σ is the variance of the error term, and $\bar{x}(\mathbf{s})$ the mean over the forecasts of M (exchangeable) ensemble members.

Using R-INLA, one can generate samples $\{a_j\}_{j=1}^r$, $\{b_j\}_{j=1}^r$, and $\{\sigma_j\}_{j=1}^r$ from the marginal posterior distributions of $a(\mathbf{s})$, $b(\mathbf{s})$, and σ. These marginal posterior samples can be employed to generate a sample of arbitrary size $N = r$ (not necessarily related to the raw ensemble size) from the predictive distribution (4.23).

The basic MNR method formally yields a univariate predictive distribution, where the model parameters vary smoothly in space. However, the discrete GMRF representation also defines the spatial field at any arbitrary location **s**, thus allowing predictions on a continuous domain.

To obtain a truly multivariate postprocessing model, Möller et al. (2016) introduce an MNR-ECC method, which combines the MNR model with the ECC approach described in Section 4.4.2. At each spatial location **s**, a discrete sample $\tilde{x}_{j1}(\mathbf{s}), \ldots, \tilde{x}_{jM}(\mathbf{s})$ of the size M of the raw forecast ensemble, for a fixed $j \in \{1, \ldots, r\}$, is drawn from the MNR predictive distribution (4.23) using quantization scheme (4.10) in Section 4.4.1, and each of these r samples of size M are reordered according to the rank order structure of the raw ensemble in the ECC manner.

The Bayesian framework used for model fitting overcomes the drawback of the standard ECC post-processed ensemble being constrained to have the same size M as the original ensemble (Möller et al., 2016). One can obtain multiple discrete samples of size M by repeatedly (r times) drawing samples from the posterior MNR predictive distribution (4.23). The ECC procedure is then applied to each of these r draws $\tilde{x}_{j1}(\mathbf{s}), \ldots, \tilde{x}_{jM}(\mathbf{s})$, $j \in \{1, \ldots, r\}$, individually, yielding r postprocessed ensembles of size M. By merging these, an ensemble of size rM (i.e., an integer multiple of the raw ensemble size) is obtained, where the number r of posterior samples can be chosen by the user.

4.5.2 TEMPORAL DEPENDENCIES

While Section 4.5.1 outlined prominent univariate spatially adaptive approaches based on NR or BMA, this section describes recently developed univariate temporally adaptive NR versions, where either the dependence between lead times or the dependence in time for a fixed lead time are considered.

Section 4.3.3 already introduced genuinely multivariate approaches that jointly model multiple forecast lead times, for example by means of a Gaussian copula. Quite recently, Dabernig, Mayr, Messner, and Zeileis (2017a) proposed a univariate temporally adaptive approach to account for differences in lead times. The method is a modification of the approach of Dabernig et al. (2017b) mentioned in Section 4.5.1, in which standardized anomalies are utilized instead of the original observations and forecasts to remove location-specific characteristics in the data, allowing a single model to be fit at all spatial locations.

To account for lead time-specific characteristics, Dabernig et al. (2017a) define standardized anomalies as well. However, the climatological mean and standard deviations used to define the anomalies are estimated from a specific generalized additive model with smooth temporal effects (instead of spatial effects as in the original approach). The estimated (temporally) standardized anomalies are then used in an NR model instead of the original data. Although the coefficients of the resulting NR model are not lead time-specific, this model allows prediction at any arbitrary lead time, even at lead time resolutions not part of typical dynamical model outputs, by means of the temporally standardized anomalies. This approach brings up the fact that stationarity across lead times is not necessarily given and should be accounted for in a postprocessing model. Removing these characteristics from forecasts and observations is one possibility to deal with different properties of forecasts at different lead times.

A different univariate approach that considers autoregressive dependencies present in the temporal evolution of temperature ensemble forecasts for a given lead time and location is proposed by Möller and Groß (2016). Their autoregressive NR model is temporally adaptive, but univariate in nature. However, Möller and Groß (2016) mention that an extension to a fully multivariate setting as described in Section 4.3 is possible and is scheduled to be developed in future research. Such a multivariate autoregressive model may consider the autocorrelation in time jointly with spatial or intervariable (or even inter-lead time) dependence by employing multivariate time series models.

Möller and Groß (2016) found that for an observation $y(t)$ at time point t, the forecast error $e(t) := y(t) - x_m(t)$ of each individual ensemble member $x_m(t)$, $m \in \{1, ..., M\}$, exhibits substantial autoregressive behavior. While many postprocessing models implicitly assume stationarity in time (and space, see Section 4.5.1), the case study of Möller and Groß (2016) reveals that this assumption is not necessarily valid. However, it is possible to account for the nonstationarity in time by removing time point-specific characteristics and employing forecasts corrected for these dependence patterns. This idea is related to the approach of Dabernig et al. (2017a) who use anomalies that are corrected for lead time-specific characteristics.

Möller and Groß (2016) construct an NR-like predictive distribution based on a corrected forecast ensemble, where the correction parameters are obtained from an autoregressive (AR) process fitted to the forecast errors of the individual ensemble members. The parametric correction of the ensemble members themselves has some connection to the approach of Pinson (2012) and other methods that produce a corrected forecast ensemble. One advantage of such an ensemble transformation approach is that the multivariate flow-dependent information present in the original ensemble can be conserved, while this information is typically lost when fitting a univariate predictive distribution based on the ensemble data.

Specifically, an $AR(p)$ process is fitted in a first step to the forecast error series of each ensemble member individually, that is, the model

$$e(t) - \mu = \sum_{j=1}^{p} \alpha_j [e(t-j) - \mu] + \varepsilon(t)$$

is assumed, where $\{\varepsilon(t)\}$ is white noise, and $\mu, \alpha_1, ..., \alpha_p$ are model parameters.
Then,

$$\hat{x}_m(t) := x_m(t) + \mu + \sum_{j=1}^{p} \alpha_j [y(t-j) - x_m(t-j) - \mu]$$

can be seen as a corrected forecast based on the forecast $x_m(t)$ as well as on past observations $y(t-j)$ and forecasts $x_m(t-j)$, $j \in \{1, ..., p\}$. The coefficients $\mu, \alpha_1, ..., \alpha_p$ are obtained by fitting an $AR(p)$ process to the observed error series $\{e(t)\}$ from a training period. This correction process is performed for each forecast ensemble member $x_1(t), \ldots, x_M(t)$ individually, yielding the AR-corrected ensemble $\hat{x}_1(t), \ldots, \hat{x}_M(t)$.

In a second step, an NR predictive distribution based on the AR-corrected forecast ensemble (called AR-NR) is constructed. As in the classical univariate NR approach, it is assumed that

$$Y(t)|x_1(t), \ldots, x_M(t) \sim \mathcal{N}\left(\tilde{\mu}(t), \tilde{\sigma}^2(t)\right).$$

Contrary to the original NR approach, the parameters $\tilde{\mu}(t)$ and $\tilde{\sigma}^2(t)$ are not estimated by score minimization, but by utilizing the autoregressive fits to the individual error series and the respective AR-corrected ensemble members. A simple estimate for the mean $\tilde{\mu}(t)$ is given by the AR-corrected ensemble mean. The variance term $\tilde{\sigma}^2(t)$ is parameterized as a linear function of both the M variances of the individual AR fits and the empirical variance of the AR-corrected ensemble. This parameterization ensures that both the longitudinal information in time (variances of AR fits) and the cross-sectional ensemble spread information (variance of corrected ensemble) is utilized to correct the dispersion of the original ensemble. In a third step, the proposed AR-NR and the standard NR predictive distribution are combined in a spread-adjusted linear pool (SLP) to further improve predictive performance.

At this point, the twofold advantage of the approach of Möller and Groß (2016) should be highlighted: The method allows construction of a corrected forecast ensemble and a full predictive probability distribution at the same time. Thus, it is possible to make use of the advantages of both concepts.

The basic AR-correction method, the AR-NR distribution and the SLP combination are implemented in the R package ensAR (Groß & Möller, 2016). As already mentioned, a multivariate AR-NR extension is to be developed by utilizing vector-autoregressive (VAR) processes to fit a multivariate time series model to multiple locations simultaneously. Modifications for other than normally distributed weather quantities are possible as well.

4.6 DISCUSSION

In this chapter, different ensemble postprocessing approaches that incorporate intervariable, spatial, and/or temporal dependencies have been reviewed. Several classes of such methods can be distinguished. Approaches yielding truly multivariate predictive distributions can be either parametric or nonparametric.

Parametric multivariate postprocessing techniques are typically specifically tailored to either intervariable (Schuhen et al., 2012, for instance) or spatial (Feldmann et al., 2015, for instance) or temporal (Pinson et al., 2009, for instance) settings, work well in low dimensions and are very likely to outperform the more general nonparametric methods in such cases. They are mostly based on the use of Gaussian copulas, which are often a convenient and reasonable tool in dependence modeling. In the context of ensemble postprocessing, Gaussian copulas may even be appropriate when there are one or more highly non-Gaussian and nonstationary weather variables with a spatially and temporally intermittent nature (such as precipitation) involved in the multivariate setting. Here, the Gaussian copula implicitly models the forecast error, and not the weather variables directly. This is exemplified by the approach of Möller et al. (2013), in which forecasts are postprocessed for each weather quantity individually using suitable univariate methods, and the Gaussian copula is employed solely to model the residual correlation remaining after the univariate postprocessing. However, if it is desired to model a vector of weather quantities in the preceding scenario directly with a copula, the choice of a Gaussian copula may be problematic and inadequate.

Nonparametric empirical copula-based multivariate postprocessing methods, which employ a reordering of univariately postprocessed ensemble forecasts according to a dependence template, can model intervariable, spatial, and temporal dependencies simultaneously and can be employed in rather high-dimensional settings. They provide a natural benchmark, as virtually no additional computational effort is needed beyond that for the univariate postprocessing, which, however, can be efficiently performed using NR or BMA, for instance. In contrast, parametric methods usually require a large number of parameters to be estimated in the multivariate model, even in settings of moderate dimension.

The nonparametric multivariate approaches based on empirical copulas essentially differ in terms of the specific selection of the dependence template. There are two prominent options how the dependence template can be chosen: either the raw ensemble forecasts can be employed, as in the ECC approach (Schefzik et al., 2013), or historical observations, as in the Schaake shuffle and modifications thereof (Clark et al., 2004; Schefzik, 2016b; Scheuerer et al., 2017).

In order to apply ECC, the underlying forecasts should reflect the actually observed dependence structures reasonably well, and the raw ensemble should be large enough, such that the dependence template is derived from sufficiently many data points. ECC-type approaches typically do not perform that well if the correlation structure is misspecified by the ensemble or if the ensemble is very small. But for all that, their advantage over Schaake shuffle-based methods is that they do not additionally require a large observational database for building the dependence template. However, the Schaake shuffle-based approaches have a somewhat broader applicability than the standard ECC technique, as they are not restricted to generate postprocessed ensembles of the same size as the raw ensemble.

Besides the parametric and nonparametric methods leading to truly multivariate postprocessed predictive distributions, approaches that yield univariate distributions, but are able to incorporate dependencies by defining smoothly varying parameters, have been discussed as well. While such parametric techniques appear occasionally in the context of temporal correlations (Dabernig et al., 2017a; Möller & Groß, 2016), they have mainly been used for modeling spatial correlations so far (Kleiber, Raftery, Baars, et al., 2011; Scheuerer & Büermann, 2014, for instance), often allowing interpolation of postprocessed forecasts to unknown out-of-sample locations. In particular, univariate spatial approaches may be Bayesian in nature, such as the MNR method of Möller et al. (2016). Such spatially or temporally adaptive models have the advantage that they can implicitly deal with nonstationary data either by defining smoothly varying coefficients adapting to local conditions or by removing location- or time-specific characteristics and employing corrected forecasts and observations.

In principle, it is possible to use univariate postprocessing approaches incorporating dependencies (instead of standard NR or BMA, for instance) in the second step of the general scheme for empirical copula-based ensemble postprocessing in Section 4.4.1. This is exemplified by the MNR-ECC approach (Möller et al., 2016), which combines the univariate MNR method accounting for spatial dependencies with the ECC notion based on a reordering of the postprocessed ensemble forecasts according to the rank order structure of the raw ensemble.

The points discussed herein may provide some guidance for choosing among postprocessing methods for incorporating dependencies, and have been confirmed in real-data case studies (Feldmann et al., 2015; Schefzik, 2016b; Scheuerer et al., 2017; Schuhen et al., 2012; Wilks, 2015) and simulations (Gräter, 2016; Lerch & Gräter, 2017). However, the comparisons in the respective papers in each case only cover a subset of the methods discussed in this chapter, and it remains to design and implement further, more comprehensive studies for more insights on the use of the different methods. Last, the predictive performance of a postprocessing approach accounting for dependencies may not only depend on the parameters discussed before (such as specific settings, dimension, (mis)specification of correlation structure, ensemble size, or data availability), but also on other criteria such as the verification metric that is considered (Thorarinsdottir & Schuhen, 2018, Chapter 6 of this book).

ACKNOWLEDGMENTS

The forecasts used in Figs. 4.1 and 4.4 have been made available by the European Center for Medium-Range Weather Forecasts, and the corresponding observations by the German Weather Service, which is gratefully acknowledged. The authors are thankful to the editors and an anonymous reviewer for providing helpful comments and suggestions, and to Sebastian Lerch for providing materials.

REFERENCES

Aas, K., Czado, C., Frigessi, A., & Bakken, H. (2009). Pair-copula constructions of multiple dependence. *Insurance: Mathematics & Economics*, *44*, 182–198.

Baran, S., & Möller, A. (2015). Joint probabilistic forecasting of wind speed and temperature using Bayesian model averaging. *Environmetrics*, *26*, 120–132.

Baran, S., & Möller, A. (2017). Bivariate ensemble model output statistics approach for joint forecasting of wind speed and temperature. *Meteorology and Atmospheric Physics*, *129*, 99–112.

Ben Bouallègue, Z., Heppelmann, T., Theis, S. E., & Pinson, P. (2016). Generation of scenarios from calibrated ensemble forecasts with a dual-ensemble copula-coupling approach. *Monthly Weather Review*, *144*, 4737–4750.

Berrocal, V. J., Raftery, A. E., & Gneiting, T. (2007). Combining spatial statistical and ensemble information in probabilistic weather forecasts. *Monthly Weather Review*, *135*, 1386–1402.

Berrocal, V. J., Raftery, A. E., & Gneiting, T. (2008). Probabilistic quantitative precipitation field forecasting using a two-stage spatial model. *Annals of Applied Statistics*, *2*, 1170–1193.

Bremnes, J. B. (2007). *Improved calibration of precipitation forecasts using ensemble techniques. Part 2: Statistical calibration methods*. Technical Report No. 04/2007, Norwegian Meteorological Institute. https://www.met.no/publikasjoner/met-report/met-report-2007/_/attachment/download/758b26ea-11eb-4808-a510-eb7df09a5642:c5d0375a363df24adfa4c59ef2867bf07ca5e09c/MET-report-04-2007.pdf (Accessed 22 January 2018).

Bröcker, J. (2012). Evaluating raw ensembles with the continuous ranked probability score. *Quarterly Journal of the Royal Meteorological Society*, *138*, 1611–1617.

Buizza, R. (2018). Ensemble forecasting and the need for calibration. In S. Vannitsem, D. S. Wilks, & J. W. Messner (Eds.), *Statistical Postprocessing of Ensemble Forecasts*. Elsevier.

Chaloulos, G., & Lygeros, J. (2007). Effect of wind correlation on aircraft conflict probability. *Journal of Guidance, Control, and Dynamics*, *30*, 1742–1752.

Chilès, J. P., & Delfiner, P. (2012). *Geostatistics: Modeling Spatial Uncertainty* (pp. 734) Hoboken: Wiley.

Clark, M. P., Gangopadhyay, S., Hay, L. E., Rajagopalan, B., & Wilby, R. L. (2004). The Schaake shuffle: A method for reconstructing space-time variability in forecasted precipitation and temperature fields. *Journal of Hydrometeorology*, *5*, 243–262.

Cressie, N. A. C. (1993). *Statistics for Spatial Data* (pp. 900). Hoboken: Wiley.

Cressie, N., & Wikle, C. K. (2011). *Statistics for Spatio-Temporal Data* (pp. 624). Hoboken: Wiley.

Dabernig, M., Mayr, G. J., Messner, J. W., & Zeileis, A. (2017a). Simultaneous ensemble postprocessing for multiple lead times with standardized anomalies. *Monthly Weather Review, 145*, 2523–2531.

Dabernig, M., Mayr, G. J., Messner, J. W., & Zeileis, A. (2017b). Spatial ensemble post-processing with standardized anomalies. *Quarterly Journal of the Royal Meteorological Society, 143*, 909–916.

Deheuvels, P. (1979). La fonction de dépendance empirique et ses propriétés. Un test non paramétrique d'indépendance. *Académie Royale de Belgique, Bulletin de la Classe des Sciences, Série 5, 65*, 274–292.

Delle Monache, L., Nipen, T., Liu, Y., Roux, G., & Stull, R. (2011). Kalman filter and analog schemes to postprocess numerical weather predictions. *Monthly Weather Review, 139*, 3554–3570.

Demarta, S., & McNeil, A. J. (2005). The *t* copula and related copulas. *International Statistical Review, 73*, 111–129.

Doblas-Reyes, F. J., Hagedorn, R., & Palmer, T. N. (2005). The rationale behind the success of multi-model ensembles in seasonal forecasting—II. Calibration and combination. *Tellus A, 57*, 234–252.

ECMWF Directorate (2012). Describing ECMWF's forecasts and forecasting system. *ECMWF Newsletter, 133*, 11–13.

Feldmann, K., Scheuerer, M., & Thorarinsdottir, T. L. (2015). Spatial postprocessing of ensemble forecasts for temperature using nonhomogeneous Gaussian regression. *Monthly Weather Review, 143*, 955–971.

Flowerdew, J. (2012). *Calibration and combination of medium-range ensemble precipitation forecasts*. United Kingdom MetOffice: Technical Report No. 567.

Flowerdew, J. (2014). Calibrating ensemble reliability whilst preserving spatial structure. *Tellus A, 66*, 22662.

Gel, Y., Raftery, A. E., & Gneiting, T. (2004). Calibrated probabilistic mesoscale weather field forecasting: The geostatistical output perturbation (GOP) method (with discussion and rejoinder). *Journal of the American Statistical Association, 99*, 575–590.

Genest, C., & Favre, A. C. (2007). Everything you always wanted to know about copula modeling but were afraid to ask. *Journal of Hydrologic Engineering, 12*, 347–368.

Gneiting, T., Raftery, A. E., Westveld, A. H., & Goldman, T. (2005). Calibrated probabilistic forecasting using ensemble model output statistics and minimum CRPS estimation. *Monthly Weather Review, 133*, 1098–1118.

Gneiting, T., Stanberry, L. I., Grimit, E. P., Held, L., & Johnson, N. A. (2008). Assessing probabilistic forecasts of multivariate quantities, with applications to ensemble predictions of surface winds (with discussion and rejoinder). *Test, 17*, 211–264.

Graf, S., & Luschgy, H. (2000). *Foundations of Quantization for Probability Distributions* (pp. 230). Berlin: Springer.

Gräter, M. (2016). Simulation study of dual ensemble copula coupling (Unpublished master's thesis). Karlsruhe Institute of Technology, Germany.

Groß, J., & Möller, A. (2016). *ensAR: Autoregressive postprocessing methods for ensemble forecasts*. R package version 0.0.0.9000, https://github.com/JuGross/ensAR. (Accessed 22 January 2018).

Hagedorn, R., Buizza, R., Hamill, T. M., Leutbecher, M., & Palmer, T. N. (2012). Comparing TIGGE multimodel forecasts with reforecast-calibrated ECMWF ensemble forecasts. *Quarterly Journal of the Royal Meteorological Society, 138*, 1814–1827.

Hamill, T. M., & Whitaker, J. S. (2006). Probabilistic quantitative precipitation forecasts based on reforecast analogs: theory and application. *Monthly Weather Review, 134*, 3209–3229.

Hemri, S., Lisniak, D., & Klein, B. (2015). Multivariate postprocessing techniques for probabilistic hydrological forecasting. *Water Resources Research, 51*, 7436–7451.

Hemri, S., Scheuerer, M., Pappenberger, F., Bogner, K., & Haiden, T. (2014). Trends in the predictive performance of raw ensemble weather forecasts. *Geophysical Research Letters, 41*, 9197–9205.

Hu, Y., Schmeits, M. J., Van Andel, S. J.Verkade, J. S., Xu, M., Solomatine, D. P., & Liang, Z. (2016). A stratified sampling approach for improved sampling from a calibrated ensemble forecast distribution. *Journal of Hydrometeorology*, *17*, 2405–2417.

Joe, H. (2014). *Dependence Modeling With Copulas* (pp. 480). Boca Raton: CRC Press.

Johnson, C., & Bowler, N. (2009). On the reliability and calibration of ensemble forecasts. *Monthly Weather Review*, *13*, 1717–1720.

Kleiber, W., Raftery, A. E., Baars, J., Gneiting, T., Mass, C., & Grimit, E. P. (2011). Locally calibrated probabilistic temperature forecasting using geostatistical model averaging and local Bayesian model averaging. *Monthly Weather Review*, *139*, 2630–2649.

Kleiber, W., Raftery, A. E., & Gneiting, T. (2011). Geostatistical model averaging for locally calibrated probabilistic quantitative precipitation forecasting. *Journal of the American Statistical Association*, *106*, 1291–1303.

Kolesárová, A., Mesiar, R., Mordelová, J., & Sempi, C. (2006). Discrete copulas. *IEEE Transactions on Fuzzy Systems*, *14*, 698–705.

Krzysztofowicz, R., & Toth, Z. (2008). Bayesian processor of ensemble (BPE): Concept and implementation. Workshop slides 4th NCEP/NWS Ensemble User Workshop, Laurel, MD. Available from: http://www.emc.ncep.noaa.gov/gmb/ens/ens2008/Krzysztofowicz_Presentation_Web.pdf (Accessed 22 January 2018)

Kurowicka, D., & Joe, H. (2011). *Dependence Modeling: Vine Copula Handbook* (pp. 368). Singapore: World Scientific.

Lerch, S., & Gräter, M. (2017). *Comparison of multivariate post-processing approaches*. Poster at the European Geosciences Union General Assembly 2017.

Lindgren, F., & Rue, H. (2015). Bayesian spatial modelling with R-INLA. *Journal of Statistical Software*, *63*, 1–25.

Lindgren, F., Rue, H., & Lindström, J. (2011). An explicit link between Gaussian fields and Gaussian Markov random fields: the stochastic partial differential equation approach (with discussion). *Journal of the Royal Statistical Society Series B*, *73*, 423–498.

Mayor, G., Suñer, J., & Torrens, J. (2005). Copula-like operations on finite settings. *IEEE Transactions on Fuzzy Systems*, *13*, 468–477.

Mayor, G., Suñer, J., & Torrens, J. (2007). Sklar's theorem in finite settings. *IEEE Transactions on Fuzzy Systems*, *15*, 410–416.

McNeil, A. J., & Nešlehová, J. (2009). Multivariate Archimedean copulas, d-monotone functions and ℓ_1-norm symmetric distributions. *Annals of Statistics*, *37*, 3059–3097.

Mesiar, R. (2005). Discrete copulas – what they are. In E. Montseny & P. Sobrevilla (Eds.), *Joint EUSFLAT-LFA 2005* (pp. 927–930). Barcelona: Universitat Politècnica de Catalunya.

Möller, A., & Groß, J. (2016). Probabilistic temperature forecasting based on an ensemble autoregressive modification. *Quarterly Journal of the Royal Meteorological Society*, *142*, 1385–1394.

Möller, A., Lenkoski, A., & Thorarinsdottir, T. L. (2013). Multivariate probabilistic forecasting using ensemble Bayesian model averaging and copulas. *Quarterly Journal of the Royal Meteorological Society*, *139*, 982–991.

Möller, A., Thorarinsdottir, T. L., Lenkoski, A., & Gneiting, T. (2016). Spatially adaptive, Bayesian estimation for probabilistic temperature forecasts. Available from: http://arxiv.org/abs/1507.05066. (Accessed 22 January 2018).

Nelsen, R. B. (2006). *An Introduction to Copulas* (pp. 272). New York: Springer.

Pinson, P. (2012). Adaptive calibration of (u, v)-wind ensemble forecasts. *Quarterly Journal of the Royal Meteorological Society*, *138*, 1273–1284.

Pinson, P. (2013). Wind energy: forecasting challenges for its operational management. *Statistical Science*, *28*, 564–585.

Pinson, P., Madsen, H., Nielsen, H. A., Papaefthymiou, G., & Klöckl, B. (2009). From probabilistic forecasts to statistical scenarios of short-term wind power production. *Wind Energy, 12*, 51–62.

Pinson, P., & Messner, J. W. (2018). Application of post-processing for renewable energy. In S. Vannitsem, D. S. Wilks, & J. W. Messner (Eds.), *Statistical postprocessing of ensemble forecasts*. Elsevier.

Raftery, A. E., Gneiting, T., Balabdaoui, F., & Polakowski, M. (2005). Using Bayesian model averaging to calibrate forecast ensembles. *Monthly Weather Review, 133*, 1155–1174.

Roulin, E., & Vannitsem, S. (2012). Postprocessing of ensemble precipitation predictions with extended logistic regression based on hindcasts. *Monthly Weather Review, 140*, 874–888.

Rue, H., Martino, S., & Chopin, N. (2009). Approximate Bayesian inference for latent Gaussian models by using integrated nested Laplace approximation (with discussion). *Journal of the Royal Statistical Society Series B, 71*, 319–392.

Rüschendorf, L. (2009). On the distributional transform, Sklar's theorem, and the empirical copula process. *Journal of Statistical Planning and Inference, 139*, 3921–3927.

Schaake, J., Demargne, J., Hartman, R., Mullusky, M., Welles, E., Wu, L., et al. (2007). Precipitation and temperature ensemble forecasts from single-valued forecasts. *Hydrology and Earth Systems Sciences Discussions, 4*, 655–717.

Schefzik, R. (2015). Multivariate discrete copulas, with applications in probabilistic weather forecasting. *Annales de l'ISUP–Publications de l'Institute de Statistique de l'Université de Paris, 59, fasc. 1–2*, 87–116.

Schefzik, R. (2016a). Combining parametric low-dimensional ensemble postprocessing with reordering methods. *Quarterly Journal of the Royal Meteorological Society, 142*, 2463–2477.

Schefzik, R. (2016b). A similarity-based implementation of the Schaake shuffle. *Monthly Weather Review, 144*, 1909–1921.

Schefzik, R. (2017). Ensemble calibration with preserved correlations: unifying and comparing ensemble copula coupling and member-by-member postprocessing. *Quarterly Journal of the Royal Meteorological Society, 143*, 999–1008.

Schefzik, R., Thorarinsdottir, T. L., & Gneiting, T. (2013). Uncertainty quantification in complex simulation models using ensemble copula coupling. *Statistical Science, 28*, 616–640.

Scheuerer, M., & Büermann, L. (2014). Spatially adaptive post-processing of ensemble forecasts for temperature. *Journal of the Royal Statistical Society Series C, 63*, 405–422.

Scheuerer, M., & Hamill, T. M. (2015). Variogram-based proper scoring rules for probabilistic forecasts of multivariate quantities. *Monthly Weather Review, 143*, 1321–1334.

Scheuerer, M., Hamill, T. M., Whitin, B., He, M., & Henkel, A. (2017). A method for preferential selection of dates in the Schaake shuffle approach to constructing spatio-temporal forecast fields of temperature and precipitation. *Water Resources Research, 53*, 3029–3046.

Scheuerer, M., & König, G. (2014). Gridded, locally calibrated, probabilistic temperature forecasts based on ensemble model output statistics. *Quarterly Journal of the Royal Meteorological Society, 140*, 2582–2590.

Schoelzel, C., & Friederichs, P. (2008). Multivariate non-normally distributed random variables in climate research – Introduction to the copula approach. *Nonlinear Processes in Geophysics, 15*, 761–772.

Schoelzel, C., & Hense, A. (2011). Probabilistic assessment of regional climate change in Southwest Germany by ensemble dressing. *Climate Dynamics, 36*, 2003–2014.

Schuhen, N., Thorarinsdottir, T. L., & Gneiting, T. (2012). Ensemble model output statistics for wind vectors. *Monthly Weather Review, 140*, 3204–3219.

Sempi, C. (2011). Copulae: some mathematical aspects. *Applied Stochastic Models in Business and Industry, 27*, 37–50.

Sklar, A. (1959). Fonctions de répartition à *n* dimensions et leurs marges. *Publications de l'Institut de Statistique de l'Université de Paris, 8*, 229–231.

Sloughter, J. M., Gneiting, T., & Raftery, A. E. (2013). Probabilistic wind vector forecasting using ensembles and Bayesian model averaging. *Monthly Weather Review, 141*, 2107–2119.

Stein, M. L. (1999). *Interpolation of Spatial Data: Some Theory for Kriging* (pp. 249). New York: Springer.

Thorarinsdottir, T. L., Scheuerer, M., & Heinz, C. (2016). Assessing the calibration of high-dimensional ensemble forecasts using rank histograms. *Journal of Computational and Graphical Statistics, 25*, 105–122.

Thorarinsdottir, T. L., & Schuhen, N. (2018). Verification: assessment of calibration and accuracy. In S. Vannitsem, D. S. Wilks, & J. W. Messner (Eds.), *Statistical Postprocessing of Ensemble Forecasts*. Elsevier.

Vannitsem, S. (2009). A unified linear Model Output Statistics scheme for both deterministic and ensemble forecasts. *Quarterly Journal of the Royal Meteorological Society, 135*, 1801–1815.

Van Schaeybroeck, B., & Vannitsem, S. (2011). Post-processing through linear regression. *Nonlinear Processes in Geophysics, 18*, 147–160.

Van Schaeybroeck, B., & Vannitsem, S. (2015). Ensemble post-processing using member-by-member approaches: theoretical aspects. *Quarterly Journal of the Royal Meteorological Society, 141*, 807–818.

Voisin, N., Pappenberger, F., Lettenmaier, D. P., Buizza, R., & Schaake, J. C. (2011). Application of a medium-range global hydrologic probabilistic forecast scheme to the Ohio River basin. *Weather and Forecasting, 26*, 425–446.

Vrac, M., & Friederichs, P. (2015). Multivariate – inter-variable, spatial, and temporal – bias correction. *Journal of Climate, 28*, 218–237.

Wilks, D. S. (2015). Multivariate ensemble Model Output Statistics using empirical copulas. *Quarterly Journal of the Royal Meteorological Society, 141*, 945–952.

Wilks, D. S. (2018). Univariate ensemble postprocessing. In S. Vannitsem, D. S. Wilks, & J. W. Messner (Eds.), *Statistical Postprocessing of Ensemble Forecasts*. Elsevier.

Wilks, D. S., & Hamill, T. M. (2007). Comparison of ensemble-MOS methods using GFS reforecasts. *Monthly Weather Review, 135*, 2379–2390.

Wood, A. W., & Schaake, J. C. (2008). Correcting errors in streamflow forecast ensemble mean and spread. *Journal of Hydrometeorology, 9*, 132–148.

POSTPROCESSING FOR EXTREME EVENTS

Petra Friederichs*, Sabrina Wahl*,†, Sebastian Buschow*

University of Bonn, Bonn, Germany* Hans-Ertel Center for Weather Research, Bonn, Germany†

CHAPTER OUTLINE

5.1 INTRODUCTION

The start of operational ensemble weather forecasting in the 1990s raised expectations that predictive guidance for extreme weather conditions would be improved (Bougeault et al., 2010). Ever since, dynamical weather prediction models have become more and more accurate. High-resolution dynamical models for short-range prediction are now operated in ensemble prediction systems (e.g., Gebhardt,

Theis, Paulat, & Ben Bouallègue, 2011). They simulate processes on scales down to the convective scale and are particularly developed for the prediction of high-impact weather related to deep moist convection. Wahl et al. (2017) show that such a convective-scale dynamical model is able to simulate 3-hourly precipitation rates with a distribution and extremal behavior similar to point observations. These models allow for a better representation of mesoscale processes, but still have problems in realistically simulating severe weather phenomena related to deep moist convection.

Ensemble prediction systems provide a vast amount of information that encompasses the conditions for extreme events. For example, positive values of the vertically integrated updraft helicity can be used as an indicator for locations of intense convection (Marsh et al., 2012; Sobash et al., 2011). Thus blending the ensemble forecast with the most advanced statistical postprocessing techniques generally leads to more accurate probabilistic guidance (e.g., Fritsch & Carbone, 2004). Mylne, Woolcock, Denholm-Price, and Darvell (2002), however, showed that calibration substantially improves the quality of ensemble probabilities for nonextreme events, but can actually degrade the skill for extreme events. A special treatment of extreme weather is hence necessary.

Legg and Mylne (2004) provide the first attempt to routinely calibrate model output from ensemble prediction systems (ECMWF-EPS) to probabilistic predictions of severe weather in the UK. The ensemble probabilities are tuned by balancing the mean forecast probability with the sample frequency of severe events. To circumvent representativity issues of the direct model output with respect to observed local weather conditions, Lalaurette (2003) propose quantification of extreme events with respect to the model climate instead of observations. They define the extreme forecast index (EFI) as a measure of deviation of the probabilistic forecast from the model's climate distribution. Positive (negative) values of the EFI denote higher probability for record breaking high (low) values of the variable of interest.

Goodwin and Wright (2010) point out that major challenges in statistical postprocessing include the limited amount of available data (which may contain no extreme events), outdated data sets (historic models or observations are a sample of the past, not the future), and inappropriate statistical models. The assumption of a normally distributed forecast error may largely underestimate the probability of extremes if the true distribution is heavy-tailed. They underline that using extreme-value theory (EVT) concentrates the analysis on the extreme values, thereby avoiding biases toward the bulk of the distribution. Given a robust estimate of the extreme-value model parameters, it is then possible to extrapolate extremes to events that are not contained in the data set.

This chapter thus explores the potential of EVT for postprocessing. The literature and methods available for applications in weather forecasting are vast, and many applications employ EVT such as flood forecasting, statistical downscaling, and many others. Friederichs (2010) uses EVT for the statistical downscaling of extreme precipitation. She uses a nonstationary peak-over-threshold (POT) approach to predict high conditional quantiles and compares theses estimates with conditional quantiles using quantile regression (QR). The differences in terms of quantile scores are small, although the uncertainty of the quantile estimates is largely reduced when using EVT.

None of the methods discussed in Wilks (2018, Chapter 3 of this book) focus on predicting extremes. Although, for example, Lerch and Thorarinsdottir (2013), Scheuerer (2014), or Baran and Nemoda (2016) use a generalized extreme-value distribution (GEV) in their EMOS (i.e., ensemble model output statistics) approach, they use it as a flexible skewed distribution for the complete sample space rather than explicitly simulating the conditional tail behavior of a distribution. As Williams, Ferro, and Kwasniok (2014) show for the Lorenz 1996 model, this procedure does not necessarily result in bad forecasts for extremes, but may fail in situations with heavy tail behavior as will be shown in

Section 5.3. Studies explicitly addressing postprocessing of ensemble predictions for weather extremes are rare. Bentzien and Friederichs (2012) investigated several parametric and nonparametric models for the calibration of ensemble precipitation forecasts. The parametric models based on a lognormal or gamma distribution perform well within the bulk of the distribution. Less skill is obtained for extremal quantiles (99%–99.9%), where both models provide less skill than QR. Using a mixture distribution with a gamma for the bulk of the distribution and generalized Pareto distribution (GPD) for the tail, the predictive skill of extremal conditional quantiles is improved and comparable to QR. The transition from gamma to GPD was set at the 95%-quantile of the gamma distribution. To avoid inconsistencies at the transition, Frigessi, Haug, and Rue (2002) propose a weighted mixture model where the transition of a light-tailed density function to a GPD tail is modeled by a smooth weighting function. For an extension of this idea see Vrac and Naveau (2007) and Naveau, Huser, Ribereau, and Hannart (2016). Bentzien and Friederichs (2012) also show that it is preferable to use a very long training period including the state of seasonal cycle as an additional covariate.

In many applications, not only local information is needed, but realizations of realistic fields of the forecast variable are required (e.g., for risk maps). Thus once an informative model for the marginal (i.e., univariate) distribution is derived, dependence in space and eventually time needs to be taken into account. Oesting, Schlather, and Friederichs (2017) provide an application of a bivariate max-stable process to postprocess spatial predictions of maximum wind gusts. The model they use is a bivariate formulation of a spatial Brown-Resnick process as described in Section 5.4.2. They formulate a joint spatial model for station observations and model forecasts of a 6-hour maximum peak wind speed. After a suitable homogenization of the marginals, the approach of Oesting et al. (2017) may be used to directly postprocess extremal model output toward station observations. In principle, the spatial extreme-value model provides a way to conditionally model observed extremes given the model simulations. However, conditional simulation of max-stable fields is still a matter of intensive research (e.g., Dombry, Engelke, & Oesting, 2016).

We first give an introduction of univariate EVT in Section 5.2. We then turn to postprocessing of extremes using forecasts from the high-resolution COSMO-DE ensemble prediction system (Gebhardt et al., 2011). Our first application is a univariate EMOS-like postprocessing for very high quantiles of quantitative precipitation in Section 5.3. We demonstrate the benefit of using EVT in postprocessing. We then briefly review extremal dependence and multivariate and spatial EVT in Section 5.4. In a second application we postprocess forecasts for spatial wind gusts using a spatial extreme-value process (Section 5.5), in order to draw attention to the potential of available methods for spatial extremes.

5.2 **EXTREME-VALUE THEORY**

"Il est impossible que l'improbable n'arrive jamais" ("It is impossible that the improbable will never occur") is a quotation of Emil Julius Gumbel (Gumbel, 1958). Gumbel expresses the aim for EVT, namely, to provide a probabilistic concept for the prediction of extreme events that, although possible, have not yet been observed, or in other words, a statistical model for extrapolating beyond the range of the data. EVT formulates a universal limit law on the asymptotic behavior of a random variable X as $x \rightarrow x_F$, where x_F is the upper endpoint of the sample space of X. The limit law holds under mild conditions, namely the von Mises condition (de Haan & Ferreira, 2006), and is related to the concept of max-stability.

There exist two procedures to define extremes in a sample. The first refers to extremes as block maxima, that is, the maximum over a subsample of length n of the data. The second works with exceedances over a sufficiently high threshold u. For a large class of distributions EVT proves that the distribution of sample maxima converges for $n \to \infty$ to a GEV, while threshold exceedances for $u \to \infty$ asymptotically follow a GPD. Both procedures are consistent with each other and related by the representation of extremes via a Poisson point process (PP). In our applications herein we work with both threshold exceedances for precipitation and block maxima for wind gusts.

In the following we provide the major findings of EVT in a nutshell. For an easily accessible introduction to EVT the reader is referred to Coles (2001). A deeper introduction also to multivariate EVT is provided, for example, in the textbook by Beirlant, Goegebeur, Segers, and Teugels (2004). There also exist several short introductions to EVT with applications in earth science (e.g., Davison & Huser, 2015; Katz, Parlange, & Naveau, 2002).

5.2.1 GENERALIZED EXTREME-VALUE DISTRIBUTION

Let $X_1, \ldots, X_n$ be identical and independent copies of the random variate X with $F_X(x) = Pr(X_i \leq x)$, and $M_n = \max(X_1, X_2, \ldots, X_n)$ the block maximum. The Fisher-Tippett-Gnedenko theorem (Fisher & Tippett, 1928; Gnedenko, 1943) states that if there exists a sequence of $a_n > 0$ and b_n such that the distribution of the normalized maximum $\tilde{M}_n = (M_n - b_n)/a_n$ for $n \to \infty$ is a nondegenerate distribution $G(z)$, then the distribution function $G(z) = \lim_{n \to \infty} Pr(\tilde{M}_n < z)$ has a form that may be expressed as

$$G(z; \mu, \sigma, \xi) = \exp\left(-\left(1 + \xi \frac{z - \mu}{\sigma}\right)_+^{-1/\xi}\right) \tag{5.1}$$

where $\mu \in \mathbb{R}$ is the location, $\sigma > 0$ the scale, and $\xi \in \mathbb{R}$ the shape parameter. The notation $_+$ indicates that $z_+ = \max(z, 0)$. For $\xi > 0$ the GEV is of Fréchet type, and one says that $F_X(x)$ is in the domain of attraction $\mathcal{D}_\xi$ of a Fréchet type GEV, whereas for $\xi < 0$ the GEV is of Weibull type. For $\xi = 0$ the GEV is of the Gumbel type with

$$\lim_{\xi \to 0} G(z; \mu, \sigma, \xi) = \exp\left(-\exp\left(-\frac{z - \mu}{\sigma}\right)\right) \tag{5.2}$$

which is often called the Fisher-Tippett Type I distribution. The three types are shown in Fig. 5.1.

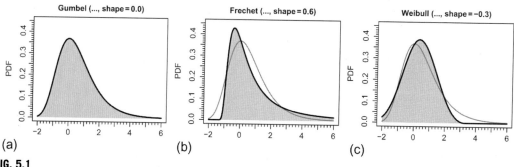

FIG. 5.1

Probability density function of a GEV (*black line with shading*) with parameters (a) $\xi = 0$ (Gumbel), (b) $\xi = 0.6$ (Fréchet), and (c) $\xi = -0.3$ (Weibull). For comparison, the *thin lines* in (b) and (c) show the Gumbel density of (a).

The domain of attraction $\mathcal{D}_\xi$ of a distribution $F_X(x)$ is given by the behavior of the survival function $\overline{F}_X(x) = 1 - F_X(x)$ as $x \rightarrow \infty$. For a Fréchet type $\mathcal{D}_\xi$ ($\xi > 0$) the right end point of the distribution is infinity, and the distribution has a rather heavy right tail. If for instance, the survival function tends to zero at a polynomial speed with

$$\overline{F}_X(x) \sim cx^{-1/\xi} \text{ for } x \rightarrow \infty, \quad c > 0 \text{ and } \xi > 0$$

then the distribution is in a Fréchet type domain of attraction (de Haan & Ferreira, 2006). The expected value only exists for $\xi < 1$, and the variance for $\xi < 0.5$. More generally, the kth moment of a GEV exists for $\xi < k^{-1}$. If $F_X(x)$ is in the domain of attraction of a Weibull type GEV (i.e., $\xi < 0$) there exists an upper endpoint x_F. Any distribution with

$$\overline{F}_X(x_F - x) \sim cx^{-1/\xi} \text{ for } x \rightarrow 0, \quad c > 0, \text{ and } \xi < 0$$

is in the domain of attraction of a Weibull type GEV. Many distributions commonly used in atmospheric science are in the domain of attraction of $\mathcal{D}_0$ (i.e., a Gumbel type GEV). The upper endpoint in this case can be finite or infinite. In the latter case, the tail distribution is light tailed, and a distribution with

$$\overline{F}_X(x) \sim e^{-x} \text{ for } x \rightarrow \infty$$

is in the domain of attraction of $\mathcal{D}_0$.

The intriguing aspect of EVT is that it provides a universal law for the asymptotic behavior of sample maxima. Max-stability is of particular interest for the definition of multivariate extremes and extremal processes (de Haan, 1984). Let $X_1, \ldots, X_n$ be independent copies of a standard Fréchet random variable X with

$$F_X(x) = \exp\{-x^{-1}\}$$

For the maximum $M_n = \max\{X_1, \ldots, X_n\}$ we obtain

$$Pr(M_n \leq x) = F_X^n(x) = \exp\{-nx^{-1}\}$$

Thus, the condition of max-stability (i.e., $F_X^n(nx) = F_X(x)$) holds. Max-stability implies that the form of the probability distribution of the sample maxima of a max-stable random variable is unchanged, while the maxima only have to be rescaled accordingly. The condition of max-stability is the major asset of EVT as it allows for an extrapolation toward unobserved extremes.

5.2.2 PEAK-OVER-THRESHOLD APPROACH

An alternative approach in EVT is POT analysis. Here an extreme is defined as the exceedance over a sufficiently high threshold u. Let X_i be independent copies of X with distribution $F_X(x)$ and upper endpoint x_F. Then $\tilde{X}_i = X_i - u$ for $X_i > u$ defines the threshold exceedance or excess $\tilde{X}$. For $u \rightarrow x_F$ the distribution of $\tilde{X}_i$ can be approximated by a GPD with distribution function

$$H(\tilde{x}; \sigma_u, \xi) = 1 - \left(1 + \xi\frac{\tilde{x}}{\sigma_u}\right)_+^{-1/\xi} \tag{5.3}$$

for some scale parameter σ_u depending on u, and shape parameter ξ independent of u. The relationship between the block maxima and the POT approaches is shown in Pickands (1975) and Balkema and De

Haan (1974) who prove that σ_u and ξ exist if and only if the respective normalization of the sample maximum exists. Then the shape parameter ξ of the GEV and the GPD are equal, and the GPD scale parameter is related to the GEV scale by $\sigma_u = \sigma + \xi(u - \mu)$. As in the block maxima approach we obtain three classes of distributions. For $\xi > 0$ the GPD is defined on $0 < \tilde{x} < \infty$ and the survival function $\overline{H}(\tilde{x}) = 1 - H(\tilde{x})$ follows a classical Pareto distribution with $\overline{H}(\tilde{x}; \sigma_u, \xi) \sim c\tilde{x}^{-1/\xi}$ and $c > 0$. For $\xi < 0$ the GPD has a finite upper endpoint $\tilde{x}_H = \sigma_u/|\xi|$. For $\xi \to 0$ the GPD reduces to the exponential distribution with

$$H(\tilde{x}; \sigma, 0) = 1 - e^{-\tilde{x}/\sigma}$$

and expectation value $E(\tilde{X}) = \sigma$.

A unifying framework for GEV and GPD is provided by the Poisson point process representation (Beirlant et al., 2004; Coles, 2001). Let X be a random variable with distribution $F_X(x)$ where $F_X(x)$ is in the domain of attraction $\mathcal{D}_\xi$ of a GEV and u a sufficiently high threshold. The number of threshold exceedances K in a finite size sample of X is thus small, and for a large sample size K approximately follows a Poisson distribution. For a sufficiently high threshold u the exceedances $X_i - u|X_i > u$ approximately follow a GPD. It can be shown that for a high threshold u the point process $(i/(n + 1),$ $(X_i - a_n)/b_n)$ can be approximated by a Poisson point process on the region $A = [t_1, t_2] \times [x, \infty)$ with the interval $[t_1, t_2] \subset [0, 1]$, $x > u$, and intensity

$$\Lambda(A) = (t_2 - t_1)\left(1 + \xi\frac{x - \mu}{\sigma}\right)_+^{-1/\xi} \tag{5.4}$$

Coles (2001) shows that the Poisson point process provides an elegant way to describe extreme-value distributions for both the block maxima and the POT approach.

5.2.3 NONSTATIONARY EXTREMES

In many applications the processes leading to extremes have characteristics that change with time. Weather extremes do not occur "out of the blue," but rather, the atmosphere sets the conditions for these extremes. For example, extreme precipitation generated by deep moist convection requires the availability of convectively available potential energy and a process that overcomes convective inhibition. Further, weather extremes are generally not well—if at all—represented by dynamical weather prediction models. Thus often it is not simply a matter of postprocessing the variable of interest but, in addition, taking into account a variety of variables (i.e., covariates). The reasoning behind ensemble postprocessing for extremes is thus that the intensity of the Poisson process, or likewise the distribution of the maxima, change with the weather condition, and that information on the weather condition is provided by the ensemble of dynamical weather forecasts. We then ask, what is the tail distribution of a variable given the respective weather forecast?

To account for this information, we extract informative covariates and assume that each extreme-value parameter $\theta \in \{\mu, \sigma, \xi\}$ can be written as $\theta = h(\boldsymbol{\beta}_\theta^T \mathbf{Y})$, respectively, where h is the inverse link function, $\mathbf{Y} = (1, Y_1, ..., Y_K)^T$ a vector of K covariates including a constant, $\boldsymbol{\beta}_\theta = (\theta_0, ..., \theta_K)^T$ a vector of regression coefficients, and superscript T denotes the transpose. Let $\mathbf{y}^{(i)}$ be a realization of the covariate at time t_i. The location parameter may be simulated to linearly depend on $\mathbf{Y}$ with $\mu^{(i)} = \mu_0 + \sum_{k=1}^{K} \mu_k y_k^{(i)}$ and $\boldsymbol{\beta}_\mu = (\mu_0, ..., \mu_K)^T$, where the link function is the identity. Likewise, the scale parameter may be assumed to follow $\sigma^{(i)} = \exp\left(\sigma_0 + \sum_{k=1}^{K} \sigma_k y_k^{(i)}\right)$, which corresponds to a logarithm as link function and ensures a positive variance. A Poisson process may then be estimated given a sufficiently high threshold using maximum likelihood estimation. For a sample of times $i = 1, \ldots, n$ the corresponding likelihood reads

$$L(\mu_0, \sigma_0, \ldots, \mu_K, \sigma_K, \xi) = \exp\left(-\frac{1}{n_p} \sum_i \left(1 + \xi \frac{u - \mu^{(i)}}{\sigma^{(i)}}\right)_+^{-1/\xi}\right)$$
$$\prod_i \left(\frac{1}{\sigma^{(i)}} \left(1 + \xi \frac{x^{(i)} - \mu^{(i)}}{\sigma^{(i)}}\right)_+^{-(1/\xi + 1)}\right)^{\delta_i} \tag{5.5}$$

The shape parameter ξ is generally kept constant over time (Coles, 2001). The time scaling parameter n_p determines the block size of the corresponding GEV distribution. The parameter δ_i is equal 1 only if the observation $x^{(i)} > u$ and 0 otherwise, thus the summation in Eq. (5.5) runs over all times i, whereas the product only considers threshold exceedances $x^{(i)} > u$. Given the nonstationary GEV parameter, the conditional quantiles $q_\tau^{(i)} = F_{\text{GEV}}^{-1}(\tau; \mu^{(i)}, \sigma^{(i)}, \xi)$ with probability τ or exceedance probabilities $p_u^{(i)} = 1 - F_{\text{GEV}}(u; \mu^{(i)}, \sigma^{(i)}, \xi)$ for a threshold u are calculated.

There are a few assumptions behind this procedure. First, the threshold is chosen as constant, thus independently of the weather situation the Poisson process is fitted to the exceedances only. An alternative approach as described in Bentzien and Friederichs (2012) would be to fit a nonstationary GPD to exceedances over a varying threshold defined as a conditional high quantile. In both cases, one has to ensure that the threshold is sufficiently high. This may be assessed via quantile-quantile plots as shown in Friederichs (2010). Further, the likelihood in Eq. (5.5) assumes conditional independence of the threshold exceedances $x^{(i)}$. Although EVT is valid for time series with short-range dependence, a declustering might be necessary in case of conditional dependence of extremes (c.f., Coles, 2001; Davison & Huser, 2015).

For block maxima the procedure would be very similar, except that the likelihood is based on the density of a GEV.

5.3 POSTPROCESSING OF UNIVARIATE EXTREMES: PRECIPITATION
5.3.1 DATA AND ENSEMBLE FORECASTS

The aim of our first application is to provide predictive distributions for extreme precipitation to occur some time between 12 UTC and 18 UTC. The target variable is, hence, the maximum of hourly precipitation accumulations between 12 UTC and 18 UTC on each day. Our observational dataset comprises hourly observations at 409 stations located across Germany.

The predictive distributions are based on ensemble forecasts taken from the operational COSMO-DE ensemble prediction system (COSMO-DE-EPS) operated by DWD (German Meteorological Service in Offenbach). The ensemble forecasts are initialized at 00 UTC. The forecast lead time is thus between 12 and 18 hours. For a description of the COSMO-DE model and the setup of the ensemble system the reader is referred to Baldauf et al. (2011), Gebhardt et al. (2011), and Peralta, Ben Bouallègue, Theis, Gebhardt, and Buchhold (2012).

Both ensemble forecasts and observations are available for a period of 6 years between January 2011 and December 2016. A large selection of variables of the COSMO-DE-EPS at the grid point nearest to each station location is available, including statistics over small and large neighborhoods around each grid point. A description of the most important variables is provided in Table 5.1.

5.3.2 APPROACHES AND VERIFICATION

We employ the nonstationary Poisson point approach (PP) as given in Section 5.2.3 to model the conditional tail distribution. Note that we pool the data for all 409 stations together, assuming thereby spatially stationary regression coefficients and conditional independence in time and space. We set $n_p = 1$ in order to model the tail distribution of the maxima with one value per day. During the optimization of the likelihood, n_p may be used for reparametrization to obtain better inference (Sharkey & Tawn, 2017). The threshold u is chosen as the climatological 99% quantile of the daily data for each month in the year, respectively. Good accordance in the quantile-quantile plots (not shown) justifies this choice.

We want to demonstrate that an explicit postprocessing of the tail distribution is beneficial for (i) reducing a bias in the high quantiles in comparison with a postprocessing that applies to the complete distribution, and (ii) reducing uncertainty in the estimates in comparison to a semiparametric QR model. Both (i) and (ii) are necessary to obtain reliable and sharp predictions of extremes.

Two reference approaches are employed in the following. The first reference model is a flexible parametric distribution for the complete range of the data as proposed in, for example, Scheuerer (2014). We chose a zero-censored GEV (cGEV) to account for zero precipitation events, and use penalized maximum likelihood estimation as suggested by Coles and Dixon (1999) to derive the parameters of the cGEV. As outlined in Section 5.2.3 and identical to the PP approach, the cGEV parameters are modeled as linear functions of the covariates. The main difference to the PP approach is that the nonstationary cGEV is not targeted on the extremes, but is fitted on all 6 hourly maxima of precipitation.

Second, we use as a semiparametric approach censored QR (c.f. Friederichs & Hense, 2007; Koenker, 2005; Wilks 2018, Chapter 3 of this book). Censored QR estimates conditional τ-quantiles using $F^{-1}(\tau|\mathbf{Y} = \mathbf{y}) = \max(0, \boldsymbol{\beta}^T \mathbf{y})$, where the coefficients $\boldsymbol{\beta}$ minimize the asymmetric cost function

$$\hat{\boldsymbol{\beta}} = \underset{\boldsymbol{\beta}}{\operatorname{argmin}} \sum_{i=1}^{N} \left(\max(0, \boldsymbol{\beta}^T \mathbf{y}^{(i)}) - x^{(i)} \right) \left(\mathbb{I}_{x^{(i)} \leq \boldsymbol{\beta}^T \mathbf{y}^{(i)}} - \tau \right) \tag{5.6}$$

over the training data set. As in Section 5.2.3 $x^{(i)}$ denotes the observations and $\mathbf{y}^{(i)}$ a vector of covariates at time i. $\mathbb{I}_v$ is an indicator function with $\mathbb{I}_v = 1$ if v is true and $\mathbb{I}_v = 0$ otherwise. The optimization of Eq. (5.6) is not straightforward, and we follow a three-step procedure proposed by Chernozhukov and Hong (2002) and described in detail in Friederichs and Hense (2007). All reference approaches use the same covariates as in the PP approach.

Our verification concentrates on selected quantiles using the quantile score. Let q_τ be the forecast quantile with probability τ, using either the inverse of the forecast distributions $F_{\text{GEV}}^{-1}(\tau)$ or $F_{\text{cGEV}}^{-1}(\tau)$ or the predicted τ quantile of QR. The quantile score (QS) is based on the cost function in Eq. (5.6) with

$$QS = \sum_{i=1}^{N} \left(q_\tau^{(i)} - x^{(i)} \right) \left(\mathbb{I}_{x^{(i)} \leq q_\tau^{(i)}} - \tau \right) \tag{5.7}$$

where $x^{(i)}$ is the ith observation of N forecast-observation pairs. The QS can be decomposed into three terms called uncertainty (UNC), reliability (REL), and resolution (RES)

$$QS = UNC - RES + REL \tag{5.8}$$

A perfect forecast has zero QS. The uncertainty is the score of the climatology, and hence does not depend on the forecast. The maximum achievable resolution equals the uncertainty, while reliability as the difference between forecast and true distribution should be close to zero. If we divide Eq. (5.8) by the uncertainty, we obtain the quantile skill score (QSS) with

$$QSS = 1 - \frac{QS}{UNC} = \frac{RES}{UNC} - \frac{REL}{UNC} \tag{5.9}$$

The left-hand side of Eq. (5.9) is the QSS using climatology as reference. Positive skill is obtained if the scaled resolution RES/UNC is larger than REL/UNC. The optimal value of RES/UNC is 1, while the optimal value of REL/UNC remains 0. Thus, the scaled resolution term can be understood as the percentage of maximal achievable resolution. On the estimation of the different score components the reader is referred to Bentzien and Friederichs (2014).

Model training and verification uses cross-validation. In order to respect the pronounced seasonal cycle in precipitation over Germany, we perform training on moving 3-month windows. To this end we withhold a month of data in 1 year, and perform training using the same month, the preceding and the following month of the other 5 years (e.g., predictions for April 2016 are based on model estimates using March to May of the years 2011–2015). Verification is performed for DJF and JJA, separately.

In order to estimate the uncertainty of the quantile estimates, we repeat the complete estimation procedure including cross-validation on 100 bootstrap samples of the 6-year dataset with replacement. To account for temporal correlations, we keep data of 5 consecutive days together when generating a bootstrapped dataset. We thus obtain 100 model estimates from each approach for each of the 409 locations and each day within the time period from 2011 to 2016. All uncertainty estimates indicated in the following are obtained via this bootstrap approach.

5.3.3 VARIABLE SELECTION

Ensemble prediction systems provide a huge amount of information, not only on the variable of interest, but on the weather conditions, physical processes, and feedbacks with local conditions. Variable selection has generally two steps, because the amount of data is huge: First, a preselection of predictors that relies on physical reasoning and, second, a statistical selection approach that chooses predictors with respect to their information content (i.e., in terms of capability to reduce the respective score function). Several methods exist for variable selection such as stepwise backward or forward selection (see e.g., Fahrmeir & Tutz, 1994, Chapter 4.1.2), or penalized regression techniques (Kyung, Gill, Ghosh, & Casella, 2010). Here, we use in a Bayesian framework the least absolute shrinkage and selection operator (LASSO) (Tibshirani, 1996). It is similar to penalized regression in the sense that it introduces a penalty to nonzero regression coefficients. To this end, the Bayesian LASSO approach assumes independent zero-mean Laplacian priors for the regression coefficients. Its scale parameter, also included as LASSO parameter, determines the strength that forces the regression coefficients toward zero. A small LASSO parameter leads to only few nonzero regression coefficients.

We use Bayesian QR (Yu & Moyeed, 2001) to select the most informative predictors. The approach follows the LASSO application in Wahl (2015), which also provides details on Bayesian QR. Fig. 5.2 displays the posterior distribution of the regression coefficients using a strong LASSO penalty. A table of the covariates and a description is given in Section 5.7. The results are based on a Monte Carlo sample of 15,000 realizations. Among all 63 covariates, only few coefficients are different from zero (i.e., all Monte

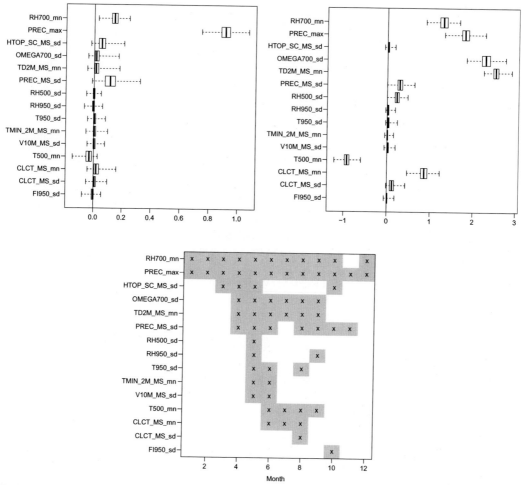

FIG. 5.2

Bayesian quantile regession with LASSO for the 99% quantile of precipitation for the months January (*upper left*) and July (*upper right*). The *boxplots* indicate the posterior distribution of the respective regression coefficients. *Lower panel* indicates selected variable (i.e., all 15,000 Monte Carlo posterior draws are either positive or negative) for each month of the year. Only covariates are displayed that are selected at least once. For a description of the variables see Table 5.1.

Carlo realizations are either above or below zero). During January the most relevant predictors are the ensemble maximum precipitation (PREC-max) and ensemble average relative humidity in 700 hPa (RH700-m), while in July, also the ensemble standard deviation of the vertical velocity in 700 hPa (OMEGA700-sd), the ensemble and small-neighborhood average dew point at 2 m height (TD2M-M-m), and the ensemble standard deviation of averaged total precipitation over a small neighborhood (PREC-M-sd) are of relevance. These are the variables that are most often selected (Fig. 5.2, bottom panel). In July, also the ensemble average temperature at 500 hPa and the average total cloud cover over a small neighborhood are relevant. In the following we use PREC-max, RH700-m, OMEGA700-sd, TD2M-M-m, and PREC-M-sd as predictor variables for all months.

5.3.4 COMPARISON OF POSTPROCESSING APPROACHES

Comparison is made in terms of the QSS for high quantiles with probabilities $\tau = 99\%, 99.5\%, 99.9\%,$ 99.95%, 99.99%. The 99.99% quantile is expected to be exceeded at one station once in 27 years. As we pool all 409 stations together the 99.99% quantile is exceeded about 90 times in our data set.

Our comparison starts with the overall QSS for a range of probability levels τ for DJF and JJA, respectively (Fig. 5.3). QR is performed for five τ-levels only, while cGEV and PP provide estimates for all τ-levels. Because the QSS in Fig. 5.3 is estimated over all 409 stations, it assess skill in time as well as space. QR and PP provide positive skill even for the 99.99% quantile. In DJF the QSS of QR and PP ranges between 50% and 60% for the high quantiles. The skill of the PP quantile forecasts is reduced for probability levels below 99%, because postprocessing using PP deliberately concentrates on levels

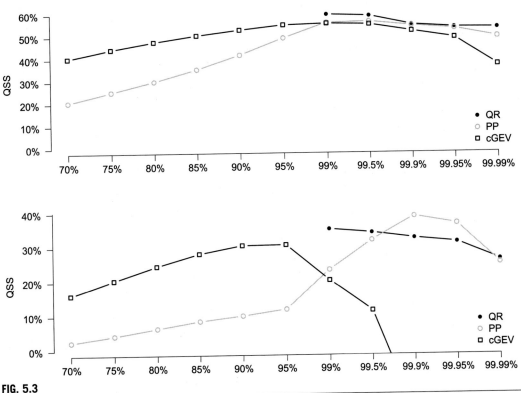

FIG. 5.3

Quantile skill scores for nonstationary censored GEV (cGEV, *light gray squares*), quantile regressions (QR, *black dots*), and nonstationary Poisson point process (PP, *gray circles*). The QSS is averaged over all days in DJF (*upper panel*) and JJA (*lower panel*).

above the climatological 99% quantile. The cGEV approach performs well during DJF for mostly all probability levels, although skill is significantly reduced to about 30% for the 99.99% quantile.

Skill during JJA is largely reduced with values between 25% and 40% for the quantiles above the 99% level. Most remarkable is the failure of the cGEV to provide skillful quantile forecasts above the 99.5% probability level. PP outperforms QR for the 99.9% and 99.95% quantiles, whereas PP and QR show similar skill for the 99.99% level. Thus, the parametric estimate of the tail distribution using a nonstationary Poisson point process provides comparable skill to a semiparametric QR approach.

The failure of our censored GEV approach in JJA is due to the inability of the GEV to represent simultaneously the bulk and the tail of the conditional distribution. Fig. 5.4 represents the cross-validated shape parameter estimates for the cGEV and the PP approach over time. During DJF the shape parameter estimates are slightly negative, and they become largely positive during JJA, particularly for the cGEV approach. The large shape parameter for cGEV in JJA results in strongly over-estimated quantile forecasts at very high τ-levels (e.g., the maximum 99.99% quantile forecast with cGEV amounts to about 2360 mm/h). Although less skillful, cGEV provides acceptable skill at high τ-levels during DJF where the tail distribution resembles an exponential or even beta distribution.

More insight about the performance of quantile forecasts is given by their reliability, in terms of the reliability component of the QS. Fig. 5.5 shows the reliability plot for the 99.99% quantile forecasts in DJF and JJA. Significant underestimation of the conditional quantiles above values of about 7 mm/h is observed for QR and PP during DJF, with a slightly worse reliability for the PP quantiles. PP provides more reliable quantile forecasts than cGEV, which underestimates the conditional quantiles for values above of 2 mm/h with increasing strength toward higher quantile forecasts. Differences in calibration between QR and PP are generally small during JJA. The small quantile forecasts are significantly over-estimated, while PP systematically underestimates quantiles in the range from 25 to 40 mm/h. The cGEV quantile forecasts for JJA are very badly calibrated as already suggested by the largely negative skill score of the higher quantiles as shown in Fig. 5.3. Very similar results are observed for the other τ-levels (not shown).

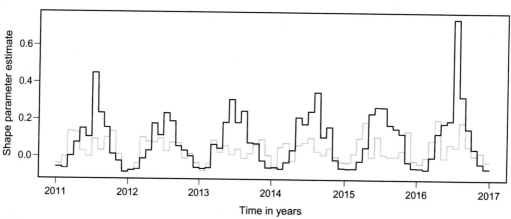

FIG. 5.4

Shape parameter estimates for nonstationary censored GEV (*black line*) and Poisson point process (*gray line*).

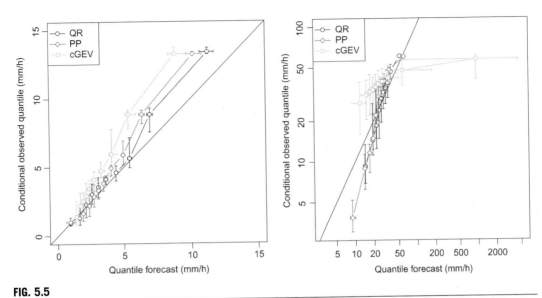

FIG. 5.5

Reliability of the QR, PP, and cGEV quantile forecasts for probability level $\tau = 99.99\%$ during DJF (*left panel*) and for QR and PP during JJA (*right panel*). Forecasts are binned into equally populated intervals. The *x*-axis represents the mean value of forecasts of the respective bin. The *y*-axis shows the observation quantiles estimated conditionally on the forecast bin. The 95% bootstrap confidence intervals of the binning and the conditional quantile estimates are indicated by *whiskers*.

A more detailed analysis of forecast performance is provided in Fig. 5.6 in terms of scaled resolution and reliability. The REL/RES diagram shows a comprehensive comparison of forecast performance, as it simultaneously involves all terms of the score decomposition. Best performance is provided by the model that lies closest to the upper left corner of the plot. Fig. 5.6 shows that QR is superior to PP in terms of both reliability and resolution during DJF. cGEV has less resolution and worse reliability than PP and QR for all quantiles. During JJA, PP is superior in terms of resolution and reliability for the 99.9% and 99.95% quantiles, while reliability is lost for the 99.99% quantile forecasts with PP. Note that we have excluded cGEV from the evaluation during JJA because the values of cGEV lie far outside the displayed range. As indicated in Fig. 5.3, cGEV has negative skill for the displayed quantiles, which ranges between −0.2 for the 99.9% quantile down to −1.4 for the 99.99% quantile. The negative score results from both a low scaled resolution around 0.1 and a poor scaled reliability component ranging between 0.35 and 1.5.

There is evidence in the uncertainty of the score components that QR forecasts are more uncertain than PP forecasts. In order to assess the uncertainty of the quantile estimates, Fig. 5.7 represents the 95% intervals of the block-bootstrap samples with respect to the predicted quantile. Uncertainty for the lower-value quantile forecasts are relatively large for the PP approach, which again is due to the fact that PP is fitted to the exceedances above the 99% climatological quantile only. Otherwise the differences in the uncertainty of the QR and PP estimates are small. However, for very large quantiles during

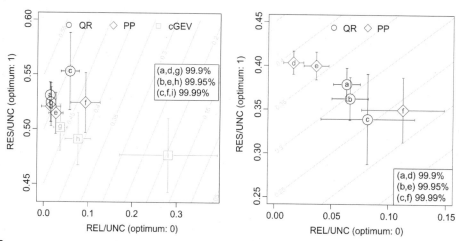

FIG. 5.6

Scaled resolution (RES/UNC) against scaled reliability (REL/UNC) for the QR and PP quantile forecasts and probability levels $\tau = 99\%$, $\tau = 99.95\%$, and $\tau = 99.99\%$, respectively. *Left panel* represents DJF and *right panel* JJA. The *whiskers* denote the 95% bootstrap uncertainty interval of 100-member bootstrap samples. The *gray contours* show lines of constant QSS.

JJA, PP estimates have a smaller uncertainty range compared with QR. Note that the uncertainty range amounts to about 50% of the predicted value of the quantile for the 99.99% probability level.

We conclude that, as long as the tail distribution is an exponential or beta distribution, a suitable conditional distribution for the bulk of the data may also provide satisfactory forecasts for conditional extremes. This is the case in the study of Williams et al. (2014), who investigate common postprocessing methods in the Lorenz 1996 model (Lorenz, 1996). They already remark that the tails of the Lorenz 1996 climatology are quite short, and the tested approaches provide reasonable performance also for extremes. However, in cases where the conditional tail distribution is heavy, a single distribution may not be able to reliably predict the complete range of the distribution, and a separate description of the tail distribution might be necessary. Although slightly more uncertain, QR provides as skilful forecasts as PP even at very high probability levels. Extrapolation is not possible with QR, whereas the good performance of the PP approach suggests that extrapolation may provide valuable guidance for conditional extremes, even above the observed levels.

5.4 EXTREME-VALUE THEORY FOR MULTIVARIATE AND SPATIAL EXTREMES

The theory of multivariate and spatial extreme values is less accessible. It is not obvious how to order multivariate observations, and the class of max-stable multivariate distributions does not consist of parametric distributions with a finite number of parameters. Thus the class of dependence structures is very large. The reader is referred to Davison, Huser, and Thibaud (2013), and Davison and Huser

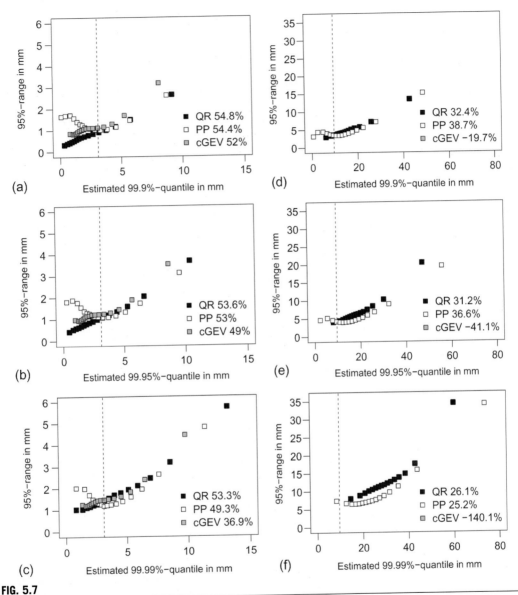

FIG. 5.7

Bootstrap uncertainty of quantile estimates as a function of quantile forecast. The forecasts are binned into equally populated intervals (centers given on x-axis). The *squares* indicate the 95% uncertainty range of a block-bootstrap sample. (a), (b), and (c) represent the quantile forecasts for $\tau = 99.9\%$, 99.95%, and 99.99% during DJF, respectively, and (d), (e), and (f) the same τ-levels during JJA. The *numbers* indicate the QSS of the quantile forecasts. The *horizontal dashed lines* indicate the climatological 99% quantile.

(2015) for a short introduction or to the textbooks of Beirlant et al. (2004) and Coles (2001). Here we only present the main concepts of multivariate EVT, extremal dependence, and spatial extremes.

5.4.1 EXTREMAL DEPENDENCE AND MULTIVARIATE EXTREME-VALUE DISTRIBUTIONS

In many applications, the dependence of extremes, either between variables, or in space or time is of great interest. The dependence is characterized by the properties of the joint tail distribution, which are relevant for extrapolation and thus also of importance for statistical prediction or postprocessing of extremes. A natural measure for the extremal dependence of a bivariate random variable $\mathbf{X} = (X_1, X_2)$ is the dependence measure χ

$$\chi = \lim_{x \to x_F} Pr(X_1 > x | X_2 > x) \tag{5.10}$$

where x_F is the upper endpoint of X_1 and X_2, and assuming that X_1 and X_2 have identical distributions. The dependence measure χ describes the probability of X_1 being extreme given that X_2 is extreme. If $\chi \neq 0$, then X_1 and X_2 are asymptotically dependent, whereas $\chi = 0$ represents asymptotic independence. The function $\chi(x) = Pr(X_1 > x | X_2 > x)$ can also be interpreted as a measure of dependence over a range of quantiles. Moreover, if the distribution of $\mathbf{X} = (X_1, X_2)$ falls in the class of bivariate extreme-value distributions (i.e., is a bivariate max-stable distribution) then $\chi(x)$ is constant for all x.

Extremes in more than one dimension are defined as component-wise maxima. Here we restrict the discussion to the bivariate case, but it may easily be extended to arbitrary dimension q. The component-wise maximum is given as

$$\mathbf{M}_n = \left(\max(X_1^{(1)}, ..., X_1^{(n)}), \max(X_2^{(1)}, ..., X_2^{(n)}) \right)$$

where n is the number of independent realizations. As in univariate EVT, $\mathbf{X}$ is in the domain of attraction of a bivariate extreme-value distribution if there exists a sequence of normalization vectors such that the normalized maxima follow a nondegenerate distribution with GEV marginals. Likewise, the concept of max-stability as formulated in Section 5.2.1 holds for bivariate or multivariate extremes.

Multivariate EVT and practice breaks down into two parts. First, one has to model the marginal distribution of each component of the multivariate extremes and, second, a formulation of the dependence structure between standardized components is needed. Thus, to study the dependence structure of a max-stable random variable, each component is standardized such that it follows a standard GEV. A very useful choice is standard Fréchet margins, although spatial extreme-value models often work with standard Gumbel margins.

Let us assume component-wise maxima with standard Fréchet margins. Then the limit distribution of bivariate maxima is a bivariate extreme-value distribution of the form

$$G(z_1, z_2) = \exp(-V(z_1, z_2)) \tag{5.11}$$

where $V(z_1, z_2)$ is called the exponent measure. The exponent measure is defined by conditions that ensure max-stability, namely, $V(z_1, \infty) = 1/z_1$ and $V(\infty, z_2) = 1/z_2$ to provide standard Fréchet margins, and $V(tz_1, tz_2) = t^{-1}V(z_1, z_2), t > 0$, as a homogeneity condition of order -1 to ensure multivariate max-stability. The exponent measure provides information on extremal dependence, such that, for example, $V(z_1, z_2) = 1/z_1 + 1/z_2$ in case of asymptotic independence and $V(z_1, z_2) = \max(1/z_1, 1/z_2)$ in case of complete dependence.

An alternative measure for extremal dependence is the extremal coefficient θ with

$$\theta = \lim_{z \to z_F} \log Pr(X_1 \leq z, X_2 \leq z) / \log Pr(X_2 \leq z)$$

It can be shown that the dependence measures χ and θ are related, namely that $\chi = 2 - \theta$. Further, if X_1 and X_2 follow a bivariate extreme-value distribution, then it can easily be shown using the homogeneity condition of order -1 that $\theta = V(1, 1)$. The extremal coefficient θ is often used to describe dependence and takes values between $\theta = 1$ for asymptotic dependence and $\theta = 2$ for asymptotic independence.

5.4.2 SPATIAL MAX-STABLE PROCESSES

Turning now to spatial extremes, we first discuss dependence in space using the extremal coefficient θ. We assume a stationary spatial random process X defined on a 2D space $S \subset \mathbb{R}^2$. For the spatial process X at locations $\mathbf{s}, \mathbf{s}' \in S$ and distance $\mathbf{h} = \mathbf{s} - \mathbf{s}'$, one may define the extremal coefficient function $\theta(\mathbf{h}) = \lim_{z \to \infty} \theta(z, \mathbf{h})$ similar to a variogram in geostatistics with

$$\theta(z, \mathbf{h}) = \frac{\log Pr(X(\mathbf{s}+\mathbf{h}) \leq z, X(\mathbf{s}) \leq z)}{\log Pr(X(\mathbf{s}) \leq z)} \tag{5.12}$$

The extremal coefficient $\theta(\mathbf{h})$ is directly related to the F-madogram (Cooley, Naveau, & Poncet, 2006) $\nu(\mathbf{h}) = E(|F_X(z, \mathbf{s} + \mathbf{h}) - F_X(z, \mathbf{s})|)$, where F_X is the marginal distribution of X, and

$$\theta(\mathbf{h}) = \frac{1 + 2\nu(\mathbf{h})}{1 - 2\nu(\mathbf{h})}$$

This relation is later used to estimate the spatial extreme-value process.

Random processes for spatial extremes are of infinite dimension. To be a valid process for spatial extremes it has to possess the max-stability property. Thus, like in multivariate EVT, there is no parametric formulation, but there exist several ways to construct valid max-stable spatial processes. Smith (1990) provides a clear introduction to this construction concept for max-stable processes, as his "rainfall-storms" interpretation is very intuitive.

Max-stable processes are generally constructed as pointwise maxima over an infinite number of stochastic processes. A spatial max-stable process is defined as a process on $S \subset \mathbb{R}^2$ with $\mathbf{s} \in S$ being the location. In our application we use a Brown-Resnick processes (Brown & Resnick, 1977; Kabluchko, Schlather , & De Haan, 2009). The Brown-Resnick processes arise as a limiting process

$$Z(\mathbf{s}) = \max_{i \in \mathbb{N}} \left\{ U_i + W_i(\mathbf{s}) - \frac{\text{Var}(W(\mathbf{s}))}{2} \right\}$$

where $0 < U_1 < U_2 < \cdots$ are the points of a Poisson process on $\mathbb{R}_+$ with intensity $e^{-u} du$, and $W_i, i \in \mathbb{N}$ independently of independent copies of a zero-mean Gaussian random field $\{W(\mathbf{s}), \mathbf{s} \in S\}$ with stationary increments and a semivariogram

$$\gamma(\mathbf{s}) = \frac{1}{2}\text{Var}(W(\mathbf{s}) - W(0))$$

The limiting process Z is stationary and max-stable with standard Gumbel margins and its law only depends on the semivariogram γ (Kabluchko et al., 2009). It is called a Brown-Resnick process associated with the semivariogram γ.

As in Oesting et al. (2017) we restrict the semivariogram to be of type

$$\gamma_\vartheta(\mathbf{h}) = \| aA(b, \zeta)\mathbf{h}\|^\alpha, \quad \mathbf{h} \in \mathbb{R}^2$$

with $\vartheta = (a, b, \zeta, \alpha)$, where $a > 0$ is a scale factor and $\alpha \in (0, 2]$. The matrix $A(b, \zeta) \in \mathbb{R}^{2 \times 2}$ with

$$A = \begin{pmatrix} \cos\zeta & \sin\zeta \\ -b\sin\zeta & b\cos\zeta \end{pmatrix}$$

allows for geometric (elliptical) anisotropy, where $b > 0$ and $\zeta \in (-\pi/4, \pi/4]$. For fitting a Brown-Resnick process to the data, we use a direct relation between the semivariogram $\gamma_\vartheta(\mathbf{h})$ and the extremal coefficient function $\theta(\mathbf{h})$ (c.f. Oesting et al., 2017).

5.5 POSTPROCESSING FOR SPATIAL EXTREMES: WIND GUSTS

The second application focuses on the postprocessing of spatial extremes. The spatial data are hourly peak wind speed observations at 121 weather stations in Germany provided by DWD. We use the 12 UTC to 18 UTC peak wind speed observations denoted as fx for the period between January 2011 and December 2016. The peak wind speed observations represent a few-seconds gusts, so their maximum over a period of 6 hours may well be described by an extreme-value distribution.

The procedure that is described in detail in the following has two steps. The first step constitutes the statistical modeling of the marginal distributions, that is, the distribution of the gust observations fx at each station location conditional on the ensemble forecasts. The second step is the description of the spatial dependence using a max-stable Brown-Resnick process as introduced in Section 5.4.2. The latter works on the residuals that are derived by using the marginal distributions of the first step to transform the gust observations fx to standard Gumbel residuals. The transformation relies on the conditional GEV using a univariate postprocessing similar to that in Section 5.3, where the predictors are again taken from the COSMO-DE-EPS.

5.5.1 POSTPROCESSING FOR MARGINAL DISTRIBUTION

The first step is similar to the approach used for precipitation in Section 5.3. The main difference is that instead of assuming a Poisson point process, we directly work on the 6 hourly maximum gusts fx at the 121 weather stations, which are assumed to follow a nonstationary GEV distribution. The ensemble maximum (VMAX-max) of the COSMO-DE-EPS gust diagnosis VMAX (Schulz, 2007) as well as the ensemble mean 10 m wind speed (VMEAN-m) are taken as predictors. We restrict our predictor selection to these two predictors, because they are available as spatial fields for the period March 2011 to February 2012.

The GEV model is formulated for the marginal of fx − VMEAN-m at time t and location s. We thus assume that fx − VMEAN-m $\sim GEV_{fx}(x; \mu(s, t), \sigma(s, t), \xi)$, with

$$\begin{aligned} \mu(s, t) &= \mu_0 + \mu_1 \text{VMAX-max}(s, t) + \mu_2 \text{VMEAN-m}(s, t) \\ \sigma(s, t) &= \exp(\sigma_o + \sigma_1 \log(\text{VMAX-max}(s, t))) \end{aligned} \tag{5.13}$$

This combination turns out to be the best model in terms of QSS for high quantiles. We further set the shape parameter $\xi = 0$, thus working with a Gumbel-type GEV. Although maximum likelihood estimates provide slightly negative ξ, a negative ξ largely underestimates large gusts. A Gumbel-type GEV provides more stable and skillful estimates for large conditional quantiles. This is in accordance with Perrin, Rootzén, and Taesler (2006) who show that the hypothesis that annual maxima of wind speed follow a Gumbel distribution is rarely rejected.

DWD issues wind gust warnings in Germany for the thresholds $u = 14, 18,$ and 25m/s. We thus calculate conditional quantiles $q_\tau(s, t) = F^{-1}_{\text{Gumbel}}(\tau; \mu(s, t), \sigma(s, t)) + \text{VMEAN-m}$ as well as exceedance probabilities for fixed thresholds $p_u = 1 - F_{\text{Gumbel}}(u - \text{VMEAN-m}; \mu(s, t), \sigma(s, t))$. For verification of the probability forecasts p_u we use the Brier score (Brier, 1950), and the Brier skill score (BSS) defined with respect to climatology.

Our estimation procedure uses cross-validation and is identical to the one used for precipitation. The seasonal cycle for wind gusts is less pronounced, thus QSS and BSS are averaged over the complete time period for each station. For a fair comparison, the climatological estimates to derive QSS and BSS have to be cross-validated, too.

Fig. 5.8 represents QSS and BSS of the nonstationary marginal model. The boxplots show the range of the QSS and BSS calculated at each station, separately. The reference forecast is the cross-validated local climatology of the respective quantile or exceedance probability. The median scores for the 50% to 95% quantiles are about 40%, while the median QSS is reduced to only about 25% for the very high 99.9% quantile. Almost all stations show positive skill even for the highest quantile. Similar results are obtained for the threshold exceedances. The median skill score for the 14 m/s threshold amounts to about 45%, and reduces to about 10% for the 25 m/s threshold. For the 25 m/s threshold skill is thus largely reduced and more than 25% of the stations have negative skill compared with climatology.

The reliability plots in Fig. 5.9 show that the nonstationary GEV distribution provides a partly reliable model for the observed maximum gusts. The reliability for the median forecasts is very good, whereas deviations are observed for the high quantile forecasts. Additional covariates improve reliability (not shown), but because the covariates are not available at all grid points, they are not included here.

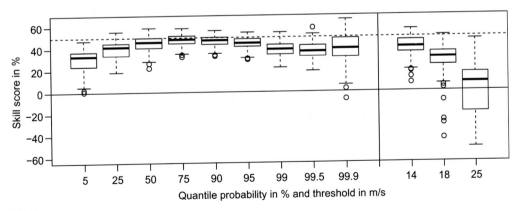

FIG. 5.8

QSS for the 5%, ..., 99.9% quantile forecasts of wind gusts (*left*) and BSS for the exceedance probabilities of thresholds 14, 18, and 25 m/s (*right*), respectively. The *boxplots* indicate the distribution of the skill scores over the stations. The GEV model is a nonstationary Gumbel distribution with covariates for $\mu \sim$ VMEAN-m + VMAX-max and $\sigma \sim$ VMAX-max.

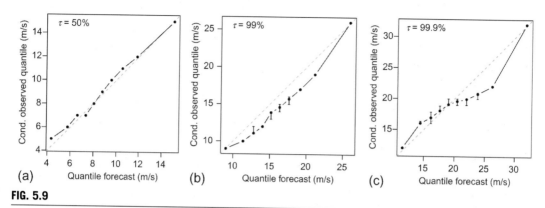

FIG. 5.9

Reliability of the wind gust quantile forecasts for the probability levels (a) 50%, (b) 99%, and (c) 99.9%. Forecasts are binned into equally populated intervals. The x-axis represents the mean value of forecasts of the respective bin. The y-axis shows the observation quantiles estimated conditionally on the forecast bin. The 95% bootstrap uncertainty of the binning and the conditional quantile estimates are indicated by *whiskers*.

The uncertainty in the quantile estimates is estimated using the same bootstrap procedure as for precipitation. Because the estimates are based on a large sample of 6 years of daily data at 121 stations, the uncertainty in the quantile estimates is very small—so small that the whiskers in Fig. 5.9(a) are invisible.

5.5.2 THE SPATIAL DEPENDENCE STRUCTURE

Let us now investigate the dependence structure of the residuals. To this end, the observations are transformed to standard Gumbel residuals using the nonstationary GEV parameters as given in Eq. (5.13). As already suggested by the deficiency of the reliability in Fig. 5.9, the residuals are not perfectly standard Gumbel (not shown).

In a next step, we estimate the bivariate extremal coefficients $\theta_{i,j}$ of the residuals at location s_i and s_j. All $\theta_{i,j}$ with respect to the Euclidean distance $|s_i - s_j|$ are displayed in Fig. 5.10. The extremal coefficients of the closest station pairs is about 1.4. The dependence of the postprocessed residuals is significant, extends to several hundreds of kilometers, and levels out to nearly 2 (i.e., independence) for very distant stations. The anisotropy of the dependence is very small (not shown). Spatial realizations of wind gusts thus would be unrealistic if the spatial dependence is ignored. The extremal coefficients $\theta_{i,j}$ are used to fit a Brown-Resnick process as described in Section 5.4.2. As indicated in Fig. 5.10, the Brown-Resnick extremal coefficient function fits the empirical $\theta_{i,j}$ well. In order to illustrate the appearance of the Brown-Resnick process, one realization of a fitted Brown-Resnick process is given in Fig. 5.10 in the right panel.

Temporal realizations of the postprocessed wind gust predictions are illustrated at Munich airport station for June 2011, and spatial predictions are discussed for June 22, 2011, in Fig. 5.11, representing the COSMO-DE-EPS gust diagnostic VMAX at 10 m height, together with the respective observations at Munich airport station for June 2011. The COSMO-DE-EPS gusts closely follow the observations. However, one notices that the observations often lie outside the range of the 20-member ensemble and

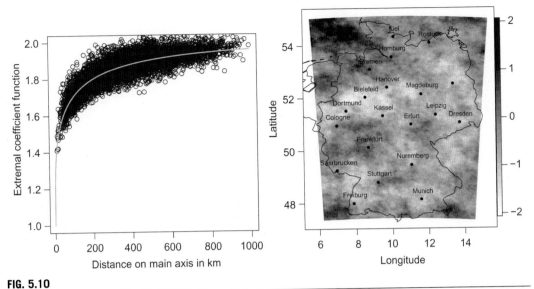

FIG. 5.10

Extremal coefficients and Brown-Resnick fit for the residuals using a nonstationary GEV. The *circles* represent the estimates of the extremal coefficient between two stations with respect to the distance in kilometers. The *gray line* indicates the extremal coefficient function of the fitted Brown-Resnick process as a function of average distance (*left panel*). A realization of the fitted Brown-Resnick process (*right panel*).

that the ensemble is largely underdispersive. This may be seen in a rank histogram (not shown), which has a distinct U-shape in which about 30% of the days the observation lies below and about 20% above the range of the ensemble.

Using the fitted Brown-Resnick process we generate realizations of the residual process and transform the residuals using the nonstationary Gumbel model. Fig. 5.11 represents an ensemble of 100 postprocessed forecasts for June 2011 at Munich airport station. The postprocessed realizations show a larger range, and although in 17% of the days, the observations lie below the range of a postprocessed ensemble, the rank histogram is much better calibrated (not shown), particularly for the moderate to higher ranks.

With the knowledge of the residual extreme-value process, that is, the fitted max-stable Brown-Resnick process, and the covariates at each grid point, it is now possible to simulate the spatial extreme-value process conditional on the COSMO-DE-EPS forecasts. We will present spatial forecasts for June 22, 2011, which represents a day with extreme weather conditions over Germany, due to a cold front associated with a low pressure system that formed around 16/17th June over the western North Atlantic (Weijenborg, Friederichs, & Hense, 2015). The maximum gust diagnosis (VMAX-max) as well as the observed gusts are represented in Fig. 5.12. COSMO-DE-EPS predicts gusts exceeding 40 m/s in the center of Germany, whereas maximum gust observations of about 36 m/s were observed in southern Germany.

Fig. 5.13 represents postprocessed realizations for June 22, 2011. We omit a complete description of the unconditional and conditional spatial wind gust process. Knowing the nonstationary GEV

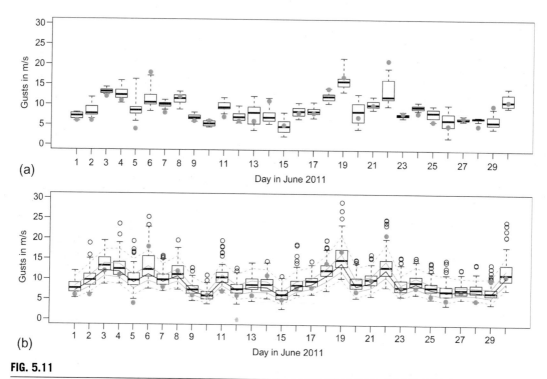

FIG. 5.11

(a) Gust diagnosis of COSMO-DE-EPS 10 m wind on June 2011 at Munich airport. The *boxplots* represent the median, the interquartile range (IQR) and the range of the ensemble. (b) Postprocessed gust forecasts using 100 realizations of the Brown-Resnick process and transformed using the nonstationary GEV. The *gray dots* show the observations. The *boxplots* represent the median, the IQR, the 1.5 IQR limits, and the outliers. The *lines* give the postprocessed median (*black solid*), the 25% and 75% quantile (*gray solid*), and the 5% and 95% quantile (*gray dashed*).

parameters and the COSMO-DE-EPS forecast at each location, we transform simulations of the Brown-Resnick spatial max-stable process to obtain realizations of the postprocessed wind gusts. Four such realizations are displayed in Fig. 5.13. The realizations show a large variety of fields, where in one realization, gusts above 35 m/s are realized in several locations, whereas others do not have gusts above 30 m/s.

The postprocessed COSMO-DE-EPS simulations provide realizations that improve the calibration of the ensemble and the spatial structure of the forecasts, that is, provide realizations that are less underdispersive, and that have a dependence structure close to the observations.

Our application represents a first attempt at using spatial EVT for ensemble postprocessing of extremes. The example leaves considerable room for improvements. First of all, the transformation of the marginals to standard Gumbel using a nonstationary GEV distribution is not perfect, and deviations in the residuals from standard Gumbel are considerable. How strongly this impacts the estimates of the spatial process is not evident. Second, the spatial process is assumed to have a stationary dependence

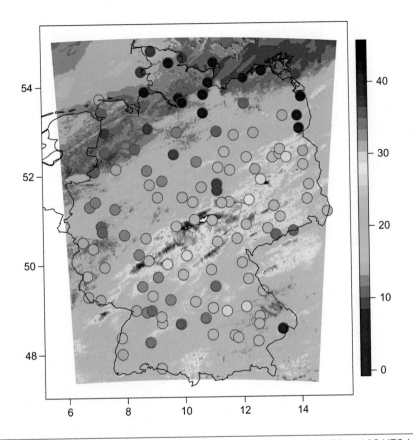

FIG. 5.12

Ensemble maximum gust diagnostic VMAX-max on June 22, 2011 between 12 UTC and 18 UTC (*shading*) and observed gusts (*dots*) in m/s (*color bar*).

structure. However, spatial dependence of the residuals might depend on the weather situation and on local conditions. And finally, it is not clear how to evaluate or validate the postprocessed spatial fields, and research is needed in developing scores for spatial extremes.

5.6 CONCLUSIONS

Extreme-value statistics is a vibrant field of research, and meteorologists have only recently started using EVT in ensemble postprocessing. The existence of a limit law similar to the central limit theorem provides statistical models for extremes, which are not available for general variables. The application of EVT is not always straightforward, and the methods become complex for multivariate or spatial extremes. However, this is the case for all nonnormal multivariate and spatial variables.

We have provided two applications, to precipitation and wind gusts, showing the potential of EVT for ensemble postprocessing. Postprocessing for precipitation employs a spatially homogeneous, univariate, nonstationary Poisson point process (PP). Because the PP is fitted to threshold exceedances

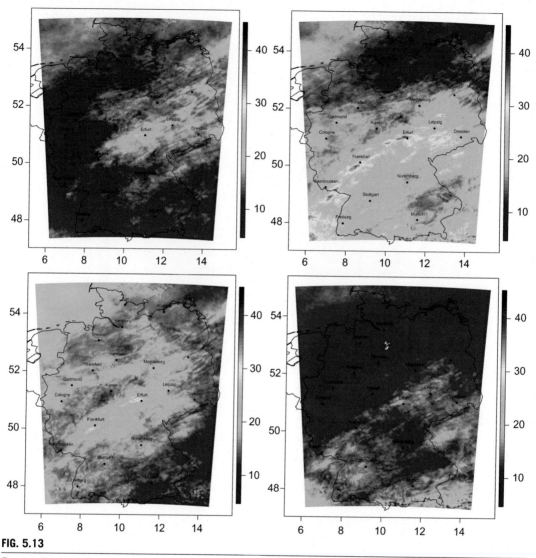

FIG. 5.13

Four realizations of gust forecasts in m/s, using the spatial Brown-Resnick process and nonstationary GEV marginals for June 22, 2011 between 12 UTC and 18 UTC.

only, it focuses on the postprocessing of extremes. For comparison, a cGEV is used to postprocess the complete distribution, and QR is applied as a method that does not imply a distributional assumption. Some attention is also given to variable selection using the LASSO approach, and to the verification of quantile forecasts.

We show that, as long as the conditional distribution is light-tailed, a suitable distribution for the bulk of the data may also provide satisfactory forecasts for conditional extremes. In cases where the conditional tail distribution is heavy, a separate description of the tail distribution seems necessary. For precipitation this is the case during the summer season, where the cGEV approach fails to provide reliable and skilful forecasts for high conditional quantiles. The skill of quantile forecasts from QR and PP is very similar, even at very high probability levels. Extrapolation is not possible with QR, whereas the good performance of the PP approach suggests that extrapolation may provide valuable guidance for conditional extremes even above the observed levels.

Spatial postprocessing was applied to maximum hourly gusts over 6-hourly forecast periods. The procedure is twofold: First describe the marginal distribution, then the dependence structure. The postprocessing of the marginals uses a spatially homogeneous, nonstationary GEV distribution. The fitted nonstationary GEV parameters are then used to derive residuals. We show that these residuals exhibit considerable spatial dependence. We then describe this spatial dependence by pairwise extremal coefficients, which are used to fit a max-stable Brown-Resnick process. The fitted Brown-Resnick process together with the conditional distribution of the marginals provides a full description of the spatial gust process, and allows for the simulation of an ensemble of postprocessed spatial gust forecasts.

Although the theory of max-stable processes is less accessible in practice, the tools to perform postprocessing for spatial extremes are available. Moreover, the EVT community provides software packages for the Statistical Programming Language R (R Core Team, 2013). The R Package `RandomFields` (Schlather, 2016) provides sophisticated software to simulate max-stable processes, and the R Package `SpatialExtremes` (Ribatet, 2015) provides tools to fit and simulate from various classes of max-stable random fields.

5.7 APPENDIX

Table 5.1 List of Selected Covariates From COSMO-DE-EPS

Acronyms	Variables	Statistics
CLCT-M-m	Total cloud cover	Ensemble mean of area mean
CLCT-M-sd	Total cloud cover	Ensemble standard deviation of area mean
FI950-sd	Geopotential at 950 hPa	Ensemble standard deviation
HTOP_SC-M-sd	Cloud top above mean sea level (shallow convection)	Ensemble standard deviation of area mean
OMEGA700-sd	Vertical velocity at 700 hPa	Ensemble standard deviation
PREC-max	Total precipitation	Ensemble maximum
PREC-M-sd	Total precipitation	Ensemble standard deviation of area mean
RH500-sd	Relative humidity at 500 hPa	Ensemble standard deviation
RH700-m	Relative humidity at 700 hPa	Ensemble mean
RH950-sd	Relative humidity at 950 hPa	Ensemble standard deviation
TMIN_2M-M-m	Minimum 2m temperature	Ensemble mean of area mean
TD2M-M-m	2m dew point temperature	Ensemble mean of area mean
T500-m	Temperature at 500 hPa	Ensemble mean

Continued

Table 5.1 List of Selected Covariates From COSMO-DE-EPS—cont'd

Acronyms	Variables	Statistics
T950-sd	Temperature at 950 hPa	Ensemble standard deviation
VMAX-max	10m gust diagnosis	Ensemble maximum
VMEAN-m	10m wind velocity	Ensemble mean

Notes: *Five different statistics are use to calculate the predictors: m indicates the mean value of 20 ensemble members, sd the standard deviation of 20 ensemble members, max the maximum value of 20 ensemble members, M the mean over a spatial neighborhood of 11 × 11 gridpoints, and SD the respective standard deviation over the neighborhood. The statistic over the neighborhood (M, SD) is calculated prior to the statistic over the ensemble (m, sd, max).*

ACKNOWLEDGMENTS

The authors would like to offer our special thanks to Reinhold Hess of the German Meteorological Service (DWD, Offenbach, Germany) for providing the COSMO-DE-EPS forecasts. The authors greatly appreciated the valuable comments on the manuscript by Marco Oesting (University of Siegen, Germany), an anonymous reviewer, and the editors.

REFERENCES

Baldauf, M., Seifert, A., Förstner, J., Majewski, D., Raschendorfer, M., & Reinhardt, T. (2011). Operational convective-scale numerical weather prediction with the COSMO model: description and sensitivities. *Monthly Weather Review, 139*, 3887–3905.

Balkema, A. A., & De Haan, L. (1974). Residual life time at great age. *The Annals of Probability, 2*, 792–804.

Baran, S., & Nemoda, D. (2016). Censored and shifted gamma distribution based EMOS model for probabilistic quantitative precipitation forecasting. *Environmetrics, 27*, 280–292.

Beirlant, J., Goegebeur, Y., Segers, J., & Teugels, J. (2004). *Statistics of Extremes: Theory and Applications.* Chichester: Wiley.

Bentzien, S., & Friederichs, P. (2012). Probabilistic quantitative precipitation forecasting using the high-resolution convection-permitting NWP model COSMO-DE. *Weather and Forecasting, 27*, 988–1002.

Bentzien, S., & Friederichs, P. (2014). Decomposition and graphical portrayal of the quantile score. *Quarterly Journal of the Royal Meteorological Society, 140*, 1924–1934.

Bougeault, P., Toth, Z., Bishop, C., Brown, B., Burridge, D., & Chen, D. H. (2010). The THORPEX interactive grand global ensemble. *Bulletin of the American Meteorological Society, 91*, 1059–1072.

Brier, G. W. (1950). Verification of forecasts expressed in terms of probability. *Monthly Weather Review, 78*, 1–3.

Brown, B. M., & Resnick, S. I. (1977). Extreme values of independent stochastic processes. *Journal of Applied Probability, 14*, 732–739.

Chernozhukov, V., & Hong, H. (2002). Three-step censored quantile regression, with an application to extramarital affairs. *Journal of the American Statistical Association, 97*, 872–882.

Coles, S. (2001). *An Introduction to Statistical Modeling of Extreme Values. Springer Series in Statistics.* London: Springer-Verlag.

Coles, S. G., & Dixon, M. J. (1999). Likelihood-based inference for extreme value models. *Extremes, 2*, 5–23.

Cooley, D., Naveau, P., & Poncet, P. (2006). Variograms for spatial max-stable random fields. *Dependence in Probability and Statistics* (pp. 373–390). New York: Springer.

Davison, A. C., & Huser, R. (2015). Statistics of extremes. *Annual Review of Statistics and Its Application, 2,* 203–235.

Davison, A. C., Huser, R., & Thibaud, E. (2013). Geostatistics of dependent and asymptotically independent extremes. *Mathematical Geosciences, 45,* 511–529.

de Haan, L. (1984). A spectral representation for max-stable processes. *Annals of Probability, 12,* 1194–1204.

de Haan, L., & Ferreira, A. (2006). *Extreme Value Theory: An Introduction.* New York: Springer Science & Business Media.

Dombry, C., Engelke, S., & Oesting, M. (2016). Exact simulation of max-stable processes. *Biometrika, 103,* 303–317.

Fahrmeir, L., & Tutz, G. (1994). *Multivariate Statistical Modelling Based on Generalized Linear Models.* New York: Springer.

Fisher, R. A., & Tippett, L. H. (1928). On the estimation of the frequency distributions of the largest or smallest member of a sample. *Proceedings of the Cambridge Philosophical Society, 24,* 180–190.

Friederichs, P. (2010). Statistical downscaling of extreme precipitation using extreme value theory. *Extremes, 13,* 109–132.

Friederichs, P., & Hense, A. (2007). Statistical downscaling of extreme precipitation events using censored quantile regression. *Monthly Weather Review, 135,* 2365–2378.

Frigessi, A., Haug, O., & Rue, H. (2002). A dynamic mixture model for unsupervised tail estimation without threshold selection. *Extremes, 5,* 219–235.

Fritsch, J. M., & Carbone, R. (2004). Improving quantitative precipitation forecasts in the warm season: a USWRP research and development strategy. *Bulletin of the American Meteorological Society, 85,* 955–965.

Gebhardt, C., Theis, S., Paulat, M., & Ben Bouallègue, Z. (2011). Uncertainties in COSMO-DE precipitation forecasts introduced by model perturbations and variation of lateral boundaries. *Atmospheric Research, 100,* 168–177.

Gnedenko, B. (1943). Sur la distribution limite du terme maximum d'une serie aléatoire. *Annals of Mathematics, 44,* 423–453.

Goodwin, P., & Wright, G. (2010). The limits of forecasting methods in anticipating rare events. *Technological Forecasting and Social Change, 77,* 355–368.

Gumbel, E. J. (1958). *Statistics of Extremes.* New York: Columbia University Press.

Kabluchko, Z., Schlather, M., & De Haan, L. (2009). Stationary max-stable fields associated to negative definite functions. *The Annals of Probability, 37,* 2042–2065.

Katz, R. W., Parlange, M. B., & Naveau, P. (2002). Statistics of extremes in hydrology. *Advances in Water Resources, 25,* 1287–1304.

Koenker, R. (2005). *Quantile Regression.* Cambridge: Cambridge University Press.

Kyung, M., Gill, J., Ghosh, M. & Casella, G. (2010). Penalized regression, standard errors, and Bayesian lasso. *Bayesian Analysis, 5,* 369–411.

Lalaurette, F. (2003). Early detection of abnormal weather conditions using a probabilistic extreme forecast index. *Quarterly Journal of the Royal Meteorological Society, 129,* 3037–3057.

Legg, T. P., & Mylne, K. R. (2004). Early warnings of severe weather from ensemble forecast information. *Weather and Forecasting, 19,* 891–906.

Lerch, S., & Thorarinsdottir, T. (2013). Comparison of non-homogeneous regression models for probabilistic wind speed forecasting. *Tellus A, 65,* 21206.

Lorenz, E. N (1996). Predictability: A problem partly solved. *Proc. Seminar on Predictability, Vol. 1.* (pp. 1–18). Berkshire, UK: ECMWF, Reading.

Marsh, P. T., Kain, J. S., Lakshmanan, V., Clark, A. J., Hitchens, N. M., & Hardy, J. (2012). A method for calibrating deterministic forecasts of rare events. *Weather and Forecasting, 27,* 531–538.

Mylne, K. R., Woolcock, C., & Denholm-Price, J. C. W.Darvell, R. J. (2002). Operational calibrated probability forecasts from the ECMWF ensemble prediction system: Implementation and verification. *Joint Session of 16th Conf on Probability and Statistics in the Atmospheric Sciences and of Symposium on Observations,*

Data Assimilation, and Probabilistic Prediction (Orlando, Florida) (pp. 113–118). American Meteorological Society.

Naveau, P., Huser, R., Ribereau, P., & Hannart, A. (2016). Modeling jointly low, moderate, and heavy rainfall intensities without a threshold selection. *Water Resources Research, 52*, 2753–2769.

Oesting, M., Schlather, M., & Friederichs, P. (2017). Statistical post-processing of forecasts for extremes using bivariate Brown-Resnick processes with an application to wind gusts. *Extremes, 20*, 309–332.

Peralta, C., Ben Bouallègue, Z., Theis, S. E., Gebhardt, C., & Buchhold, M. (2012). Accounting for initial condition uncertainties in COSMO-DE-EPS. *Journal of Geophysical Research, 117*, D7.

Perrin, O., Rootzén, H., & Taesler, R. (2006). A discussion of statistical methods used to estimate extreme wind speeds. *Theoretical and Applied Climatology, 85*, 203–215.

Pickands, J. (1975). Statistical inference using extreme order statistics. *The Annals of Statistics, 3*, 119–131.

R Core Team. (2013). *R: A Language and Environment for Statistical Computing*. Vienna, Austria: R Foundation for Statistical Computing.

Ribatet, M. (2015). *SpatialExtremes: modelling spatial extremes*. R package version 2.02.

Scheuerer, M. (2014). Probabilistic quantitative precipitation forecasting using ensemble model output statistics. *Quarterly Journal of the Royal Meteorological Society, 140*, 1086–1096.

Schlather, M., Malinowski, A., Oesting, M., Boecker, D., Strokorb, K., & Engelke, S. (2016). *RandomFields: simulation and analysis of random fields*. R package version 3.1.8.

Schulz, J. -P. (2007). Revision of the turbulent gust diagnostics in the COSMO-model. *COSMO Newsletter, 8*, 17–22. http://www.cosmo-model.org.

Sharkey, P., & Tawn, J. A. (2017). A Poisson process reparameterisation for Bayesian inference for extremes. *Extremes, 20*, 239–263.

Smith, R. L. (1990). *Max-stable processes and spatial extremes* 205, Unpublished manuscript.

Sobash, R. A., Kain, J. S., Bright, D. R., Dean, A. R., Coniglio, M. C., & Weiss, S. J. (2011). Probabilistic forecast guidance for severe thunderstorms based on the identification of extreme phenomena in convection-allowing model forecasts. *Weather and Forecasting, 26*, 714–728.

Tibshirani, R. (1996). Regression shrinkage and selection via the lasso. *Journal of the Royal Statistical Society: Series B (Methodological), 58*, 267–288.

Vrac, M., & Naveau, P. (2007). Stochastic downscaling of precipitation: from dry events to heavy rainfalls. *Water Resources Research, 43*, W07402.

Wahl, S. (2015). Uncertainty in mesoscale numerical weather prediction: probabilistic forecasting of precipitation (Ph.D. dissertation), Universitäts-und Landesbibliothek Bonn (available online http://hss.ulb.uni-bonn.de/2015/4190/4190.htm).

Wahl, S., Bollmeyer, C., Crewell, S., Figura, C., Friederichs, P., & Hense, A. (2017). A novel convective-scale regional reanalysis COSMO-REA2: improving the representation of precipitation. *Meteorologische Zeitschrift, 26*, 345–361.

Weijenborg, C., Friederichs, P., & Hense, A. (2015). Organisation of potential vorticity during severe convection. *Tellus A, 67*, 25705.

Wilks, D. S. (2018). Univariate ensemble postprocessing. In S. Vannitsem, D. S. Wilks, & J. W. Messner (Eds.), *Statistical Postprocessing of Ensemble Forecasts*. Elsevier.

Williams, R., Ferro, C., & Kwasniok, F. (2014). A comparison of ensemble post-processing methods for extreme events. *Quarterly Journal of the Royal Meteorological Society, 140*, 1112–1120.

Yu, K., & Moyeed, R. A. (2001). Bayesian quantile regression. *Statistics & Probability Letters, 54*, 437–447.

VERIFICATION: ASSESSMENT OF CALIBRATION AND ACCURACY

6

Thordis L. Thorarinsdottir, Nina Schuhen

Norwegian Computing Center, Oslo, Norway

CHAPTER OUTLINE

6.1 INTRODUCTION

In a discussion article on the application of mathematics in meteorology, Bigelow (1905) describes the fundamentals of modeling in a timeless manner:

> There are three processes that are generally essential for the complete development of any branch of science, and they must be accurately applied before the subject can be considered to be satisfactorily explained. The first is the discovery of a mathematical analysis, the second is the discussion of numerous observations, and the third is a correct application of the mathematics to the observations, including a demonstration that these are in agreement.

Statistical Postprocessing of Ensemble Forecasts. https://doi.org/10.1016/B978-0-12-812372-0.00006-6

The topic of this chapter is methods for carrying out the last item on Bigelow's list, that is, methods to demonstrate the agreement between a model and a set of observations. Ensemble prediction systems and statistically postprocessed ensemble forecasts provide probabilistic predictions of future weather. Verification methods applied to these systems should thus be equipped to handle both the verification of the best prediction derived from the ensemble and the verification of the associated prediction uncertainty.

Murphy (1993) argues that a general prediction system should strive to perform well on three types of goodness: There should be consistency between the forecaster's judgment and the forecast, there should be correspondence between the forecast and the observation, and the forecast should be informative for the user. Similarly, Gneiting, Balabdaoui, and Raftery (2007) state that the goal of probabilistic forecasting should be to maximize the sharpness of the predictive distribution subject to calibration. Here, calibration refers to the statistical consistency between the forecast and the observation, while sharpness refers to the concentration of the forecast uncertainty; the sharper the forecast, the higher information value will it provide, as long as it is also calibrated. The prediction goal of Gneiting et al. (2007) is thus equivalent to Murphy's second and third types of goodness.

We focus on verification methods for probabilistic predictions of continuous variables in one or more dimensions under the general framework described by Murphy (1993) and Gneiting et al. (2007). Specifically, we denote an observation in d dimensions by $y = (y_1, \ldots, y_d) \in \Omega^d$ for $d = 1$, $2, \ldots$, where Ω denotes either the real axis $\mathbb{R}$, the nonnegative real axis $\mathbb{R}_{\geq 0}$, the positive real axis $\mathbb{R}_{>0}$, or an interval on $\mathbb{R}$. A probabilistic forecast for y given by a distribution function with support on Ω^d is denoted by $F \in \mathcal{F}$ for some appropriate class of distributions $\mathcal{F}$, with the density denoted by f if it exists. For ensemble forecasts, we will alternatively use the notation $\mathbf{x} = \{x_1, \ldots, x_K\}$ to describe the K ensemble members or F for the associated empirical distribution function. Verification methods for deterministic predictions and other types of variables are discussed, for example, in Wilks (2011, Chapter 8) and Jolliffe and Stephenson (2012).

This chapter is organized as follows. Diagnostic tools for checking calibration are discussed in Section 6.2. Section 6.3 describes methods that assess the accuracy of forecasts where each forecast is issued a numerical score based on the event that materializes. Scoring rules apply to individual events while divergence functions compare the empirical distribution of a series of events with a predictive distribution. The scores may focus on certain aspects of the forecast, such as the tails, and it is important also to assess the uncertainty in the scores. The properties of various univariate scores are compared in a simulation study. While the methods in Section 6.3 provide a decision-theoretically coherent approach to model evaluation and model ranking, they may hide key information about the model performance such as the direction of bias. Additional evaluation may thus be needed to better understand the performance of a single model. Approaches for this are discussed in Section 6.4. The chapter then closes with a summary in Section 6.5.

6.2 CALIBRATION

Calibration, or reliability, is the most fundamental aspect of forecast skill for probabilistic forecasts as it is a necessary condition for the optimal use and value of the forecast. Calibration refers to the statistical compatibility between the forecast and the observation; the forecast is calibrated if the observation cannot be distinguished from a random draw from the predictive distribution.

6.2.1 UNIVARIATE CALIBRATION

Several alternative notions of univariate calibration exist for a single forecast (Gneiting et al., 2007; Tsyplakov, 2013) and a group of forecasts (Strähl & Ziegel, 2017). We focus on the so-called *probabilistic calibration* as suggested by Dawid (1984); F is probabilistically calibrated if the *probability integral transform* (PIT) $F(Y)$, the value of the predictive cumulative distribution function for the random observation Y, is uniformly distributed. If F has a discrete component, a randomized version of the PIT given by

$$\lim_{y\uparrow Y} F(y) + V\left(F(Y) - \lim_{y\uparrow Y} F(y)\right)$$

with $V \sim \mathcal{U}([0,1])$ may be used, see Gneiting and Ranjan (2013). Here, we use $y\uparrow Y$ to denote that the limit is taken as y approaches Y from below.

Assume our test set consists of n observations $y_1, \ldots, y_n$. For a forecasting method issuing continuous univariate predictive distributions $F_1, \ldots, F_n$, calibration can be assessed empirically by plotting the histogram of the PIT values

$$F_1(y_1), \ldots, F_n(y_n).$$

A forecasting method that is calibrated on average will return a uniform histogram, a $\cap$-shape indicates overdispersion and a $\cup$-shape indicates underdispersion, while a systematic bias results in a triangular-shaped histogram. Examples of miscalibration are shown in Fig. 6.1, including a biased forecast (panel a), an underdispersive forecast (panel b), an overdispersive forecast (panel c), and an example of a multiply misspecified forecast where the left tail is too light, the main bulk of the distribution lacks mass and the right tail is too heavy (panel d).

The discrete equivalent of the PIT histogram, which applies to ensemble forecasts, is the verification rank histogram (Anderson, 1996; Hamill & Colucci, 1997). It shows the distribution of the ranks of the observations within the corresponding ensembles and has the same interpretation as the PIT histogram.

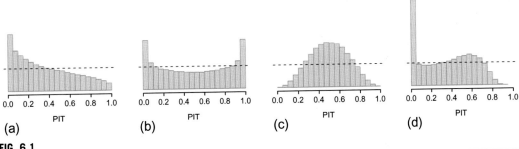

(a)   **(b)**   **(c)**   **(d)**

FIG. 6.1

Probability integral transform (PIT) histograms for 100,000 simulated standard Gaussian $\mathcal{N}(0,1)$ observations and various misspecified forecasts: (a) biased $\mathcal{N}(0.5,1)$ forecasts, (b) underdispersive $\mathcal{N}(0,0.75^2)$ forecasts, (c) overdispersive $\mathcal{N}(0,2^2)$ forecasts, and (d) multiply misspecified generalized extreme value GEV(0, 1, 0.5) forecasts. The theoretically optimal histograms are indicated with *dashed lines*.

The information provided by a rank histogram may also be summarized numerically by the reliability index (RI), which is defined as

$$RI = \sum_{i=1}^{I} \left| \zeta_i - \frac{1}{I} \right|$$

where I is the number of (equally sized) bins in the histogram and ζ_i is the observed relative frequency in bin $i = 1, ..., I$. The RI thus measures the departure of the rank histogram from uniformity (Delle Monache, Hacker, Zhou, Deng, & Stull, 2006).

6.2.2 MULTIVARIATE CALIBRATION

For assessing the calibration of multivariate forecasts, Gneiting, Stanberry, Grimit, Held, and Johnson (2008) formalized a general two-step framework. Let $S = \{x_1, ..., x_K, y\}$ denote a set of $K + 1$ points in Ω^d comprising an ensemble forecast with K members and the corresponding observation y. The rank of y in S, $\text{rank}_S(y)$, is calculated in two steps,

(i) apply a prerank function $\rho_S : \Omega^d \to \mathbb{R}_{\geq 0}$ to calculate the prerank $\rho_S(u)$ of every $u \in S$ resulting in a univariate value for each u;

(ii) set the rank of the observation y equal to the rank of $\rho_S(y)$ in $\{\rho_S(x_1), ..., \rho_S(x_K), \rho_S(y)\}$,

$$\text{rank}_S(y) = \sum_{v \in S} \mathbb{1}\{\rho_S(v) \leq \rho_S(y)\}$$

where $\mathbb{1}$ denotes the indicator function and ties are resolved at random.

Here, we focus on four different approaches that follow this general two-step framework. Further approaches are discussed in Gneiting et al. (2008), Ziegel and Gneiting (2014), and Wilks (2017). The difference between our four approaches lies in the definition of the prerank function ρ_S in step (i). The *multivariate ranking* of Gneiting et al. (2008) is defined using the prerank function

$$\rho_S^m(u) = \sum_{v \in S} \mathbb{1}\{v \preceq u\} \tag{6.1}$$

where $v \preceq u$ if and only if $v_i \leq u_i$ in all components $i = 1, ..., d$. Gneiting et al. (2008) further consider an optional initial step in the ranking procedure in which the data is normalized in each component before the ranking. The *average ranking* proposed by Thorarinsdottir, Scheuerer, and Heinz (2016) provides a similar ascending rank structure and is given by the average over the univariate ranks. That is, let

$$\text{rank}_S(u, i) = \sum_{v \in S} \mathbb{1}\{v_i \leq u_i\}$$

denote the standard univariate rank of the ith component of u among the values in S. The multivariate average rank is then defined using the prerank function

$$\rho_S^a(u) = \frac{1}{d}\sum_{i=1}^{d} \text{rank}_S(u, i) \tag{6.2}$$

Two further approaches assess the centrality of the observation within the ensemble. Under *minimum spanning tree ranking*, the prerank function $\rho_S^{mst}(u)$ is given by the length of the minimum spanning tree of the set $S \setminus u$, that is, the set S without the element u (Smith & Hansen, 2004; Wilks, 2004). Here, a spanning tree of the set $S\setminus u$ is a collection of $K - 1$ edges such that all points in $S\setminus u$ are used, with no closed loops. The spanning tree with the smallest length is then the minimum spanning tree (Kruskal, 1956); it may, for example, be calculated using the R package vegan (Oksanen et al., 2017; R Core Team, 2016).

Alternatively, the *band-depth ranking* proposed by Thorarinsdottir et al. (2016) uses a prerank function that calculates the proportion of components of $u \in S$ inside bands defined by pairs of points from S. It can be written as

$$\rho_S^{\text{bd}}(u) = \frac{1}{d} \sum_{i=1}^{d} \left[\text{rank}_S(u,i)[(K+1) - \text{rank}_S(u,i)] + [\text{rank}_S(u,i) - 1] \sum_{v \in S} \mathbb{1}\{v_i = u_i\} \right] \qquad (6.3)$$

If $u_i \neq v_i$ with probability 1 for all $u, v \in S$ with $u \neq v$ and $i = 1, \ldots, d$ the formula in Eq. (6.3) may be simplified to

$$\rho_S^{\text{bd}}(u) = \frac{1}{d} \sum_{i=1}^{d} [(K+1) - \text{rank}_S(u,i)][\text{rank}_S(u,i) - 1] \qquad (6.4)$$

This implies that the formula in Eq. (6.3) should be used for forecasts with a discrete component, for example, precipitation forecasts. The band depth in Eq. (6.3) is equivalent to the simplicial depth proposed by Liu (1990) and thus also to the simplicial depth ranking proposed by Mirzargar and Anderson (2017), see López-Pintado and Romo (2009) and Thorarinsdottir et al. (2016).

While all four methods return a uniform rank histogram for a calibrated forecast, the interpretation of the histogram shape for a misspecified forecast varies between the methods as demonstrated in the following example.

6.2.3 EXAMPLE: COMPARING MULTIVARIATE RANKING METHODS

The four multivariate ranking methods are compared in Fig. 6.2 for several different settings where $y \in \mathbb{R}^d$ can be thought of as a temporal trajectory of a real-valued variable observed at $d = 10$ equidistant time points $t = 1, \ldots, 10$. In the first two examples (rows 1 and 2), y is a realization of a zero-mean Gaussian AR(1) (autoregressive) process Y with a covariance function given by

$$\text{Cov}(Y_i, Y_j) = \exp(-|i-j|/\tau), \quad \tau > 0. \qquad (6.5)$$

The process Y thus has standard Gaussian marginal distributions while the parameter τ controls how fast correlations decay with time lag. We set $\tau = 3$ for Y and consider ensemble forecasts with 50 members of the same type, but with a different parameter value τ. That is, we set $\tau = 1.5$ in row 1 (too strong correlation) and $\tau = 5$ in row 2 (too weak correlation). It follows from this construction that a univariate calibration test at a fixed time point would not detect any miscalibration in the forecasts.

While all four methods are able to detect the misspecification in the correlation structure, the resulting histograms vary in shape. The shape of the average rank histograms and the band-depth rank histograms offer a similar interpretation as that of the univariate rank histograms in Fig. 6.1 with a ∪-shape when the correlation is too strong (underdispersion across components) and a ∩-shape when the correlation is too weak (overdispersion across components). In these 10-dimensional examples, the prerank ordering of the multivariate rank histograms (Eq. 6.1) is only able to detect miscalibration related

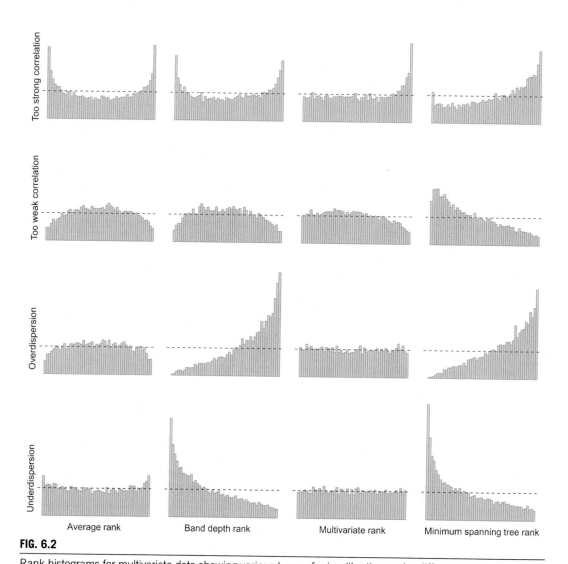

FIG. 6.2

Rank histograms for multivariate data showing various types of miscalibration under different ranking methods: average ranking (*first column*), band-depth ranking (*second column*), multivariate ranking (*third column*), and minimum spanning tree ranking (*fourth column*); 10,000 simulated observations of dimension 10 are compared with ensemble forecasts with 50 members. In the *top two rows*, the observations are realizations of a zero-mean Gaussian AR(1) process with the covariance function in Eq. (6.5) where $\tau = 3$. The forecasts follow the same model with $\tau = 1.5$ (*first row*) and $\tau = 5$ (*second row*). In the *bottom two rows*, the observations are i.i.d. standard Gaussian variables while the forecasts have variance 1.25^2 (*third row*) and 0.85^2 (*fourth row*). The theoretically optimal histograms are indicated with *dashed lines*.

to the highest ranks (see also the discussion in Pinson & Girard, 2012 and Thorarinsdottir et al., 2016). Under minimum spanning tree ranking, too many observations have high ranks when the correlation in the forecasts is too strong and the opposite holds for the example with too weak correlation in the forecasts.

In the latter two examples in Fig. 6.2 (rows 3 and 4), both observations and forecasts are i.i.d. variables in 10 dimensions. However, the marginal distributions of the ensemble forecasts are misspecified. The observations follow a standard Gaussian distribution, the forecasts in row 3 have a standard deviation of 1.25 (overdispersion) and the forecasts in row 4 have a standard deviation of 0.85 (underdispersion). The shape of the average rank histograms is exactly that of their univariate counterparts in Fig. 6.1, indicating that this ranking method cannot distinguish between miscalibration in the marginals and the higher-order structure. For the two ranking methods based on centrality, the marginal overdispersion results in too many high ranks while the marginal underdispersion results in too many low ranks. For this dimensionality, the multivariate ranking is unable to detect the miscalibration.

Further comparison of the four ranking methods is provided in Thorarinsdottir et al. (2016) and Wilks (2017). In general, it is a challenging task to represent and compare a multifaceted higher-order structure with a single value. As the different methods vary in their strengths and weaknesses, it is recommended that several of these methods be applied when assessing multivariate calibration. The multivariate ranking of Gneiting et al. (2008), for instance, does not satisfy affine invariance (Mirzargar & Anderson, 2017) while lower-dimensional positive and negative biases may cancel out under average ranking (Thorarinsdottir et al., 2016).

Furthermore, a prior assessment of the marginal calibration may increase the information value in the multivariate rank histograms and ease the interpretation of the resulting shapes. As the multivariate methods perform a simultaneous assessment of the marginal and the higher-order calibration, a specific nonuniform shape may represent multiple types of misspecifications. For example, depth-based approaches such as the band-depth ranking and the minimum spanning tree ranking are not able to distinguish between underdispersive and biased forecasts (Mirzargar & Anderson, 2017).

6.3 ACCURACY

In this section, we discuss methods for assessing forecast accuracy that are appropriate for ranking and comparing competing forecasting methods. Alternative assessment techniques that may provide additional insights for understanding the performance and errors of a single forecasting model, but are not appropriate for forecast ranking are discussed in Section 6.4.

6.3.1 UNIVARIATE ASSESSMENT

Scoring rules assess the accuracy of probabilistic forecasts by assigning a numerical penalty to each forecast-observation pair. Specifically, a scoring rule is a mapping

$$S \colon \mathcal{F} \times \Omega^d \to \mathbb{R} \cup \{\infty\} \tag{6.6}$$

where for every $F \in \mathcal{F}$ the map $y \mapsto S(F, y)$ is quasiintegrable. In our notation, a smaller penalty indicates a better prediction. A scoring rule is *proper* relative to the class $\mathcal{F}$ if

$$\mathbb{E}_G S(G, Y) \leq \mathbb{E}_G S(F, Y) \tag{6.7}$$

for all probability distributions $F, G \in \mathcal{F}$, that is, if the expected score for a random observation Y is optimized if the true distribution of Y (G) is issued as the forecast. The scoring rule is *strictly proper* relative to the class $\mathcal{F}$ if Eq. (6.7) holds with equality only if $F = G$. Propriety will encourage honesty and prevent hedging, which coincides with Murphy's first type of goodness (Murphy, 1993). That is, the scores cannot be hedged by a willful divergence of the forecast from the true distribution to improve the perceived performance, see for example the discussion in Section 1 of Gneiting (2011).

Competing forecasting methods are verified based on a proper scoring rule by comparing their mean scores over an out-of-sample test set. The method with the smallest mean score is preferred. Formal tests of the null hypothesis of equal predictive performance can also be employed, see Section 6.3.7. While average scores are directly comparable if they refer to the same set of forecast situations, this may no longer hold for distinct sets of forecast cases, for instance due to spatial and temporal variability in the predictability of weather. For ease of interpretability and to address this issue, verification results are sometimes represented as a *skill score* of the form

$$S_n^{\text{skill}}(A) = \frac{\frac{1}{n}\sum_{i=1}^{n}S(F_i^A, y_i) - \frac{1}{n}\sum_{i=1}^{n}S(F_i^{\text{ref}}, y_i)}{\frac{1}{n}\sum_{i=1}^{n}S(F_i^{\text{perf}}, y_i) - \frac{1}{n}\sum_{i=1}^{n}S(F_i^{\text{ref}}, y_i)} \qquad (6.8)$$

for the forecasting method A where F^{ref} denotes the forecast from a reference method, F^{perf} denotes the perfect forecast, and n is the size of the test set. The skill score is standardized such that it takes the value 1 for an optimal forecast and the value 0 for the reference forecast. Negative values thus indicate that the forecasting method A is of a lesser quality than the reference forecast. However, it is vital to select the reference forecast with care (Murphy, 1974, 1992) as skill scores of the form of Eq. (6.8) may be improper even if the underlying scoring rule S is proper (Gneiting & Raftery, 2007; Murphy, 1973a).

The most popular proper scoring rules for univariate real-valued quantities are the *ignorance* (or *logarithmic*) *score* (IGN) and the continuous ranked probability score, see Gneiting and Raftery (2007) for a more comprehensive list. IGN is defined as

$$\text{IGN}(F, y) = -\log f(y) \qquad (6.9)$$

where f denotes the density of F (Good, 1952). It thus applies to absolutely continuous distributions only and cannot be applied directly to ensemble forecasts. For a large enough ensemble, the density of the ensemble forecast may potentially be approximated using, for example, kernel density estimation or by fitting a parametric distribution. Alternatively, IGN may be replaced by the *Dawid-Sebastiani* (DS) *score* (Dawid & Sebastiani, 1999),

$$\text{DS}(F, y) = \log \sigma_F^2 + \frac{(y - \mu_F)^2}{\sigma_F^2} \qquad (6.10)$$

where μ_F denotes the mean value of F and σ_F^2 its variance. While the proper DS score equals IGN for a Gaussian predictive distribution F, it only requires the estimation of the ensemble mean and variance.

The *continuous ranked probability score* (CRPS) (Matheson & Winkler, 1976) is of particular interest in that it simultaneously assesses both calibration and sharpness, and thus all three types of goodness discussed by Murphy (1993). The CRPS applies to probability distributions with a finite mean and has three equivalent definitions (Gneiting & Raftery, 2007; Gneiting & Ranjan, 2011; Hersbach, 2000; Laio & Tamea, 2007),

$$\text{CRPS}(F, y) = \mathbb{E}_F |X - y| - \frac{1}{2} \mathbb{E}_F \mathbb{E}_F |X - X'| \tag{6.11}$$

$$= \int_{-\infty}^{+\infty} (F(x) - \mathbb{1}\{y \le x\})^2 dx \tag{6.12}$$

$$= \int_0^1 (F^{-1}(\tau) - y)(\mathbb{1}\{y \le F^{-1}(\tau)\} - \tau) d\tau \tag{6.13}$$

Here, X and X' denote two independent random variables with distribution F, $\mathbb{1}\{y \le x\}$ denotes the indicator function that is equal to 1 if $y \le x$ and 0 otherwise, and $F^{-1}(\tau) = \inf\{x \in \mathbb{R} : \tau \le F(x)\}$ is the quantile function of F.

It follows directly from Eqs. (6.12), (6.13) that the CRPS is tightly linked to other proper scores that focus on specific parts of the predictive distribution. The form in Eq. (6.12) can be interpreted as the integral over the *Brier score* (Brier, 1950), which assesses the predictive probability of threshold exceedance. The Brier score is usually written in the form

$$\text{BS}(F, y|u) = (p_u - \mathbb{1}\{y \ge u\})^2 \tag{6.14}$$

for a threshold u with $p_u = 1 - F(u)$. Similarly, the integrand in Eq. (6.13) equals the *quantile score* (Friederichs & Hense, 2007; Gneiting & Raftery, 2007),

$$\text{QS}(F, y|q) = (F^{-1}(q) - y)(\mathbb{1}\{y \le F^{-1}(q)\} - q) \tag{6.15}$$

which assesses the predicted quantile $F^{-1}(q)$ for a probability level $q \in (0, 1)$.

When the predictive distribution F is given by a finite ensemble $\{x_1, \ldots, x_K\}$, the CRPS representation in Eq. (6.11) is equal to

$$\text{CRPS}(F, y) = \frac{1}{K} \sum_{k=1}^K |x_k - y| - \frac{1}{2K^2} \sum_{k=1}^K \sum_{l=1}^K |x_k - x_l| \tag{6.16}$$

see Grimit, Gneiting, Berrocal, and Johnson (2006). For small ensembles, Ferro, Richardson, and Weigel (2008) propose a *fair* approximation given by

$$\text{CRPS}(F, y) \approx \frac{1}{K} \sum_{k=1}^K |x_k - y| - \frac{1}{2K(K-1)} \sum_{k=1}^K \sum_{l=1}^K |x_k - x_l| \tag{6.17}$$

For large ensembles, a more computationally efficient calculation is based on the generalized quantile function (Laio & Tamea, 2007). Let $x_{(1)} \le \cdots \le x_{(K)}$ denote the order statistics of $x_1, \ldots, x_K$. Then

$$\text{CRPS}(F, y) = \frac{2}{K^2} \sum_{i=1}^K (x_{(i)} - y) \left(K \mathbb{1}\{y < x_{(i)}\} - i + \frac{1}{2} \right) \tag{6.18}$$

see also Murphy (1970). The formula in Eq. (6.18) is implemented in the R package `scoringRules` together with exact formulas for a large class of parametric families of distributions (see Table 6.1 and Jordan, Krüger, & Lerch, 2017).

Table 6.1 Parametric Families of Distributions for Which the CRPS Is Implemented in the R **Package** scoringRules **(Jordan et al., 2017)**

Dist. on $\mathbb{R}$	Dist. on $\mathbb{R}_{>0}$	Dist. on Intervals	Discrete Dist.
Gaussian	Exponential	Generalized extreme value	Poisson
t	Gamma	Generalized Pareto	Neg. binomial
Logistic	Log-Gaussian	Trunc. Gaussian	
Laplace	Log-logistic	Trunc. t	
Two-piece Gaussian	Log-Laplace	Trunc. logistic	
Two-piece exponential		Trunc. exponential	
Mixture of Gaussians		Uniform	
		Beta	

Notes: *The truncated families can be defined with or without a point mass at the support boundaries.*

When the forecasting model is estimated using a Bayesian analysis, the predictive distribution F is commonly given by the posterior predictive distribution under the model. Here, F is rarely known in closed form and is, instead, approximated by a large sample that is often obtained using Markov chain Monte Carlo techniques. However, such techniques may yield highly correlated samples, which complicates the employment of approximation formulas as those for the CRPS shown herein. Optimal approximations for both IGN and CRPS when the distribution F is the posterior predictive distribution from a Bayesian analysis are discussed in Krüger, Lerch, Thorarinsdottir, and Gneiting (2016).

The quality of a deterministic forecast x is typically assessed by applying a *scoring function $s(x, y)$*, that assigns a numerical score based on x and the corresponding observation y. As in the case of proper scoring rules, competing forecasting methods are compared and ranked in terms of the mean scores over the cases in a test set. Popular scoring functions include the squared error, $s(x, y) = (x - y)^2$, and the absolute error, $s(x, y) = |x - y|$.

A scoring function can be applied to a probabilistic prediction $F \in \mathcal{F}$ if it is *consistent* for a functional T relative to the class $\mathcal{F}$ in the sense that

$$\mathbb{E}_{F}s(T(F),Y) \leq \mathbb{E}_{F}s(x,Y) \tag{6.19}$$

for all $x \in \Omega$ and $F \in \mathcal{F}$. A consistent scoring function becomes a proper scoring rule if the functional T in Eq. (6.19) is used as the derived deterministic prediction based on F. That is, if $S(F, y) = s(T(F), y)$. The squared error proper scoring rule is given by

$$\mathrm{SE}(F,y) = (\mathrm{mean}(F) - y)^2 \tag{6.20}$$

where $\mathrm{mean}(F)$ denotes the mean value of F, and the absolute error proper scoring rule becomes

$$\mathrm{AE}(F,y) = |\mathrm{med}(F) - y| \tag{6.21}$$

where $\mathrm{med}(F)$ denotes the median of F.

One appealing property of scoring rules that derive from scoring functions is thus the possibility of comparing deterministic and probabilistic forecasts. See Gneiting (2011) for an extensive discussion of the use of scoring functions to evaluate probabilistic predictions.

6.3.2 SIMULATION STUDY: COMPARING UNIVARIATE SCORING RULES

The purpose of this simulation study is to demonstrate a coherent approach to using proper scores and rank or PIT histograms in practice, while highlighting some of the difficulties that might arise when working with limited data sets. In particular, we investigate how different scoring rules rank forecasts according to their skill, and how these results differ with the amount of available data.

We start by generating two sets of observation data, drawn randomly from the same fixed "true" distribution. The first set consists of 100 values, which will serve as verifying observations, while the second set, the training data, consists of 300 values for each of the 100 observations. Our goal is to issue forecasts matching the observations, based on the information contained in the training data. For the first part of the simulation study, the true distribution is normal, with a random mean $\mu \sim \mathcal{N}(25,1)$ and fixed standard deviation $\sigma = 3$. In the second part, the truth is a Gumbel distribution, with the mean following a $\mathcal{N}(25,1)$ distribution and the scale parameter fixed to 3, see Table 6.2.

Using a method-of-moments approach, we estimate four competing forecast distributions for each observation, which are listed in Table 6.3. The distribution parameters are calculated by plugging the sample mean and sample standard deviation from the training data into the equations for mean and variance. For the noncentral t-distribution, the degrees of freedom are obtained numerically by a root-finding algorithm described in Brent (1973), while restricting them to $\nu \geq 3$, ensuring that both mean and variance exist. As a fifth forecaster, we use the true distribution, from which the observations

Table 6.2 Observation-Generating Distributions Used in the Simulation Study

	Distribution $F(Y)$	$\mathbb{E}(Y)$	Var(Y)
Part 1	Normal	$\mathcal{N}(\mu,\sigma^2)$	$\mu \sim \mathcal{N}(25,1)$	$\sigma^2 = 9$
Part 2	Gumbel	$G(\mu,\sigma)$	$\mu + \sigma \cdot \gamma \sim \mathcal{N}(25,1)$	$\frac{\pi^2}{6}\sigma^2 = \frac{3\pi^2}{2}$

Notes: *The expected values are random variables following a normal distribution, while the scale parameters are fixed. γ denotes the Euler-Mascheroni constant.*

Table 6.3 Forecasters Used in Both Parts of the Simulation Study, and Their Expected Values and Variances as Functions of the Distribution Parameters

Distribution $F(Y)$		$\mathbb{E}(Y)$	Var(Y)
Normal	$\mathcal{N}(\mu,\sigma^2)$	μ	σ^2
Noncentral t	$t(\nu,\mu)$	$\mu\sqrt{\frac{\nu}{2}}\dfrac{\Gamma\left(\frac{\nu-1}{2}\right)}{\Gamma\left(\frac{\nu}{2}\right)}$, if $\nu > 1$	$\dfrac{\nu(1+\mu^2)}{\nu-2} - \dfrac{\mu^2\nu}{2}\left(\dfrac{\Gamma\left(\frac{\nu-1}{2}\right)}{\Gamma\left(\frac{\nu}{2}\right)}\right)^2$, if $\nu > 2$
Lognormal	$\ln \mathcal{N}(\mu,\sigma^2)$	$\exp\left(\mu+\frac{\sigma^2}{2}\right)$	$(\exp(\sigma^2)-1)\exp(2\mu+\sigma^2)$
Gumbel	$G(\mu,\sigma)$	$\mu + \sigma \cdot \gamma$	$\dfrac{\pi^2}{6}\sigma^2$

Note: *γ denotes the Euler-Mascheroni constant.*

are generated. An ensemble of 50 members is drawn randomly from each of the forecast distributions, which is then paired with the observations.

The performance of the five forecasters is evaluated using the absolute error, the squared error, the ignorance score, the CRPS, and the PIT histogram. We also produced rank histograms, but they turned out to be almost identical to the PIT histograms. As we encountered variations in the scores depending on the initial random seed, the whole process is repeated 10 times with different initial seeds, so that the final number of forecast-observation pairs comes to 1000.

In order to understand the true ranking of the five forecasting methods in terms of skill, we reproduce the simulation study with 10 times 100,000 forecasts. For the case of a normal true distribution, Fig. 6.3 shows the mean absolute error, mean CRPS and mean ignorance score, along with 95% bootstrap confidence intervals (see Section 6.3.7) computed from 1000 bootstrap samples. We have omitted the squared error from this plot, as its values are on a much larger scale than the other scores. Looking at the results for the small sample size in the top row, all scores assign the lowest mean value, and therefore the highest skill, to the normal distribution with the true parameters. However, if no knowledge about the true distribution is available, as in a real forecast setting, the absolute error and the CRPS

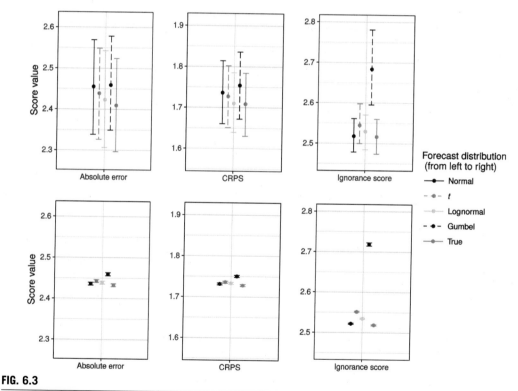

FIG. 6.3

Top row: Mean absolute error, CRPS and ignorance score, and the 95% bootstrap confidence interval for the five forecast distributions, if the true distribution is normal. Scores are based on 1000 forecast-observation pairs. *Bottom row:* Same as above, but scores are based on 1 million forecast-observation pairs.

would prefer the lognormal distribution over all other forecasters, while the ignorance score judges the normal distribution with estimated parameters to be the best.

The bottom panel of Fig. 6.3 shows the results from running the same study with the larger sample size, which changes the order in which we would expect the forecasters to rank. Here, all scores correctly find the Gumbel distribution, which has a completely different shape and tail behavior than the truth, to be the worst forecast, and the two forecasts based on normal distributions to be the best. This contradicts the results in the top panel, where only the ignorance score ranked the forecasters in the same order as we would expect.

Due to assigning large penalties to outliers, the ignorance score is able to discriminate between the shapes of the forecast distributions, and shows a significant difference at the 95% level between the Gumbel and the normal, lognormal, and true distributions. The relatively poor performance of the noncentral t-distribution can probably be explained by the fact that, while this distribution approximates a normal distribution if the degrees of freedom are large, the asymptotic distribution will have a standard deviation of 1, which does not match the given standard deviation of 3 in this example.

Judging from Fig. 6.4, which shows PIT histograms for the small-sample study with a normal true distribution, we cannot make any statements about the forecast ranking, except that the Gumbel distribution forecast is clearly uncalibrated. Only when looking at the large sample equivalent in Fig. 6.5 do we see that the normal and the true forecasters are the only ones not suffering from miscalibration. A formal chi-squared test (see Section 6.3.7) rejects the assumption of uniformity for the Gumbel

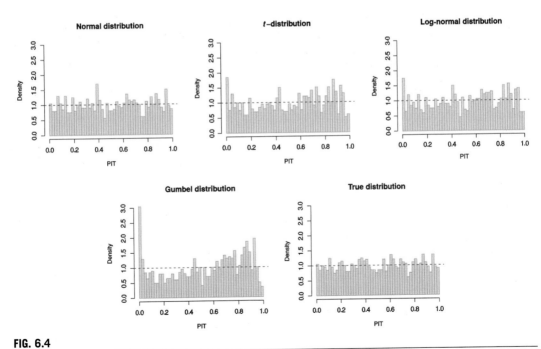

FIG. 6.4

PIT histograms for the five forecast distributions, if the true distribution is normal, based on 1000 forecast-observation pairs.

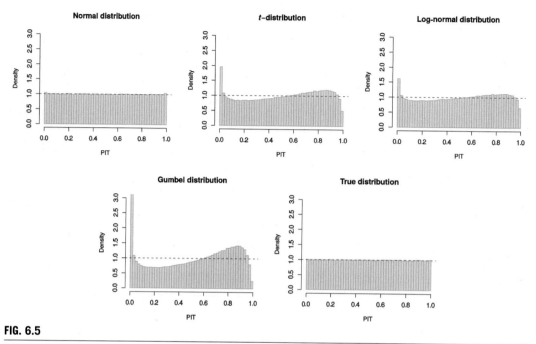

FIG. 6.5

PIT histograms for the five forecast distributions, if the true distribution is normal, based on 1 million forecast-observation pairs.

distribution and even the t and lognormal distributions (at a level of 5%) in the small-sample case, and for all distributions apart from the true one in the large sample case.

Fig. 6.6 illustrates one example forecast, for which the scores are plotted as functions of the verifying observation, in this case a sample value from a $N(27.16, 9)$ distribution. While the score minima largely coincide for the true and the t-distribution, it becomes clear from the shape of the ignorance score why it is much better at identifying the Gumbel distribution as inferior: because of the lack of symmetry, Gumbel forecasts will receive a much higher penalty if the observation lies left of the distribution mode than if it lies on the right.

For the second part of the simulation study, we used a Gumbel distribution as truth, where the mean is distributed as $N(25, 1)$ and the scale parameter is 3. The same kinds of forecasts are produced again: normal, noncentral t, lognormal, and Gumbel distributions, based on the sample means and variances of the training data. In Fig. 6.7, the outcome of the study is shown for a small sample size (top row) and a very large sample size (bottom row). As previously, all scores agree on the forecast ranking when the sample is large. The Gumbel distribution with estimated parameters and the true Gumbel distribution are assigned the lowest scores, while the normal forecaster now has the lowest skill.

However, the rankings look different in the top panel, where the true distribution is only ranked the third best by the absolute error and the CRPS, behind the estimated Gumbel and noncentral t-distributions. The ignorance score again is the only score able to reproduce the forecast ranking we expect from the bottom panel. This is, of course, concerning and hints at the fact that even for a data set of apparently sufficient size, such as the 1000 50-member ensembles used here, the scores do not necessarily provide robust and proper results.

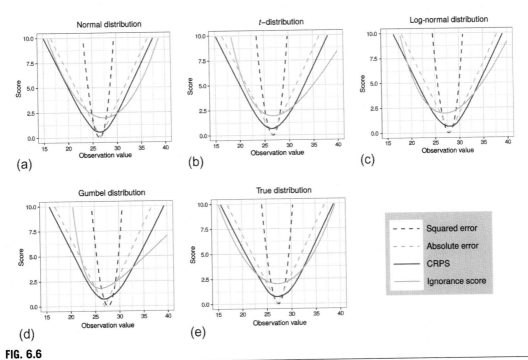

FIG. 6.6

Squared error, absolute error, CRPS, and ignorance score as functions of the verifying observation, for one forecast case in the simulation study: (a) normal distribution forecast, (b) noncentral t-distribution forecast, (c) lognormal distribution forecast, (d) Gumbel distribution forecast, and (e) forecast based on the true normal distribution.

Again we cannot really judge the degree of forecast calibration by just looking at the small-sample PIT histograms in Fig. 6.8, except for the clearly uncalibrated normal distribution. A case could be made that the histogram for the true distribution looks slightly flatter than the other ones, but not with great certainty. It becomes clear, however, from Fig. 6.9, that the forecasts based on noncentral t and lognormal distributions also suffer from multiple types of miscalibration. These findings are confirmed by a chi-squared test, which rejects the uniformity hypothesis for all except the Gumbel distributions in Fig. 6.8 and all except the true distribution in Fig. 6.9.

Picking an example forecast from the data set, Fig. 6.10 shows that the ignorance score for the two Gumbel distribution forecasters is again nonsymmetric, and therefore minimizes at a different value compared with the CRPS. In general, the ignorance score takes its minimum value at the mode of the distribution, and the CRPS at the median.

We can gather from this simulation study that even proper scores can behave very differently, depending on the size of the underlying data set, and are not necessarily able to rank competing forecasters according to their actual skill. Therefore, we suggest always using a combination of scoring rules to get a maximum amount of information about the performance of a particular model or forecaster. The ignorance score is more sensitive to the shape of a distribution and thus is suitable to check if a chosen distribution actually fits the data. The CRPS is very useful for comparing models when the forecasts do not take the form of a standard probability distribution, or if for a given data set such a distribution cannot be perfectly specified.

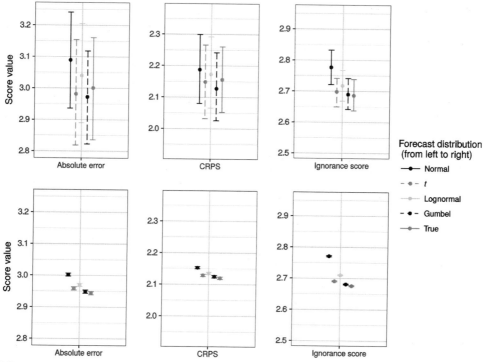

FIG. 6.7

Top row: Mean absolute error, CRPS and ignorance score, and the 95% bootstrap confidence interval for the five forecast distributions, if the true distribution is a Gumbel distribution. Scores are based on 1000 forecast-observation pairs. *Bottom row*: Same as above, but scores are based on 1 million forecast-observation pairs.

These results also have implications for the ongoing discussion of whether to use maximum likelihood methods or minimize the CRPS to estimate model parameters (Gneiting, Raftery, Westveld, & Goldman, 2005), in that there might not be a definitive answer. Depending on the forecast situation and model choice, it could be preferable to switch between the two approaches. A case can be made for performing a thorough exploratory analysis of the data at hand before fitting any distributions, to find one that matches the data best. If it is difficult to select one distribution over the other, the simpler model should be preferred.

In all circumstances, the ranking of forecasters should not be solely based on the mean score, even if the sample size seems to be sufficiently large, but confidence intervals should be given, for example, by applying bootstrapping techniques. We found that even for 1 million data points, differences between the forecast scores were often not significant at the 5% level.

6.3.3 ASSESSING EXTREME EVENTS

Forecasts specifically aimed at predicting extreme events can be assessed in a standard manner, for example, by using the scoring rules discussed in Section 6.3.1 (Friederichs & Thorarinsdottir, 2012).

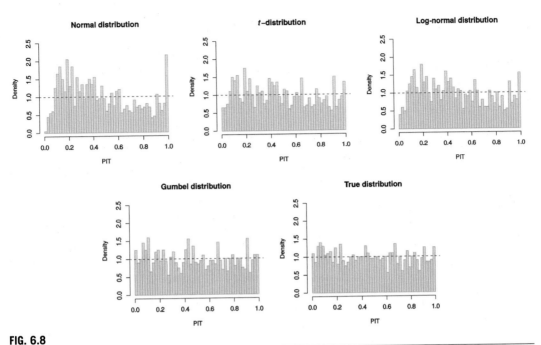

FIG. 6.8

PIT histograms for the five forecast distributions, if the true distribution is a Gumbel distribution, based on 1000 forecast-observation pairs.

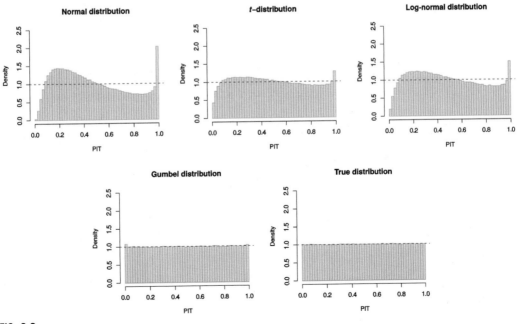

FIG. 6.9

PIT histograms for the five forecast distributions, if the true distribution is a Gumbel distribution, based on 1 million forecast-observation pairs.

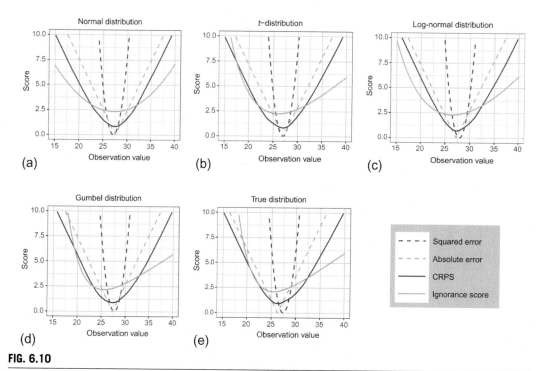

FIG. 6.10

Squared error, absolute error, CRPS, and ignorance score as functions of the verifying observation, for one forecast case in the simulation study: (a) normal distribution forecast, (b) noncentral t-distribution forecast, (c) lognormal distribution forecast, (d) Gumbel distribution forecast, and (e) forecast based on the true Gumbel distribution.

However, the restriction of conventional forecast evaluation to subsets of extreme observations by selecting the extreme observations after-the-fact while discarding the nonextreme ones, and to proceed with standard evaluation tools, will invalidate their theoretical properties and encourage hedging strategies (Lerch, Thorarinsdottir, Ravazzolo, & Gneiting, 2017).

Specifically, Gneiting and Ranjan (2011) show that a proper scoring rule S is rendered improper if the product with a nonconstant weight function w is formed, where w depends on the observed value y. That is, consider the weighted scoring rule

$$S_0(F,y) = w(y)S(F,y). \qquad (6.22)$$

Then if Y has density g, the expected score $\mathbb{E}_g S_0(F,Y)$ is minimized by the predictive distribution F with density

$$f(y) = \frac{w(y)g(y)}{\int w(z)g(z)\mathrm{d}z} \qquad (6.23)$$

which is proportional to the product of the weight function w and the true density g. In particular, if $w(y) = \mathbb{1}\{y \geq u\}$ for some high threshold value u, then S_0 corresponds to evaluating F only on observed values exceeding u under the scoring rule S.

Instead, one can apply proper *weighted scoring rules* that are tailored to emphasize specific regions of interest. Diks, Panchenko, and Van Dijk (2011) propose two weighted versions of the ignorance score that correct for the result in Eq. (6.23). The *conditional likelihood* (CL) score is given by

$$CL(F,y) = -w(y) \log \left(\frac{f(y)}{\int_\Omega w(z)f(z)\mathrm{d}z} \right)$$

and the *censored likelihood* (CSL) score is defined as

$$CSL(F,y) = -w(y) \log f(y) - (1-w(y)) \log \left(1 - \int_\Omega w(z)f(z)\mathrm{d}z \right)$$

Here, w is a weight function such that $0 \le w(y) \le 1$ and $\int w(y)f(y)\mathrm{d}y > 0$ for all potential predictive distributions $F \in \mathcal{F}$. When $w(y) \equiv 1$, both the CL and the CSL score reduce to the unweighted ignorance score in Eq. (6.9).

Gneiting and Ranjan (2011) propose the *threshold-weighted continuous ranked probability score* (twCRPS), defined as

$$\mathrm{twCRPS}(F,y) = \int_\Omega w(z)(F(z) - \mathbb{1}\{y \le z\})^2 \mathrm{d}z$$

where, again, w is a nonnegative weight function, see also Matheson and Winkler (1976). When $w(y) \equiv 1$, the twCRPS reduces to the unweighted CRPS in Eq. (6.12) while $w(y) = \mathbb{1}\{y = u\}$ equals the Brier score in Eq. (6.14). More generally, the twCRPS puts emphasis on a particular part of the forecast distribution F as specified by w. For focusing on the upper tail of F, Gneiting and Ranjan (2011) consider both indicator weight functions of the type $w(y) = \mathbb{1}\{y \ge u\}$ and nonvanishing weight functions such as $w(y) = \Phi(y|u,\sigma^2)$ where Φ denotes the cumulative distribution function of the Gaussian distribution with mean u and variance σ^2. Corresponding weight functions for the lower tail of F are given by $w(y) = \mathbb{1}\{y \le u\}$ and $w(y) = 1 - \Phi(y|u,\sigma^2)$ for some low threshold value u.

Nonstationarity in the mean climate, for example, due to spatial heterogeneity, may render it difficult to define a common threshold value u over a large number of forecast cases. Here, it may be more natural to define a weight function in quantile space using the CRPS representation in Eq. (6.13),

$$\mathrm{twCRPS}(F,y) = \int_0^1 w(\tau)(F^{-1}(\tau) - y)(\mathbb{1}\{y \le F^{-1}(\tau)\} - \tau)\mathrm{d}\tau$$

where w is a nonnegative weight function on the unit interval (Gneiting & Ranjan, 2011; Matheson & Winkler, 1976). Setting $w(\tau) \equiv 1$ retrieves the unweighted CRPS in Eq. (6.13) while this definition of twCRPS with $w(\tau) = \mathbb{1}\{\tau = q\}$ equals the quantile score in Eq. (6.15). Examples of more general weight functions for this setting include $w(\tau) = \mathbb{1}\{\tau \ge q\}$ and $w(\tau) = \tau^2$ for the upper tail, and $w(\tau) = \mathbb{1}\{\tau \le q\}$ and $w(\tau) = (1-\tau)^2$ for the lower tail, with appropriate threshold values q, see also Gneiting and Ranjan (2011).

Lerch et al. (2017) find that there are limited benefits in using weighted scoring rules compared with using standard, unweighted scoring rules when testing for equal predictive performance. However, the application of weight functions as described here may facilitate interpretation of the forecast skill.

6.3.4 EXAMPLE: PROPER AND NONPROPER VERIFICATION OF EXTREMES

In the following, we illustrate that the use of nonproper methods to verify and compare competing forecasts for extremes can lead to a distortion of the results and possibly false inference. Taking the same setting as the first part of the simulation study in Section 6.3.2, we generate sets of observation and training data from a normal distribution with standard deviation 3 and the mean a random value from a $\mathcal{N}(25,1)$ distribution.

Four of the forecasting methods in Section 6.3.2 are compared: a normal distribution with estimated parameters based on the training data, a Gumbel distribution with estimated parameters, a normal distribution with the true parameters, and a Gumbel distribution with the true means as location parameter and scale parameter $\sigma = 3$. The forecasters' performance for extremes, which we consider to be values greater or equal to the 97.5% quantile of the observations u, will be measured using the threshold-weighted CRPS with three different weight functions and the unweighted CRPS, where the cases are restricted to observations above the threshold. The weight functions considered are variations on the indicator function:

$$w_1(y) = \mathbb{1}\{y \geq u\}$$
$$w_2(y) = 1 + \mathbb{1}\{y \geq u\}$$
$$w_3(y) = 1 + \mathbb{1}\{y \geq u\} \cdot u$$

Mean scores and 95% confidence intervals, calculated by numerical integration based on the small sample data set from Section 6.3.2, are shown in Fig. 6.11 for the threshold-weighted CRPS and the CRPS with restricted observations, along with the unweighted CRPS. The results for the twCRPS with weight function w_1 are omitted, as they are equal to 0 for all forecasters.

However, just by adding 1 to the indicator function, we obtain meaningful scores with weight function w_2, showing the Gumbel distribution with fixed parameters to be the least skillful forecast, while the two normal distribution forecasters are of significantly better quality. The twCRPS with weight function w_3 and the unweighted CRPS lead to similar conclusions, although the differences between the scores are sometimes not significant. In contrast to the other scores, the CRPS based on the restricted data set clearly shows the Gumbel distribution with fixed parameters to be the preferred forecaster.

Although the fixed Gumbel parameters and shape are obviously wrong, this is no surprise, as this distribution was purposely chosen because it has a heavy tail. Fig. 6.12 shows predictive densities for one example from the data set. If we restrict the evaluation to the area above the chosen threshold, represented by the black vertical line, the Gumbel distribution with fixed parameters is indeed the seemingly best forecast, as it assigns the highest probabilities to extreme values. The two normal distributions and the Gumbel distribution with estimated parameters, which tries to approximate the true normal distribution, have a very similar tail behavior, explaining their similar performance in terms of all scores.

We come to the same conclusion as Lerch et al. (2017), that conditioning a data set on extremal observations can result in preferring a forecaster who predicts extremes with inflated probabilities. When evaluating forecasts for a certain range of values, proper methods such as the threshold-weighted CRPS should be used, where the whole data set is considered.

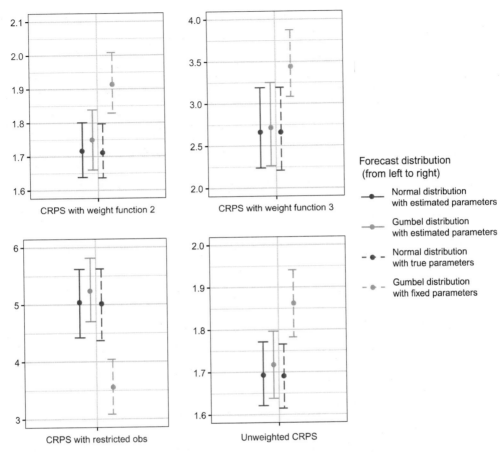

FIG. 6.11

Mean scores and 95% bootstrap confidence interval for the four versions of the CRPS. *Top row*: twCRPS with weight functions w_2 and w_3. *Bottom row*: CRPS restricted to observations above the threshold u and unweighted CRPS.

6.3.5 MULTIVARIATE ASSESSMENT

Two general approaches can be employed to assess multivariate forecasts with scoring rules: Use specialized multivariate scores, or reduce the multivariate forecast to a univariate quantity and subsequently apply the univariate scores discussed previously. For the latter approach, the appropriate univariate quantities depend on the context. Multivariate forecasts of single weather quantities are usually in the form of temporal trajectories, spatial fields, or space-time fields. Here it can, for instance, be useful to assess the predictive performance of derived quantities such as maxima, minima, and accumulated totals, all of which depend on accurate modeling of both marginal and higher order structures. See, for example, Feldmann, Scheuerer, and Thorarinsdottir (2015) for an assessment of spatial forecast fields for temperature.

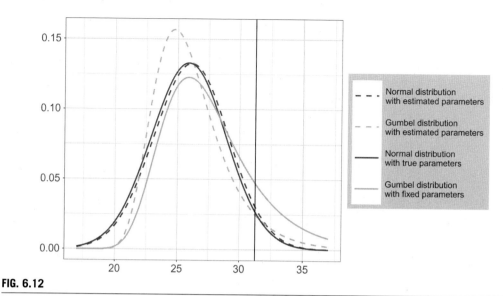

FIG. 6.12

Example predictive densities given by the four competing forecasters. The *black vertical line* shows the threshold *u*, above which observations are considered to be extreme.

Scores that directly assess multivariate forecasts are rather scarce and, as noted by Gneiting and Katzfuss (2014), there is a need to develop further decision-theoretically principled methods for multivariate assessment. The univariate Dawid-Sebastiani score in Eq. (6.10) can be applied in a multivariate setting with

$$DS(F,y) = \log \det \Sigma_F + (y - \mu_F)^\top \Sigma_F^{-1}(y - \mu_F) \qquad (6.24)$$

where μ_F is the mean vector and Σ_F the covariance matrix of the predictive distribution with $\det \Sigma_F$ denoting the determinant of Σ_F (Dawid & Sebastiani, 1999). However, note that unless the sample size is much larger than the dimension of the multivariate quantity, sampling errors can affect the calculation of $\det \Sigma_F$ and Σ_F^{-1} (see e.g., Table 2 in Feldmann et al., 2015). Similarly, if the multivariate predictive density is available, the ignorance score in Eq. (6.9) can be employed (Roulston & Smith, 2002).

Gneiting and Raftery (2007) propose the *energy score* (ES) as a multivariate generalization of the CRPS. It is given by

$$ES(F,y) = \mathbb{E}_F \| X - y \| - \frac{1}{2} \mathbb{E}_F \mathbb{E}_F \| X - X' \| \qquad (6.25)$$

where X and X' are two independent random vectors distributed according to F and $\| \cdot \|$ is the Euclidean norm. For ensemble forecasts, the natural analog of the formulas in Eqs. (6.16), (6.17) apply. If the multivariate observation space Ω^d consists of weather variables on varying scales, the margins should be standardized before computing the joint energy score for these variables (Schefzik, Thorarinsdottir, & Gneiting, 2013). This can be done using the marginal means and standard deviations of the observations in the test set. The energy score has been developed with low-dimensional quantities in mind and it may lose discriminatory power in higher dimensions (Pinson, 2013).

Scheuerer and Hamill (2015) propose a multivariate scoring rule that considers pairwise differences of the components of the multivariate quantity. In its general form, the *variogram score (VS) of order p* is given by

$$\mathrm{VS}_p(F,y) = \sum_{i=1}^{d}\sum_{j=1}^{d}\omega_{ij}\left(|y_i - y_j|^p - \mathbb{E}_F|X_i - X_j|^p\right)^2 \tag{6.26}$$

where y_i and y_j are the ith and the jth component of the observation, X_i and X_j are the ith and the jth component of a random vector X that is distributed according to F, and ω_{ij} are nonnegative weights. Scheuerer and Hamill (2015) compare different choices of the order p and find that the best results in terms of discriminative power are obtained with $p = 0.5$. Furthermore, they recommend using weights proportional to the inverse distance between the components unless a prior knowledge regarding the correlation structure is available.

A comparison of the three multivariate scores in Eqs. (6.24)–(6.26) is provided in Scheuerer and Hamill (2015). The authors conclude by recommending the use of multiple scores as they complement each other in their strengths and weaknesses. The variogram score is generally able to distinguish between correct and misspecified correlation structures, but it has certain limitations resulting from the fact that it is proper but not strictly proper. Some of these limitations can be addressed by also using the energy score that is more sensitive to misspecifications in the predictive mean and less affected by finite representations of the predictive distribution. While the latter is an issue for the Dawid-Sebastiani score, it performs well for continuous predictive distributions, in particular for multivariate Gaussian models (Wei, Balabdaoui, & Held, 2017).

6.3.6 DIVERGENCE FUNCTIONS

In some cases, in particular in climate modeling, it is of interest to compare the predictive distribution F against the true distribution of the observations, which is commonly approximated by the *empirical distribution function* of the available observations $y_1, \ldots, y_n$,

$$\hat{G}_n(x) = \frac{1}{n}\sum_{i=1}^{n}\mathbb{1}\{y_i \leq x\}. \tag{6.27}$$

The two distributions, F and $\hat{G}_n$, can be compared using a *divergence*

$$D: \mathcal{F} \times \mathcal{F} \to \mathbb{R}_{\geq 0} \tag{6.28}$$

where $D(F, F) = 0$.

Assume that the observations $y_1, \ldots, y_n$ forming the empirical distribution function $\hat{G}_n$ are independent with distribution $G \in \mathcal{F}$. A propriety condition for divergences corresponding to that for scoring rules (Eq. 6.7) states that the divergence D is *n-proper* for a positive integer n if

$$\mathbb{E}_G D(G, \hat{G}_n) \leq \mathbb{E}_G D(F, \hat{G}_n) \tag{6.29}$$

and *asymptotically proper* if

$$\lim_{n\to\infty}\mathbb{E}_G D(G, \hat{G}_n) \leq \lim_{n\to\infty}\mathbb{E}_G D(F, \hat{G}_n) \tag{6.30}$$

for all probability distributions $F, G \in \mathcal{F}$ (Thorarinsdottir, Gneiting, & Gissibl, 2013). While the condition in Eq. (6.30) is fulfilled by a large class of divergences, only score divergences have been shown to fulfill Eq. (6.29) for all integers n. A divergence D is a *score divergence* if there exists a proper scoring rule S such that $D(F,G) = \mathbb{E}_G S(F,Y) - \mathbb{E}_G S(G,Y)$.

A score divergence that assesses the full distributions is the *integrated quadratic divergence* (IQD)

$$IQD(F,G) = \int_{-\infty}^{+\infty} (F(x) - G(x))^2 dx \qquad (6.31)$$

which is the score divergence of the continuous ranked probability score (Eq. 6.12). Alternative score divergences that assess specific properties of the predictive distribution include the *mean value divergence* (MVD),

$$MVD(F,G) = (\text{mean}(F) - \text{mean}(G))^2 \qquad (6.32)$$

which is the divergence associated with the squared error scoring rule (Eq. 6.20), and the *Brier divergence* (BD) associated with the Brier score (Eq. 6.14),

$$BD(F,G|u) = (G(u) - F(u))^2 \qquad (6.33)$$

for some threshold u.

Fig. 6.13 provides a comparison of the score divergences in Eqs. (6.31)–(6.33) for two simple settings where the observation distribution is given by a standard normal distribution and all the forecast distributions are also normal distributions but with varying parameters. In the left plot, the variance is correctly specified while the forecast mean value varies. In the right plot, the forecast mean values equal that of the observation distribution while the standard deviation varies. We compare the IQD, the MVD, and the BD with thresholds $u = 0.67$ and $u = 1.64$, which equal the 75% and the 95% quantiles of the observation distribution, respectively. The divergences are more sensitive to forecast errors in the mean than the spread. In particular, the MVD is, naturally, not able to detect errors in the forecast spread. Furthermore, integrating over the BD for all possible thresholds u and obtaining the IQD yields

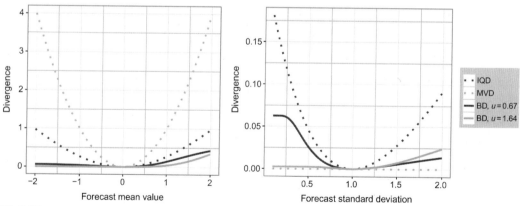

FIG. 6.13

Comparison of expected score divergence values for a standard normal observation distribution and normal forecast distributions with varying mean values (*left*) or standard deviations (*right*).

a better discrimination than investigating the differences for individual quantiles. The right plot also shows that the model ranking obtained under the BD strongly depends on the threshold u.

While every proper scoring rule is associated with a score divergence, not all score divergences are practical for use in the setting where the empirical distribution function $\hat{G}_n$ is used. One example is the Kullback-Leibler divergence, which is the score divergence of the ignorance score in Eq. (6.9). The Kullback-Leibler divergence becomes ill-defined if the forecast distribution F has positive mass anywhere where the observation distribution G has mass zero. When G is replaced by $\hat{G}_n$ and, especially, if the sample size n is relatively small, such issues might occur. One option to circumvent the issue is to treat the data as categorical and bin it in b bins prior to the evaluation. That is, identify the probability distribution F with a probability vector $(f_1, ..., f_b)$ and, similarly, G with a probability vector $(g_1, ..., g_b)$. The *Kullback-Leibler divergence* is then given by

$$\mathrm{KLD}(F,G) = \sum_{i=1}^{b} f_i \log \frac{f_i}{g_i}$$

see also the discussion in Thorarinsdottir et al. (2013).

Historically, much of the forecast evaluation literature has focused on the evaluation of probabilistic forecasts against deterministic observations and an in-depth discussion of optimal theoretical and/or practical properties of divergences is lacking. Applied studies commonly employ divergences that are asymptotically proper rather than n-proper for all positive integer n, see for example, Palmer (2012) and Perkins, Pitman, Holbrook, and McAneney (2007).

6.3.7 TESTING EQUAL PREDICTIVE PERFORMANCE

As demonstrated in the simulation study in Section 6.3.2, the estimation of the mean score over a test set may be associated with a large uncertainty. A simple bootstrapping procedure over the individual scores may be used to assess the uncertainty in the mean score, see for example, Friederichs and Thorarinsdottir (2012). Assume we have n score values $S(F_1, y_1), ..., S(F_n, y_n)$. By repeatedly resampling vectors of length n (with replacement) and calculating the mean of each sample, we obtain an estimate of the variability in the mean score. Note that some care is needed if the forecast errors, and thus the resulting scores, are correlated. A comprehensive overview over bootstrapping methods for dependent data is given in Lahiri (2003).

Formal statistical tests can be applied to test equal predictive performance of two competing methods under a proper scoring rule. The most commonly applied test is the *Diebold-Mariano test* (Diebold & Mariano, 1995), which applies in the time series setting. Consider two competing forecasting methods F and G that for each time step $t = 1, ..., n$ issue forecasts F_t and G_t, respectively, for an observation y_{t+k} that lies k time steps ahead. The mean scores under a scoring rule S are given by

$$\bar{S}_n^F = \frac{1}{n}\sum_{t=1}^{n} S(F_t, y_{t+k}) \quad \text{and} \quad \bar{S}_n^G = \frac{1}{n}\sum_{t=1}^{n} S(G_t, y_{t+k})$$

The Diebold-Mariano test uses the test statistic

$$t_n = \sqrt{n}\frac{\bar{S}_n^F - \bar{S}_n^G}{\hat{\sigma}_n} \tag{6.34}$$

where $\hat{\sigma}_n^2$ is an estimator of the asymptotic variance of the score difference. Under the null hypothesis of equal predictive performance and standard regularity conditions, the test statistic t_n in Eq. (6.34) is asymptotically standard normal (Diebold & Mariano, 1995). When the null hypothesis is rejected in a two-sided test, F is preferred if t_n is negative and G is preferred if t_n is positive.

Diebold and Mariano (1995) note that for ideal k-step-ahead forecasts, the forecast errors are at most $(k-1)$-dependent. An estimator for the asymptotic variance $\hat{\sigma}_n^2$ based on this assumption is given by

$$\hat{\sigma}_n^2 = \begin{cases} \hat{\gamma}_0 & \text{if } k = 1 \\ \hat{\gamma}_0 + 2\sum_{j=1}^{k-1}\hat{\gamma}_j, & \text{if } k \geq 2 \end{cases} \tag{6.35}$$

where $\hat{\gamma}_j$ denotes the lag j sample autocorrelation of the sequence $\{S(F_i, y_{i+k}) - S(G_i, y_{i+k})\}_{i=1}^n$ for $j = 0$, 1, 2, ... (Gneiting & Ranjan, 2011). Alternative estimators are discussed in Diks et al. (2011) and Lerch et al. (2017).

In the spatial setting, Hering and Genton (2011) propose the *spatial prediction comparison test*, which accounts for spatial correlation in the score values without imposing assumptions on the underlying data or the resulting score differential field. This test is implemented in the R package SpatialVx (Gilleland, 2017). Weighted scoring rules and their connection to hypothesis testing are discussed in Holzmann and Klar (2017).

A simple test for the uniformity of a rank or PIT histogram is the chi-squared test. It tests if the histogram values can be considered samples from a uniform distribution and therefore if any deviations of uniformity are random or systematic (Wilks, 2004, 2011). The chi-squared statistic based on n cases and K ensemble members is

$$\chi^2 = \sum_{i=1}^{K+1} \frac{(m_i - f)^2}{f} \tag{6.36}$$

with m_i denoting the actual number of counts for bin i and $f = \frac{n}{K+1}$ the expected number of counts for a uniform distribution. We can reject the null hypothesis of the histogram being uniform if this statistic exceeds the quantile of the chi-squared distribution with K degrees of freedom at the chosen level of significance.

In its general form, however, the chi-squared test only applies to independent data, which is not the case in many forecast settings due to, for example, temporal or spatial correlation between forecast data points. Some methods to address this effect are proposed in Wilks (2004). If the goal is to not only test for uniformity, but also for the other deficiencies in calibration shown in Section 6.2.1, Elmore (2005) and Jolliffe and Primo (2008) present alternatives that are more flexible and appropriate. Wei et al. (2017) propose calibration tests for multivariate Gaussian forecasts based on the Dawid-Sebastiani score in Eq. (6.24).

6.4 UNDERSTANDING MODEL PERFORMANCE

When assessing the performance of an individual model, for example, to identify weaknesses and test potential improvements, it might be useful to look at tools that do not necessarily follow the principles of propriety described in Section 6.3. For instance, it can be useful to investigate the forecast bias to better understand the potential sources of forecast errors even if competing forecasting models should not be ranked based on mean bias as it is not a proper score (Gneiting & Raftery, 2007). Here, we discuss a few tools that may be used to provide a better understanding of the performance of an individual forecasting model, even though ranking of competing forecasters should not be based on these tools.

One of the most popular measures used by national weather services is the anomaly correlation coefficient (ACC), a valuable tool to track the gain in forecast skill over time (Jolliffe & Stephenson, 2012). The ACC quantifies the correlation between forecast anomalies and the anomalies of the observation, typically an analysis. Anomalies are defined as the difference between the forecast or analysis and the climatology for a given time and location. Usually, the climatology is based on the model climate, calculated from the range of values predicted by the dynamical forecast model over a long time period.

For a deterministic forecast f_i, valid at time i, with a corresponding analysis a_i and climate statistic c_i, there are two equivalent definitions for the ACC (e.g., Miyakoda, Hembree, Strickler, & Shulman, 1972):

$$\text{ACC} = \frac{\displaystyle\sum_{i=1}^{N}(f_i - c_i) \cdot (a_i - c_i) - \sum_{i=1}^{N}(f_i - c_i) \cdot \sum_{i=1}^{N}(a_i - c_i)}{\sqrt{\displaystyle\sum_{i=1}^{N}(f_i - c_i)^2 - \left(\sum_{i=1}^{N}(f_i - c_i)\right)^2} \cdot \sqrt{\displaystyle\sum_{i=1}^{N}(a_i - c_i)^2 - \left(\sum_{i=1}^{N}(a_i - c_i)\right)^2}}$$

$$= \frac{\displaystyle\sum_{i=1}^{N}(f_i' - \bar{f}')(a_i' - \bar{a}')}{\sqrt{\displaystyle\sum_{i=1}^{N}(f_i' - \bar{f}')^2 \sum_{i=1}^{N}(a_i' - \bar{a}')^2}}$$

Here, $f_i' = f_i - c_i$ is the forecast anomaly and $a_i' = a_i - c_i$ the anomaly of the analysis, with respective sums $\bar{f}' = \sum_{i=1}^{N}(f_i - c_i)$ and $\bar{a}' = \sum_{i=1}^{N}(a_i - c_i)$. The ACC is a preferred evaluation tool for gridded forecasts and spatial fields, as these are usually compared with an analysis or a similar gridded observation product.

However, there are certain limitations and pitfalls one has to be aware of when using this measure. Due to it being a correlation coefficient, the ACC does not give any information about forecast biases and errors in scale, so that it can overestimate the forecast skill (Murphy & Epstein, 1989). As such, it should always be used in conjunction with an estimate of the actual bias, or applied to previously bias-corrected data.

It has been established empirically that an anomaly correlation of 0.6 corresponds to a limit in usefulness for a medium-range forecast. Murphy and Epstein (1989) warn, however, that the ACC is an upper limit of the actual skill and that the ACC should be seen as a measure of potential skill. Naturally, the ACC relies to a large extent on the underlying climatology used to compute the anomalies.

When evaluating forecast skill with proper scores, it is often useful to compute separate indicators for the degree of calibration and the sharpness of the forecast. The well-known and widely used decomposition of the Brier score by Murphy (1973b) separates the score value in three parts, quantifying reliability, resolution, and uncertainty.

Consider a forecast sample of size N, where probability forecasts $p_u = 1 - F(u)$ are computed for exceeding a threshold u and binary observations take the form $o = \mathbb{1}\{y \geq u\}$. If the forecasts take K unique values, with n_k denoting the number of forecasts within the category k and $p_{u,k}$ the probability forecast associated with category k, then the Brier score can be written as

$$\text{BS}(F, y | u) = \frac{1}{N} \sum_{k=1}^{K} n_k (p_{u,k} - \bar{o}_k)^2 - \frac{1}{N} \sum_{k=1}^{K} n_k (\bar{o}_k - \bar{o})^2 + \bar{o}(1 - \bar{o}) \tag{6.37}$$

where $\bar{o}_k$ is the event frequency for each of the forecast values and $\bar{o} = \frac{1}{N} \sum_{i=1}^{N} o_i$ the climatological event frequency, computed from the sample. The first part of the sum in Eq. (6.37) relates to the reliability or calibration, the second, having a negative effect on the total score, to the resolution or sharpness, and the last part is the climatological uncertainty of the event.

This representation of the Brier score relies on the number of discrete forecast values K being relatively small. If p_u takes continuous values, care must be taken when binning the forecast into categories, so as not to introduce biases (Bröcker, 2008; Stephenson, Coelho, & Jolliffe, 2008). Several analog decompositions have been proposed for other scores, such as the CRPS (Hersbach, 2000), the quantile score (Bentzien & Friederichs, 2014), and the ignorance score (Weijs, van Nooijen, & van de Giesen, 2010). Bröcker (2009) shows that any proper score can be decomposed analogously to Eq. (6.37). Recently, Siegert (2017) formulated a general framework allowing for the decomposition of arbitrary scores.

While it is common and advisable to look at a model's performance in certain weather situations or for certain periods of time, it is important to be aware of Simpson's paradox (Simpson, 1951). It describes the phenomenon that a certain effect appearing in several subsamples may not be found in a combination of these samples, or that the larger sample may even show the complete opposite effect.

For example, a forecast model can have superior skill over all four seasons, compared with another model, but still be worse when assessed over the whole year. Hamill and Juras (2006) showed this to be true for a synthetic data set of temperature forecasts on two islands. In this case, the climatologies of the two islands were so different that the values of performance measures were misleadingly improved. Fricker, Ferro, and Stephenson (2013) found that this spurious skill does not affect proper scores derived from scoring rules, but care should be taken when using scores derived from a contingency table that are not proper, and skill scores in general.

In general, it is recommended to use statistical significance testing in order to evaluate potential model improvements. Differences in scores are often very small and it is hard to judge if they are caused by genuine improvement or chaotic error growth. Geer (2016) investigate a version of the Student's t-test modified for multiple models and taking account of autocorrelation in the scores. They also found that in order to detect an improvement of 0.5%, at least 400 forecast fields on a global grid would be required. This confirms our findings from Section 6.3.2 that it is essential to carefully consider the experiment sample size in order to generate meaningful and robust results.

6.5 SUMMARY

In this chapter, a variety of methods to assess different aspects of forecast goodness were presented and discussed. Calibration errors can be diagnosed with the help of histograms, in both univariate and multivariate settings. It is recommended to use multiple such diagnostics, especially in the multivariate case, as different tools highlight different types of miscalibration.

Scoring rules provide information about the accuracy of a forecast and are valuable tools for comparing forecasting methods. In this context, only proper scores should be used, as they ensure that the forecast based on the best knowledge will receive the best score. There are many such scores available, with the CRPS and the ignorance score being among the most popular. However, only looking at the mean of one such score can be misleading, even if the underlying sample seems to be of sufficient size. Therefore, it is crucial to also provide information about the error of a mean score, and to base decisions about model preference on the evaluation of multiple scoring rules, if possible. If we do not want to compare models, but rather understand the behavior of a model, it can be helpful to use measures that are not necessarily proper. Especially skill scores and the ACC are widely used.

By adding appropriate weight functions to the CRPS and the ignorance score, it is possible to evaluate extreme event forecasts in a proper way. These weight functions can be designed to emphasize, for example, different parts of the climatological distribution. Scores for multivariate quantities not only give information about the calibration and sharpness of the forecast, but also assess the correct representation of the covariance structure between locations, forecast times, or variables. However, some of them have limitations and do not work well if the number of dimensions is large.

Given the multitude of available evaluation tools and scores, which are constantly growing due to new research and applications, it is essential to be aware of their properties and how to choose a suitable measure. To make sure that all aspects of a forecast's performance are addressed, a number of scores should be calculated and a quantification of the associated uncertainty given.

REFERENCES

Anderson, J. (1996). A method for producing and evaluating probabilistic forecasts from ensemble model integrations. *Journal of Climate, 9*, 1518–1530.

Bentzien, S., & Friederichs, P. (2014). Decomposition and graphical portrayal of the quantile score. *Quarterly Journal of the Royal Meteorological Society, 140*, 1924–1934.

Bigelow, F. (1905). Application of mathematics in meteorology. *Monthly Weather Review, 33*, 90–90.

Brent, R. (1973). *Algorithms for Minimization Without Derivatives*. Englewood Cliffs: Prentice-Hall.

Brier, G. (1950). Verification of forecasts expressed in terms of probability. *Monthly Weather Review, 78*, 1–3.

Bröcker, J. (2008). Some remarks on the reliability of categorical probability forecasts. *Monthly Weather Review, 136*, 4488–4502.

Bröcker, J. (2009). Reliability, sufficiency, and the decomposition of proper scores. *Quarterly Journal of the Royal Meteorological Society, 135*, 1512–1519.

Dawid, A. (1984). Statistical theory: the prequential approach (with discussion and rejoinder). *Journal of the Royal Statistical Society Ser. A, 147*, 278–292.

Dawid, A., & Sebastiani, P. (1999). Coherent dispersion criteria for optimal experimental design. *Annals of Statistics, 27*, 65–81.

Delle Monache, L., Hacker, J. P., Zhou, Y., Deng, X., & Stull, R. B. (2006). Probabilistic aspects of meteorological and ozone regional ensemble forecasts. *Journal of Geophysical Research: Atmospheres, 111*, D24307.

Diebold, F., & Mariano, R. (1995). Comparing predictive accuracy. *Journal of Business & Economic Statistics, 13*, 253–263.

Diks, C., Panchenko, V., & Van Dijk, D. (2011). Likelihood-based scoring rules for comparing density forecasts in tails. *Journal of Econometrics, 163*, 215–230.

Elmore, K. (2005). Alternatives to the chi-square test for evaluating rank histograms from ensemble forecasts. *Weather and Forecasting, 20*, 789–795.

Feldmann, K., Scheuerer, M., & Thorarinsdottir, T. (2015). Spatial postprocessing of ensemble forecasts for temperature using nonhomogeneous Gaussian regression. *Monthly Weather Review, 143*, 955–971.

Ferro, C., Richardson, D., & Weigel, A. (2008). On the effect of ensemble size on the discrete and continuous ranked probability scores. *Meteorological Applications, 15*, 19–24.

Fricker, T., Ferro, C., & Stephenson, D. (2013). Three recommendations for evaluating climate predictions. *Meteorological Applications, 20*, 246–255.

Friederichs, P., & Hense, A. (2007). Statistical downscaling of extreme precipitation events using censored quantile regression. *Monthly Weather Review, 135*, 2365–2378.

Friederichs, P., & Thorarinsdottir, T. (2012). Forecast verification for extreme value distributions with an application to probabilistic peak wind prediction. *Environmetrics, 23*, 579–594.

Geer, A. J. (2016). Significance of changes in medium-range forecast scores. *Tellus Ser. A, 68*, 30229.

Gilleland, E. (2017). Spatialvx: spatial forecast verification. *R package version 6-1*.

Gneiting, T. (2011). Making and evaluating point forecasts. *Journal of the American Statistical Association, 106*, 746–762.

Gneiting, T., Balabdaoui, F., & Raftery, A. (2007). Probabilistic forecasts, calibration and sharpness. *Journal of the Royal Statistical Society Ser. B, 69*, 243–268.

Gneiting, T., & Katzfuss, M. (2014). Probabilistic forecasting. *Annual Review of Statistics and Its Application, 1*, 125–151.

Gneiting, T., & Raftery, A. (2007). Strictly proper scoring rules, prediction, and estimation. *Journal of the American Statistical Association, 102*, 359–378.

Gneiting, T., Raftery, A., Westveld, A., & Goldman, T. (2005). Calibrated probabilistic forecasting using ensemble model output statistics and minimum CRPS estimation. *Monthly Weather Review, 133*, 1098–1118.

Gneiting, T., & Ranjan, R. (2011). Comparing density forecasts using threshold-and quantile-weighted scoring rules. *Journal of Business & Economic Statistics, 29*, 411–422.

Gneiting, T., & Ranjan, R. (2013). Combining predictive distributions. *Electronic Journal of Statistics, 7*, 1747–1782.

Gneiting, T., Stanberry, L., Grimit, E., Held, L., & Johnson, N. (2008). Assessing probabilistic forecasts of multivariate quantities, with applications to ensemble predictions of surface winds (with discussion and rejoinder). *Test, 17*, 211–264.

Good, I. (1952). Rational decisions. *Journal of the Royal Statistical Society Ser. B, 14*, 107–114.

Grimit, E., Gneiting, T., Berrocal, V., & Johnson, N. (2006). The continuous ranked probability score for circular variables and its application to mesoscale forecast ensemble verification. *Quarterly Journal of the Royal Meteorological Society, 132*, 2925–2942.

Hamill, T. M., & Colucci, S. (1997). Verification of Eta-RSM short-range ensemble forecasts. *Monthly Weather Review, 125*, 1312–1327.

Hamill, T. M., & Juras, J. (2006). Measuring forecast skill: is it real skill or is it the varying climatology? *Quarterly Journal of the Royal Meteorological Society, 132*, 2905–2923.

Hering, A., & Genton, M. (2011). Comparing spatial predictions. *Technometrics, 53*, 414–425.

Hersbach, H. (2000). Decomposition of the continuous ranked probability score for ensemble prediction systems. *Weather and Forecasting, 15*, 559–570.

Holzmann, H., & Klar, B. (2017). Focusing on regions of interest in forecast evaluation. *The Annals of Applied Statistics, 11*, 2404–2431.

Jolliffe, I., & Primo, C. (2008). Evaluating rank histograms using decompositions of the chi-square test statistic. *Monthly Weather Review, 136*, 2133–2139.

Jolliffe, I., & Stephenson, D. (Eds.), (2012). *Forecast Verification: A Practitioner's Guide in Atmospheric Science.* Chichester, UK: John Wiley & Sons.

Jordan, A., Krüger, F., & Lerch, S. (2017). *Evaluating probabilistic forecasts with the R package scoring Rules.* https://arxiv.org/abs/1709.04743 (Accessed 26 January 2018).

Krüger, F., Lerch, S., Thorarinsdottir, T. L., & Gneiting, T. (2016). *Probabilistic forecasting and comparative model assessment based on Markov Chain Monte Carlo output.* https://arxiv.org/pdf/1608.06802.pdf (Accessed 26 January 2018).

Kruskal, J. (1956). On the shortest spanning subtree of a graph and the traveling salesman problem. *Proceedings of the American Mathematical Society, 7,* 48–50.

Lahiri, S. (2003). *Resampling Methods for Dependent Data.* New York: Springer.

Laio, F., & Tamea, S. (2007). Verification tools for probabilistic forecasts of continuous hydrological variables. *Hydrology and Earth System Sciences Discussions, 11,* 1267–1277.

Lerch, S., Thorarinsdottir, T., Ravazzolo, F., & Gneiting, T. (2017). Forecaster's dilemma: extreme events and forecast evaluation. *Statistical Science, 32,* 106–127.

Liu, R. (1990). On a notion of data depth based on random simplices. *The Annals of Statistics, 18,* 405–414.

López-Pintado, S., & Romo, J. (2009). On the concept of depth for functional data. *Journal of the American Statistical Association, 104,* 718–734.

Matheson, J., & Winkler, R. (1976). Scoring rules for continuous probability distributions. *Management Science, 22,* 1087–1096.

Mirzargar, M., & Anderson, J. (2017). On evaluation of ensemble forecast calibration using the concept of data depth. *Monthly Weather Review, 145,* 1679–1690.

Miyakoda, K., Hembree, G., Strickler, R., & Shulman, I. (1972). Cumulative results of extended forecast experiments I. Model performance for winter cases. *Monthly Weather Review, 100,* 836–855.

Murphy, A. (1970). The ranked probability score and the probability score: a comparison. *Monthly Weather Review, 98,* 917–924.

Murphy, A. (1973a). Hedging and skill scores for probability forecasts. *Journal of Applied Meteorology, 12,* 215–223.

Murphy, A. (1973b). A new vector partition of the probability score. *Journal of Applied Meteorology, 12,* 595–600.

Murphy, A. (1974). A sample skill score for probability forecasts. *Monthly Weather Review, 102,* 48–55.

Murphy, A. (1992). Climatology, persistence, and their linear combination as standards of reference in skill scores. *Weather and Forecasting, 7,* 692–698.

Murphy, A. (1993). What is a good forecast? An essay on the nature of goodness in weather forecasting. *Weather and Forecasting, 8,* 281–293.

Murphy, A., & Epstein, E. (1989). Skill scores and correlation coefficients in model verification. *Monthly Weather Review, 117,* 572–582.

Oksanen, J., Blanchet, F., Friendly, M., Kindt, R., Legendre, P., & McGlinn, D. (2017). *Vegan: community ecology package.*

Palmer, T. (2012). Towards the probabilistic Earth-system simulator: a vision for the future of climate and weather prediction. *Quarterly Journal of the Royal Meteorological Society, 138,* 841–861.

Perkins, S., Pitman, A., Holbrook, N., & McAneney, J. (2007). Evaluation of the AR4 climate models' simulated daily maximum temperature, minimum temperature, and precipitation over Australia using probability density functions. *Journal of Climate, 20,* 4356–4376.

Pinson, P. (2013). Wind energy: forecasting challenges for its operational management. *Statistical Science, 28,* 564–585.

Pinson, P., & Girard, R. (2012). Evaluating the quality of scenarios of short-term wind power generation. *Applied Energy, 96,* 12–20.

Core Team, R. (2016). *R: A language and environment for statistical computing.* Vienna, Austria: R Foundation for Statistical Computing.

Roulston, M., & Smith, L. (2002). Evaluating probabilistic forecasts using information theory. *Monthly Weather Review, 130*, 1653–1660.

Schefzik, R., Thorarinsdottir, T., & Gneiting, T. (2013). Uncertainty quantification in complex simulation models using ensemble copula coupling. *Statistical Science, 28*, 616–640.

Scheuerer, M., & Hamill, T. M. (2015). Variogram-based proper scoring rules for probabilistic forecasts of multivariate quantities. *Monthly Weather Review, 143*, 1321–1334.

Siegert, S. (2017). Simplifying and generalising Murphy's Brier score decomposition. *Quarterly Journal of the Royal Meteorological Society, 143*, 1178–1183.

Simpson, E. (1951). The interpretation of interaction in contingency tables. *Journal of the Royal Statistical Society Ser B, 13*, 238–241.

Smith, L., & Hansen, J. (2004). Extending the limits of ensemble forecast verification with the minimum spanning tree. *Monthly Weather Review, 132*, 1522–1528.

Stephenson, D., Coelho, C. A. S., & Jolliffe, I. (2008). Two extra components in the Brier score decomposition. *Weather and Forecasting, 23*, 752–757.

Strähl, C., & Ziegel, J. (2017). Cross-calibration of probabilistic forecasts. *Electronic Journal of Statistics, 11*, 608–639.

Thorarinsdottir, T., Gneiting, T., & Gissibl, N. (2013). Using proper divergence functions to evaluate climate models. *SIAM/ASA Journal on Uncertainty Quantification, 1*, 522–534.

Thorarinsdottir, T., Scheuerer, M., & Heinz, C. (2016). Assessing the calibration of high-dimensional ensemble forecasts using rank histograms. *Journal of Computational and Graphical Statistics, 25*, 105–122.

Tsyplakov, A. (2013). *Evaluation of probabilistic forecasts: Proper scoring rules and moments.* http://ssrn.com/abstract=2236605 (Accessed 26 January 2018).

Wei, W., Balabdaoui, F., & Held, L. (2017). Calibration tests for multivariate Gaussian forecasts. *Journal of Multivariate Analysis, 154*, 216–233.

Weijs, S., & van Nooijen, R.van de Giesen, N. (2010). Kullback-Leibler divergence as a forecast skill score with classic reliability-resolution-uncertainty decomposition. *Monthly Weather Review, 138*, 3387–3399.

Wilks, D. (2004). The minimum spanning tree histogram as verification tool for multidimensional ensemble forecasts. *Monthly Weather Review, 132*, 1329–1340.

Wilks, D. (2011). Statistical Methods in the Atmospheric Sciences. Oxford: Elsevier Academic Press.

Wilks, D. (2017). On assessing calibration of multivariate ensemble forecasts. *Quarterly Journal of the Royal Meteorological Society, 143*, 164–172.

Ziegel, J., & Gneiting, T. (2014). Copula calibration. *Electronic Journal of Statistics, 8*, 2619–2638.

CHAPTER

PRACTICAL ASPECTS OF STATISTICAL POSTPROCESSING

7

Thomas M. Hamill

NOAA Earth System Research Lab, Physical Sciences Division, Boulder, CO, United States

CHAPTER OUTLINE

7.1 INTRODUCTION

Those involved with statistical model development commonly spend much of their time dealing with the practical aspects behind testing a research hypothesis. What data should be used? Are the input data of consistent quality, or must the researcher perform quality control? Are the training data so limited that no existing method produces acceptable quality guidance? Are they so voluminous as to be challenging to store and disseminate or to speedily train a model? Do the training data change in their statistical characteristics over time? How do I quickly obtain code for existing methods to use as standards of comparison? A researcher may wish to focus on the scientific aspects of the problem, but find they cannot do so until due diligence has been paid to these other issues. Such issues will not go away, but it is possible to anticipate and surmount common obstacles, individually and as a community.

Statistical Postprocessing of Ensemble Forecasts. **https://doi.org/10.1016/B978-0-12-812372-0.00007-8**

187

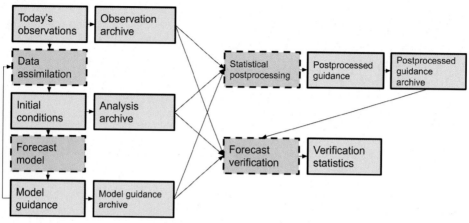

FIG. 7.1

Diagram of many of the typical components and data stores of an end-to-end weather prediction system, and the propagation of data through the system. Boxes with solid borders are data stores, and boxes with dashed borders are components of the prediction system.

Fig. 7.1 illustrates a typical weather prediction system with its components and data stores, illustrating the dependency of statistical postprocessing on previously produced data. These components commonly include a data assimilation system (Daley, 1991; Kalnay, 2003) that statistically adjusts prior dynamical forecasts to newly available observations. Its purpose is to generate accurate and dynamically balanced gridded analyses of the state of the environment suitable for the initialization of a prediction system. The forecast model (or more commonly now, an ensemble prediction system; see Chapter 2) approximates the physical laws governing the evolution of the environmental state (Durran, 2010; Warner, 2011) and simulates the evolution from the initial states. The statistical postprocessing algorithm is commonly trained using archives of forecast, observation, and/or analysis data.

The diagram in Fig. 7.1 simplifies the actual data flow. For example, statistical postprocessing often has two distinct phases: the training of a model and the application of that model to adjust to today's real-time guidance. For some variables such as precipitation, the analyses used in the statistical training may (Lespinas, Fortin, Roy, Rasmussen, & Stadnyk, 2015) or may not (Zhang et al., 2016) utilize prior model forecast guidance, as suggested in the diagram. Further, the postprocessed guidance is not necessarily the end of the product chain; it may also provide inputs to other prediction systems. For example, a hydrologic prediction system intended to produce streamflow forecasts may ingest postprocessed meteorological guidance, synthesize it with observations of the land and snow state, and then generate ensembles of hydrologic predictions that in turn may require their own statistical postprocessing (Hemri, 2018, Chapter 8 in this book; Schaake, Hamill, Buizza, & Clark, 2007).

Because of these data dependencies, the quality of the postprocessed guidance depends on more than just the sophistication of the statistical algorithm. Suppose a statistical postprocessing algorithm is trained against analysis data, regarding these as proxies for the true state. The ultimate accuracy of the postprocessed guidance thus depends upon the accuracy, bias, and temporal consistency of these analyses. Further, the postprocessing algorithm is statistically modeling the discrepancies between prior forecasts and the verification data. What should be done if the characteristics of the forecast discrepancies change in time due to something other than short-term weather variability? Perhaps the forecast model has different bias characteristics in the warm season relative to the cool season, or under El Niño versus La Niña conditions, or perhaps the forecast model was upgraded to a new version during the training period, and the old and new versions have different error characteristics. Understanding these issues and addressing them may be essential to providing the high-quality post-processed guidance desired by forecast users.

This chapter now delves more deeply into these issues and some possible ways to ameliorate them. Section 7.2 provides an example of how the classical and thorny "bias-variance tradeoff" manifests itself in statistical postprocessing; this tradeoff underlies the discussion of many of the algorithmic and data choices that follow. Section 7.3 then returns to discuss challenges with the training data, both forecast and observed/analyzed data. Section 7.4 discusses future directions to mitigate these challenges. Section 7.5 then provides a case study, discussing the tradeoffs that were made in developing a product of common interest, the probability of precipitation from multimodel ensemble guidance. Finally, in Section 7.6 we turn to a different problem: how do we accelerate progress in statistical post-processing as a community? Different investigators commonly develop methods in isolation from each other, which may make testing a hypothesis (is the proposed method better than other recently developed methods?) quite difficult. There is a way forward, providing we are willing to participate in the co-development of a community infrastructure and test data sets.

7.2 THE BIAS-VARIANCE TRADEOFF

Readers are referred to applied statistics texts such as Hastie and Tibshirani (1990, Fig. 2.2) or Hastie et al. (2001, Section 2.9) for more discussion on this subject. The bias-variance tradeoff is intimately related to a statistical concept called "overfitting." For example, this is discussed in Wilks (2011), Section 7.4). Wikipedia (2016) describes the bias-variance tradeoff this way:

> "The bias-variance tradeoff is a central problem in supervised learning.[1] Ideally, one wants to choose a model that both accurately captures the regularities in its training data, but also generalizes well to unseen data. Unfortunately, it is typically impossible to do both simultaneously. High-variance learning methods may be able to represent their training set well, but are at risk of overfitting to noisy or unrepresentative training data. In contrast, algorithms with high bias typically produce simpler models that don't tend to overfit, but may underfit their training data, failing to capture important regularities."

Let's construct a simple, synthetic observation and forecast training data set to illustrate the problem that occurs with a commonly applied statistical postprocessing algorithm, a "decaying-average bias correction" (Cui, Toth, Zhu, & Hou, 2012). Today's forecast bias is estimated as a linear combination of the most recent forecast minus observation and a previous bias estimate. This simple postprocessing approach is appealing for its minimal data storage requirements. Our theoretical construct is as follows. The true state of a univariate system at date/time t, y_t^{true} is sought. In this synthetic construct, the true

---

[1] Supervised learning is the machine learning task of inferring a function from labeled training data.

state, unknown for purposes of model training, is always exactly zero. What is available is a time series of forecasts, all for the same lead time (say, perhaps a 3-day ahead forecast) from dates/times $t0$ to tf, $\mathbf{x} = [x_{t0}, \ldots, x_{tf}]$. Past observations $\mathbf{y} = [y_{t0}, \ldots, y_{tf-1}]$ are also available. The observations are generated from the truth plus random noise: $y_t^o = y_t^{true} + e_t^o$, $e_t^o \sim N(0\frac{1}{9})$, that is, the observations at time t are normally distributed with zero mean (the true state) and random error with a variance of 1/9. Forecast-error characteristics, unknown to the data analyst but known to us here, are constructed with random, seasonally dependent, and serially correlated systematic errors. The true seasonally dependent bias is $B_t = \cos(2\pi J(t)/365)$, where $J(t)$ is the Julian day of the year minus one; that is, the bias varies over the year from 1 to -1 in a cosine-shaped function, too warm at the beginning and end of the calendar year and too cold in the middle. The forecast's daily random error $e_t^f \sim N(0, 1)$, that is, the innovation variance (Wilks, 2011, Section 9.3.1) is nine times larger than the observation variance. Finally, the time series of synthetic forecasts are simulated with a first-order autoregressive model (ibid): $x_t - B_t = k(x_{t-1} - B_{t-1}) + e_t^f$, where here $k = 0.5$.

The decaying-average bias correction assumes that estimated forecast bias for day t, $\hat{B}_t$, can be estimated as a linear combination of the previous day's bias estimate and the most recent deviation of the forecast from the observation:

$$\hat{B}_t = (1 - \alpha)\hat{B}_{t-1} + \alpha(x_{t-1} - y_{t-1}^o) \tag{7.1}$$

Here α is a user-defined parameter that indicates how much weight to apply to the most recent deviation of the observation from the forecast. When α is small, the bias tends toward being estimated as a long-term mean of the difference between forecasts and observations. When α is large, the most recent data is weighted heavily, and estimated bias may vary a lot from one day to the next.

Fig. 7.2 illustrates 100 independent Monte Carlo simulations of the estimated bias started from different initial random numbers and using different random observation errors; data is shown only after 60 days of spin-up. The four panels show the simulations for four increasing values of α. Each simulation's estimated bias is shown with a light gray line. The mean of these bias estimates is shown with the dashed black line; this is unavailable in practice, as nature provides but one realization. The true bias, again unknown to the data analyst, is denoted by the heavy black line. For small α (Fig. 7.2a), there is less variance in the 100 Monte Carlo estimates of the bias. However, because the algorithm thereby provides heavier weight to past data, and because the true bias for the past data is seasonally dependent, there are systematic errors in those bias estimates; the maximum amplitude of the bias is typically under-estimated and lags the true bias. The tradeoff made for this value of α has resulted in comparatively low variance among the bias estimates, but high systematic error with respect to the true underlying bias. It is akin to a regression analysis with too few predictors (underfitting). For large α (Fig. 7.2d), much weight is provided to the most recent forecast deviation from the observations. The several recent observations are implicitly assigned heavier weight while the long-term mean is assigned less weight. This is akin to a regression analysis with too many predictors. The bias estimates change rapidly with each new daily update, and there is a much greater variety of bias estimates over the 100 independent simulations. The tradeoff made for this α has resulted in lower bias on average, but there is high sampling variability.

In practical weather prediction, we have only a single set of observational data to work with, not 100 replicates, so the dashed lines in Fig. 7.2 are never achieved. Were a data analyst to use this algorithm without modification, she would be faced with making a choice, adjusting α to find an

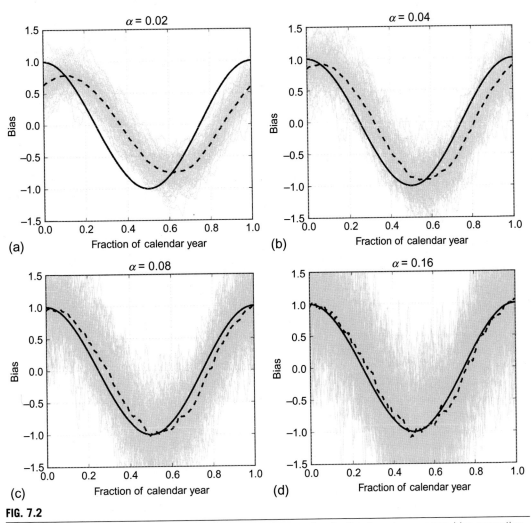

FIG. 7.2

Illustration of the bias-variance tradeoff in statistical postprocessing for the decaying- average bias-correction algorithm. Thin gray lines denote individual Monte Carlo bias estimates using the decaying-average bias-correction algorithm. Dashed black line indicates the mean of the 100 Monte Carlo bias estimates. Solid black line indicates the true underlying bias. Panels (a), (b), (c), and (d) show decaying average weights $\alpha = 0.02$, 0.04, 0.08, and 0.16, respectively.

acceptable compromise between the bias and the variance based on seeing only a single one of the 100 thin gray lines in each of the panels of Fig. 7.2. If she had developed some intuition that the biases were seasonally varying, she might choose to test the value of incorporating other predictors in a more sophisticated regression analysis, predictors such as $\cos(2\pi J(t)/365)$ and $\sin(2\pi J(t)/365)$. Why not do this? The decaying-average bias correction has one very appealing characteristic: very little data need be archived. Once the current forecast and observation have been used to update the bias, it can effectively be discarded for purposes of training. A longer time series of data would need to be stored to apply the more appropriate regression analysis and improve the bias estimates. While data storage was insignificant in this simple synthetic problem, if the method was applied to many variables on a high-resolution grid over a large area, the data storage demands might require the analyst's attention.

7.3 TRAINING-DATA ISSUES FOR STATISTICAL POSTPROCESSING

Consider now the characteristics of an ideal training data set, ideal not in the sense of providing perfect forecasts, but rather ideal in that it serves nearly all the needs of the statistician.

- *The training data should span a long period of time, thereby providing samples across the range of possible future environmental conditions.* This would provide enough samples to quantitatively estimate the probability of even relatively unusual events at each geographic location. Forecast errors are likely to be at least somewhat related to the local geographic peculiarities, including characteristics such as the terrain height, the terrain orientation, the vegetation, land-use, and soil type. With voluminous training data, models could be developed that incorporate any necessary additional predictors without overfitting.
- *Training data should be generated from the same ensemble prediction system in the training period as used for real-time predictions.* This makes the error characteristics of the forecasts more consistent over time.
- *Real-time and retrospective forecast ensembles would have many members.* This permits the atmospheric uncertainty to be quantitatively estimated with modest sampling variability.
- *Error characteristics would not change radically over time*; the forecast errors from simulations 10 or 20 years in the past would be similar to those of today.
- *Past analyses or observations used as predictand data would cover the same period as the forecasts.*
- *Past analyses or observations would be unbiased and of uniformly high quality.*
- *Observation or analysis data would be available for all the locations where postprocessed guidance is desired.*

Unfortunately, less-than-ideal training data is the norm. Consider now the issues with predictor data (typically the ensemble forecasts) followed by the issues associated with predictand (observation, analysis) data.

7.3.1 CHALLENGES IN DEVELOPING IDEAL PREDICTOR TRAINING DATA

The ideal predictor data set can be computationally expensive to generate and archive, and may even be practically impossible to achieve perfectly. Let's presume that an operational implementation occurs every year, and that many past years of forecast data are desired. Computational costs will scale linearly with the number of ensemble members and the number of past "reforecast" cases; 20 years of reforecasts will be 10 times more expensive than 2 years. Ideally, the model forecast data would be archived at the native model resolution, but if a forecast upgrade has double the horizontal resolution and twice as fine a temporal resolution, eight times more reforecast data must be stored when the model is changed, an increasing data-storage burden as the system is upgraded. If the statistical model development is occurring on another computing system, there are additional issues of data transfer to, and storage on, the computer system used for statistical development. While this may not be excessively burdensome if the statistical modeler is developing a regional postprocessing system for one or two variables, it becomes an increasingly important issue to deal with if the system is intended to produce statistical adjustments for a wide number of variables over a large geographic region.

The ideal forecast data set would also generate the reforecasts' initial conditions using a consistent data assimilation system, the same one as used for the generation of the real-time forecasts' initial conditions. Most operational centers use a computationally expensive four-dimensional variational data assimilation technique (4D-Var; Courtier, Thépaut, & Hollingsworth, 1994; Kalnay, 2003), an ensemble Kalman filter (Evensen, 2014; Hamill, 2006), or hybridizations of the two (e.g., Buehner, Morneau, & Charette, 2013; Kleist & Ide, 2015). Generating multiyear or even multidecadal reanalyses to provide reforecast initial conditions may use computational resources that could otherwise be used for increasing the real-time prediction system's resolution or its ensemble size. The additional postprocessed skill added by utilizing the extra training data must be evaluated relative to the additional skill generated from using a higher-resolution, more sophisticated real-time prediction system.

Perhaps to save the computational expense of regenerating reanalyses, the developers of a prediction system may choose to initialize reforecasts using a previously generated reanalysis based on an older version of the forecast model and assimilation system. This was the choice that was made with the recent NCEP Global Ensemble Forecast System (GEFS) reforecasts (Hamill et al., 2013). Prior to 2011, initial conditions were generated from the NCEP Climate Forecast System reanalysis (Saha et al., 2010). Subsequent to this, the forecast initial conditions were generated from the real-time data assimilation system, which underwent various changes that affected initial condition characteristics. Fig. 7.3, from Hamill (2017), shows that the character of short-range temperature and dew point analyses changed over that period with respect to an unchanging reanalysis developed at the European Centre for Medium-Range Weather Forecasts (ECMWF; Dee et al., 2011). Forecasts inherit this initial-condition bias to some extent, so the statistical character of the reforecasts was not homogeneous before versus after 2011. The practical impact of this is degraded statistically postprocessed products after 2011 if they were trained with forecast data prior to 2011.

Even if the computational and storage resources are set aside for the generation of multidecadal reanalyses and reforecasts that are consistent with the operational prediction system, the observing system that provides input to the reanalysis system may have changed dramatically over the reanalysis period. In the past few decades, assimilation systems have begun to assimilate more and more satellite data, including microwave radiances (McNally et al., 2006), infrared radiometer data (Collard & McNally, 2009), cloud-drift winds estimated from high-resolution satellite imagery time series (Velden et al., 2005), aircraft temperatures (Benjamin et al., 2010), scatterometer estimates of ocean surface winds (Bi, Jung, Morgan, & Le Marshall, 2011) and radio occultations (Anthes et al., 2008). These have increased the accuracy of analyses and reanalyses in recent years. Because of these changes, even with current state-of-the-art assimilation methods, it is not possible to generate a retrospective forecast for a date in the distant past with expected errors as small as they are for current forecasts (Dee et al., 2011, Fig. 1).

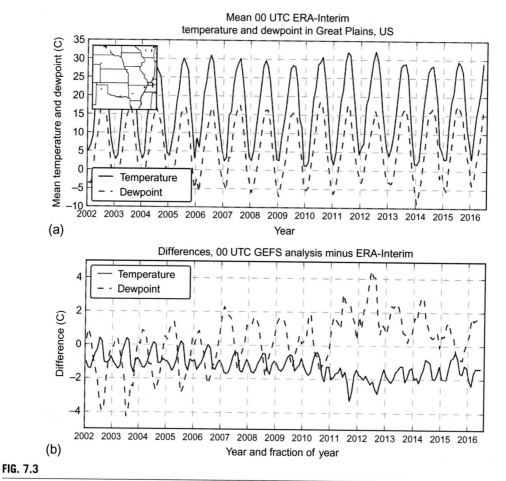

FIG. 7.3

(a) Time series of mean temperatures at 00 UTC from ERA-Interim reanalyses for area covered in map inset (central US). (b) Time series of mean differences at 00 UTC between the temperature of the GEFS initial analysis and the ERA-Interim analysis for temperature (*solid curve*) and dew point (*dashed curve*).

7.3.2 CHALLENGES IN GATHERING/DEVELOPING IDEAL PREDICTAND TRAINING DATA

Training against gridded analyses is often desired, for many users need gridded postprocessed guidance, and this is a straightforward way to achieve this. Unfortunately, some of the characteristics of the ideal analyses outlined herein are difficult to achieve. First, a long time series of analyses can be computationally expensive if generated with modern data assimilation methods such as 4D-Var, the EnKF, or hybridizations. It also requires synthesis of all available observations and massive storage of the resulting data. This may make reanalysis generation impractical for some prediction centers. Were the statistician to use the operational analyses produced in real time, these analyses would likely vary in quality and bias, reflecting both the changing nature of the observing system and the changes in the data assimilation and forecast system.

Why should one expect the analysis bias to vary over time? Presumably the analyses or reanalyses are generated by adjustment of first-guess (background) forecasts to newly available observations. Then the observations and the background should be unbiased in order for the assimilation procedure to produce the unbiased analyses desired for postprocessing. While technology for adjusting the observations to reduce bias (e.g., Auligné, McNally, & Dee, 2007) is now common, and while approaches to adjusting the background to be unbiased have been proposed (e.g., Dee, 2005) if not widely used, complete removal of bias from data assimilation information sources is still problematic. Hence many analyses should be expected to have bias.

Illustrations of analysis bias are shown in Figs. 7.4 and 7.5. Fig. 7.4 shows the time-averaged spread (standard deviation about the multianalysis mean) of 2-m surface temperatures among four different prediction centers. Spreads are calculated for each day and then averaged over the year. Analyses were interpolated to a 1-degree grid before display and were taken from the TIGGE archive (Bougeault et al., 2009; Swinbank et al., 2016). Time-averaged analysis spreads exceeding 1 °C are common, with many regions, especially in mountainous and polar regions, with much greater spread. If we examine the time series of analyses at a particular location (Fig. 7.5), here in the central Amazon river basin, we see that

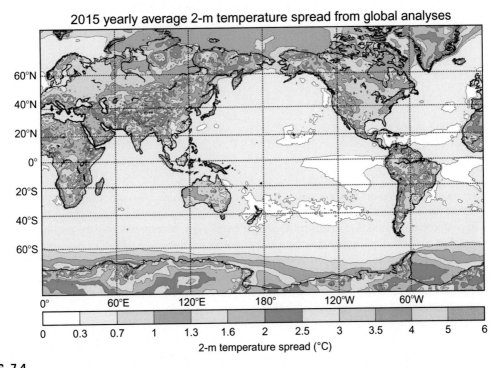

FIG. 7.4

2015's yearly average of the daily spread of 00 UTC 2-m temperature analyses. Data for each analysis system was extracted on a common 1-degree grid via European Centre for Medium-Range Weather Forecasts (ECMWF) TIGGE data portal (Bougeault et al., 2009). Analysis systems used here were National Centers for Environmental Prediction (NCEP), Canadian Meteorological Centre (CMC), the UK Met Office, and ECMWF.

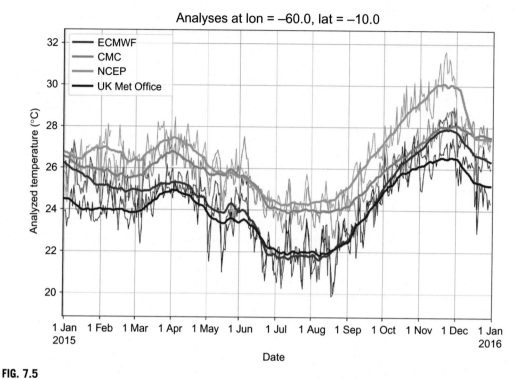

FIG. 7.5

Raw (thin lines) and +/− 15 day smoothed *(thick lines)* time series of 2-m surface temperature analyses at 00 UTC from four different global data assimilation systems for a location in the Amazon river basin.

the differences are not random; some analysis systems are systematically colder, others systematically warmer than the average. Choosing one (e.g., NCEP) and training against this analysis is thus likely to result in postprocessed guidance at this location that has a warm bias (presuming the multicenter mean is more realistic). Note that differences between analyses for upper-air variables may not be as pronounced (Park, Buizza, & Leutbecher, 2008), as near-surface variables are especially challenging to predict, given that many of the relevant processes (boundary layer, surface layer, land surface, cloud microphysics) are treated through parameterizations, that is, approximations of the sub-grid scale effects upon the resolved scales (Stensrud, 2007).

Given the challenges with training against analysis data, even if gridded products are preferred, might it be better to directly use station data? Training against observations provides more site-specific, downscaled information specifically at the observation site (Vannitsem & Hagedorn, 2011). However, if information is also desired at other nearby locations, spatial modeling is necessary. Several such techniques have been developed. These include the procedure of Glahn, Gilbert, Cosgrove, Ruth, and Sheets (2009), where postprocessing is first performed at stations and then interpolated to a grid. Scheuerer and Büermann (2014) proposed a strategy, further developed by Dabernig, Mayr, Messner, and Zeileis (2017) and Stauffer, Umlauf, Messner, Mayr, and Zeileis (2017) where climatological characteristics are interpolated to the grid and removed from both forecasts and observations, so that all locations within a region can be postprocessed simultaneously.

While these methods avoid training against analyses that contain bias, there are disadvantages to such approaches. For example, in situ observation locations are commonly sparse over bodies of water and in mountainous and less-populated regions. Given that commonly desired variables like temperature, wind speed, and precipitation amount may vary with elevation and vary from land to ocean, such statistical interpolations from observation locations to the output grid may produce lower-quality grids in such regions than are desirable. Analyses produced through data assimilation procedures, despite their contamination by bias, will often have useful information in areas devoid of in situ observations. This is because they use other sources of data such as satellites and radars, and because the model first-guess field (background) is effectively a repository of information accumulated from the assimilation of earlier observations.

7.4 PROPOSED REMEDIES FOR PRACTICAL ISSUES IN STATISTICAL POSTPROCESSING

7.4.1 IMPROVING THE APPROACHES FOR GENERATING REFORECASTS

Let's assume for the moment that the director of a prediction center has made a determination that postprocessing is an important step in the production of forecasts, and that training sample size is of sufficient importance that some computational resources must be set aside for reforecasts. Let's also assume for the time being that reanalyses that are similar in quality to the real-time analyses have already been generated. The director of the prediction center has perhaps indicated that the number of reforecasts that can be generated without unduly affecting the implementation of other model improvements is not extravagant, perhaps limited to running the ensemble system retrospectively spanning 4 years of training data with, say, 5 members. With these limits established, other configurations are possible. Two years could be spanned with 10-member reforecasts at the same computational expense. This would decrease the range of weather scenarios covered, but ensemble spread estimates for each case would be improved. Also, 20 years of reforecast data could be generated by generating a 5-member reforecast every fifth day (Hamill, Whitaker, & Wei, 2004). A regular, every-nth-day sub-sampling procedure may be nearly optimal for some variables but not for others. Suppose the most important intended application of the reforecast data set was the statistical postprocessing of heavy precipitation. In such a situation, weather-dependent procedures for determining the dates to generate reforecasts might improve the postprocessing of heavy precipitation. For example, the probability that one should generate a reforecast for a particular past date could depend on, say, the likelihood that a prior-generation reforecast was predicting much heavier than average precipitation amounts in areas of particular interest.[2] When determining a list of case days to reforecast, a day with 20% probability of heavy precipitation would be twice as likely to be selected as a day with 10% probability. Were reforecasts to be selected based on probability of heavy precipitation occurring somewhere in a large region

---

[2]What should *not* be done is to select cases on the basis of observed or analyzed heavy precipitation. In such a situation, the training data would be biased toward the occurrence of heavy precipitation events, and the postprocessing technique applied to the real-time forecast would likely over-predict the precipitation.

(say, the contiguous U.S.), one would expect that at any specific point, there would still be many samples of more ordinary weather, and the accuracy postprocessing algorithm would not be degraded for more common events.

Are there general principles that might guide reforecast configuration? Such decisions should be informed by the intended application(s). If the primary application is for sub-seasonal forecasting where the forecast is more affected by boundary conditions such as sea-surface temperature and soil moisture than by the initial atmospheric state, then a reforecast data set spanning a wider range of climate states is desirable; 20 years every fifth day is preferable to the most recent 4 years every day. If shorter-term probabilistic precipitation postprocessing is of greater interest, a weather-dependent sampling strategy that generates reforecasts on days where heavy precipitation is more likely may be desirable.

Are there principles to determine the tradeoff of length of reforecast versus ensemble size? Again, this may depend on the intended application. For products like the extreme forecast index (Lalaurette, 2003; Petroliagis & Pinson, 2014) that use the reforecasts to determine how unusual today's forecast is relative to the ensemble reforecast climatology, ECMWF's experience (Vitart et al., 2014) has shown that the product performance is improved with more members. On the other hand, for many statistical postprocessing applications, it can be more helpful to have a greater number of individual weather events than a larger ensemble. The primary skill improvement in postprocessing is more commonly a result of correcting errors in the mean state than adjusting the spread, and a wider range of weather scenarios permits more appropriate state-dependent corrections.

How might one address the changing statistical quality of the reforecasts over time due to observing network changes? Past experience has shown (Uppala et al., 2005, Fig. 14, vs. Hamill et al., 2013, Fig. 1; also Dee et al., 2014, Fig. 1) that the more advanced the data assimilation and prediction system, the higher the overall quality and more uniform the statistical characteristics of past versus current forecasts. Hence, regular production of reanalyses with the most up-to-date system is the most straightforward way to address this, however impractical it may be computationally. Assuming one does not have regularly generated reanalyses of uniformly high quality, other options might include weighting the training samples to be inversely proportional to their expected error variance, as is done for example in weighted least-squares regression. If reanalyses are not available, but some computational resources have been set aside for reforecasts, perhaps judicious use of a different reanalysis for initialization may prove useful. In the recent past, prediction centers without their own reanalyses have explored creating reforecasts through a modified initialization with another center's reanalyses, adjusted near the surface to reflect the climatology of the own center's land-surface scheme (Boisserie, Decharme, Descamps, & Arbogast, 2016; Lin et al., 2016).

Should a prediction center not have reanalyses readily available, but believe them to be necessary, their generation may be the most expensive and time-consuming part of the reforecasting process. Suppose we want to generate a 5-member reforecast every third day to +30 days lead. Every third day we have thus generated 150 member-days of reforecasts. Say now that an 80-member ensemble data assimilation approach is used, stepping the 80 members forward 6 h, producing an updated analysis, and repeating the process. The computational expense of merely generating the background ensemble of forecasts for the data assimilation over the same 3-day period is $80 \times 3 = 240$ member-days. The computational expense of the analysis update step is roughly the same order of magnitude. Because of the large computational expense and labor involved in reanalysis generation, reanalyses are thus typically generated once or twice a decade at some prediction centers (e.g., ECMWF, the U.S. NWS, and Japan Meteorological Agency) or not at all for many others.

In the end, is the computation of reforecasts, and perhaps reanalyses, a necessary precursor for effective statistical postprocessing? It likely depends on the intended application. Previous experience (e.g., Hamill, Whitaker, & Mullen, 2006; Scheuerer & Hamill, 2015) have shown that for rare events such as heavy precipitation, the enlarged sample size afforded by many reforecasts improves the postprocessing skill significantly. Another application that is greatly improved by reforecasts is the postprocessing of subseasonal forecasts. At these leads, the noise due to chaos and model error is large and the detectable signal is small; large samples are helpful to extract the small amount of signal among the bath of noise (Ou, Charles, & Collins, 2016). Reforecasts also provide the large sample sizes that can be important for validation of rare, extreme events, such as heavy precipitation events leading to floods. For other applications, such as short-term temperature calibration (Hagedorn, Buizza, Hamill, Leutbecher, & Palmer, 2012) or basic probability forecasting of nonzero precipitation amounts (Hamill et al., 2017), it may be possible to work around some of the issues related to the short training data set, as discussed in the next subsection.

7.4.2 CIRCUMVENTING COMMON CHALLENGES POSED BY SHORTER TRAINING DATA SETS

Suppose lengthy reforecasts are a practical impossibility and one must make do with a much shorter time series of forecasts for training. What procedures may be practical in such a circumstance? For some variables such as surface temperature, past experience shows that some benefit can be obtained with simple approaches such as the decaying-average bias correction discussed in Section 7.2. This is because bias commonly has a large systematic component, especially at short leads, and hence yesterday's forecast bias provides useful predictive information on today's bias. For other variables of interest such as precipitation, the past few days or weeks or even months may not provide a wide enough variety of precipitation events to achieve major improvements, especially if the postprocessing method is applied independently from one location to the next. The limited sample size requires consideration of other approaches.

An obvious candidate is the supplementation of training data using information from surrounding regions (e.g., Allen & Erickson, 2001; Mass, Baars, Wedam, Grimit, & Steed, 2008). Fig. 7.6 provides evidence for why one should be judicious with such approaches. Here, GEFS reforecast data (Hamill et al., 2013) and climatology-calibrated precipitation analyses (CCPA; Hou et al., 2014) were used for the Dec-Jan-Feb 2002–2015 period to populate cumulative distribution functions (CDFs) of 24-h accumulated precipitation at two nearby locations along the Oregon-California border in the northwest U.S. Suppose one were one to apply a quantile-mapping procedure (Hopson & Webster, 2010; Maraun, 2013; Voisin, Schaake, & Lettenmaier, 2010) to address the conditional bias of the forecasts. For example, perhaps the coastal location's training data is supplemented with the training data from the inland location. The conditional forecast biases of moderate precipitation at these two locations are opposite in sign. Along the coast, precipitation is under-forecast, while slightly inland it is over-forecast. Applying quantile mapping to the coastal location using the inland supplemental training data would likely produce a worse adjusted forecast than were the data kept separate.

Perhaps the concept of using supplemental training data is conceptually sound, provided one is careful with what supplemental data are used. A more advanced selection procedure for supplemental data was recently demonstrated in Hamill, Scheuerer, and Bates (2015) and Hamill et al. (2017), and a similar approach was discussed in Lerch and Baran (2017). For each grid point where a postprocessed precipitation forecast was desired, a number of supplemental locations were determined based on

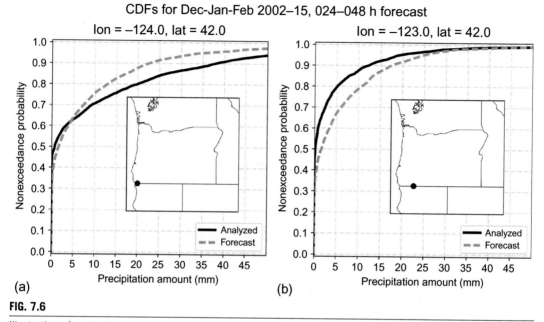

FIG. 7.6

Illustration of regionally dependent differences between forecast and analyzed precipitation, here at two locations in the western U.S., (a) along the U.S. west coast and (b) inland. CDFs from CCPA analyses are shown in the heavy black lines, while GEFS-member forecast CDFs are shown in with heavy dashed gray lines. The two locations are denoted by the two black dots in the inset maps.

similarities of climatology and geographical characteristics such as terrain height and hillslope orientation, with the presumption that many precipitation biases are related to the simplified representation of the terrain characteristics in the numerical model. Training data at the original location was supplemented with the data at these additional locations, with subsequent improvement to the postprocessed guidance.

Another issue that should be considered when working with short training data sets is the possibility of seasonally dependent bias, as with the example of Fig. 7.2. Suppose, for practical considerations, one must use only the last month or two of forecast and observed/analyzed data for training. Focusing on the statistical postprocessing of precipitation, consider forecast and analyzed CDFs again from a reforecast data set spanning four months (Fig. 7.7). The differences between forecast and observed at moderate amounts for this location in southwest Iowa (U.S.) change from a relatively neutral bias at higher precipitation amounts in February to a slight under-forecast in April and a more noticeable under-forecast in June. If the model hasn't changed in more than a year, the most straightforward approach to deal with this is to use additional training data from the same season but from the previous year.

Suppose a postprocessing application requires many years of retrospective forecasts. Consistent reforecasts are unavailable, but an archive of past forecast guidance from the operational model is available. Optimistically, perhaps only the mean bias changes with model version. In this case, an indicator (or "dummy") variable (Neter, Wasserman, & Kutner, 1990, Chapter 10) may suffice to permit

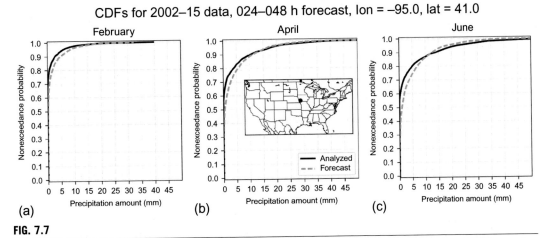

FIG. 7.7

Cumulative distribution functions of 24-h accumulated CCPA-analyzed *(heavy black curve)* and GEFS-member reforecast *(dashed gray curve)* precipitation for three months for a location in SW Iowa of the U.S. Curves were generated using 2002–15 data. (a) February, (b) April, and (c) June.

use of multiple model versions of the training data were regression-type approaches to be used for post-processing. If the regression relationship changes in other ways, leading to the need for a larger set of predictors and interactions for each model version, one must be aware of the potential for overfitting.

7.4.3 SUBSTANDARD ANALYSIS DATA

If training against analyses is an imperative for postprocessing, and if the analyses have systematic errors as previously discussed, one possible but time-consuming approach for remediating these errors is to improve the data assimilation system that generates the analyses. Meteorological statisticians in the U.S. National Weather Service (NWS) have requested such an improvement from the NWS data-assimilation system developers. The NWS wishes to perform postprocessing against high-resolution gridded analyses. The current NWS high-resolution analysis system (e.g., De Pondeca et al., 2011) produce analyses with bias and higher-than-ideal errors, especially in the mountainous western U.S. Hence, the NWS is providing resources for the improvement of this analysis system. Unfortunately, the analysis variables that are of greatest interest (surface temperature, wind, precipitation, and so forth) are often the most difficult to improve. The accuracy of estimating these variables depends on faithfully depicting how the atmosphere interacts with the land surface, with all its heterogeneities, its physical complexity, and its poorly observed soil state. Further, ubiquitous and significant errors in the depiction of clouds in prediction systems may contaminate the model estimates of downward solar radiation that largely determine the resulting surface sensible and latent heat fluxes back to the atmosphere.

Another reason that high-quality analyses suitable for postprocessing are challenging to generate is that the postprocessing requirements for analysis data may be somewhat different than requirements for forecast initialization. For postprocessing, accuracy, lack of bias, and relevant spatial detail are of paramount importance. For forecast initialization, an analysis that leads to an accurate and stable forecast is paramount. Introducing, say, the use of the actual terrain heights in the system rather than a smoothed version may produce somewhat more realistic analyses, but radically poorer predictions.

While the direct improvement of gridded analyses is desirable, major improvements will take time to achieve, and some useful analysis data may be needed right away. Here are some suggested guidelines for use of analysis data: (a) if multiple analyses are available (e.g., Fig. 7.5), then consider training and verification against some linear combination of the available analyses. The underlying hypothesis is that the different systems may have somewhat independent biases, and a (possibly weighted) mean will yield a more accurate estimate than any one individually. (b) Consider approaches that leverage station data but implicitly produce a gridded product, as in Glahn et al. (2009), Kleiber et al. (2011a); Kleiber, Raftery, and Gneiting (2011b), Scheuerer and König (2014), Scheuerer and Möller (2015), and Stauffer et al. (2017).

7.5 CASE STUDY: POSTPROCESSING TO GENERATE HIGH RESOLUTION PROBABILITY-OF-PRECIPITATION FROM GLOBAL MULTIMODEL ENSEMBLES

A practical example of a challenging problem in statistical postprocessing is now presented, illustrating some of the tradeoffs discussed herein and the choices made in the development of a pre-operational postprocessed product in the U.S.

Several years ago, the U.S. NWS instituted a statistical postprocessing initiative, the "National Blend of Models" or more simply the "National Blend." Many of the worded weather forecasts produced by the NWS are automatically generated from gridded fields of temperature, winds, precipitation, and so forth. Forecasters at several dozen NWS offices typically provide manual modifications to centrally produced model guidance grids. When one views the synthesized product nationally, abrupt discontinuities are sometimes evident at the boundaries between two weather forecast offices' areas of responsibility. The National Blend intends to provide statistically postprocessed guidance from multimodel ensembles to the forecasters, of such quality and reliability that the need for manual editing is much less necessary. The intent is to improve forecast consistency as well as quality.

The following case study illustrates a statistical postprocessing method in development for 12-h probability of accumulated precipitation (POP12) in the National Blend. In the U.S., nonzero precipitation is defined as the event of ≥ 0.254 mm, here during the 12-h period. The ultimate desired guidance is a 2.5-km grid of POP12 over the contiguous U.S. (CONUS) and adjacent coastal waters, with likely future extension to a full probabilistic quantitative precipitation forecast. For the initial technique development described here, POP12 is produced and validated on a ⅛-degree grid, which is equivalent to approximately 10.6-km grid spacing at 40°N latitude. The data inputs consist of global deterministic and ensemble forecast guidance and ⅛-degree CCPA (Hou et al., 2014). CCPA has been available since 2002, making it useful for determining a precipitation climatology as well as postprocessing model training and validation. Unfortunately, CCPA data do not cover adjacent coastal waters, one of the tradeoffs made in this application.

In this initial development stage, only two ensemble systems are used for medium-range POP12 forecasts, the U.S. National Centers for Environmental Prediction (NCEP) Global Ensemble Forecast System (Zhou et al., 2017), and the Canadian Meteorological Center (CMC) Global Ensemble Prediction System (Gagnon et al., 2014, 2015). Hereafter, these are referred to as the "NCEP" and "CMC" ensembles. Each ensemble system provides 20 ensemble-member forecasts at a resolution coarser than ⅛ degree. Single deterministic control forecasts are also used from each center. In the future, the U.S. Navy global ensemble will be used as well, although these data are not part of this study. For more on the multicenter U.S. and Canadian ensemble, see also Candille (2009).

The postprocessing method demonstrated here is the approach described in Hamill et al. (2017), which provides greater detail on the methodology and the rationale for its use. The methodology combines a variety of established algorithms, chosen for their suitability to short training data sets and multimodel ensembles. The approach is also readily extensible to full probabilistic quantitative precipitation forecasting in the future. The algorithmic approach includes five general steps: (1) Populate forecast and analysis CDFs of precipitation using the last 60 days of data. To increase the training sample size, CDFs at a particular grid point for a particular ensemble member were populated not only with training data for that grid point, but also from data at that grid point's predefined supplemental locations (ibid). (2) Quantile-map each ensemble member using the forecast and analyzed CDFs. This ameliorates conditional bias and applies an implicit statistical downscaling. (3) Dress each quantile-mapped ensemble member with random noise to correct for remaining problems with underdispersion. (4) Generate probabilities from weighted, dressed ensemble members. (5) Smooth the resulting POP field.

The first step in the real-time data processing is populating the forecast and analyzed CDFs with the prior 60 days' data. As noted earlier, postprocessing of precipitation can be very difficult with small training sample sizes, but this is ameliorated here (e.g., Fig. 2 from Hamill, Hagedorn, & Whitaker, 2008) by supplementation of the training data with data from other locations with similar terrain and precipitation characteristics. Specifically, for each output ⅛-degree grid point, a set of other grid points, or "supplemental locations" are identified, and then the forecast and analyzed CDFs for this grid point are populated with data from the original grid point and with data from the supplemental locations. Such an approach is often preferable to enlarging sample size by combining training data from radically different seasons that often have different conditional biases (Fig. 7.7). Also, should the prediction system change versions and have different biases for different system versions, older model data will be aged out by the end of 60 days as well, limiting the potential duration of degraded product quality.

Fig. 7.8 shows predefined POP12 supplemental locations for several selected grid points in the U.S. during the month of April. Though supplemental locations are shown only for six grid points in this figure, supplemental locations are defined for every ⅛-degree grid point in the CONUS and the Columbia river basin of Canada. The supplemental locations were chosen based on similarity of the CCPA precipitation climatology during the 2002–15 period, terrain height and aspect, and physical separation between grid points. The rationale was that the model's location-dependent systematic errors of precipitation were related to the precipitation climatology in part, and also to the smoothed representation of the terrain relief in the coarser-resolution numerical models (Fig. 7.6). Supplemental locations were prevented from being too close to each other so that these samples had more independent error characteristics. For more information on the supplemental location algorithm, see Hamill et al. (2017).

The next step in the real-time processing is to apply quantile mapping to each ensemble member using the CDFs generated in the previous step. Quantile mapping is illustrated in Fig. 7.9. Presuming a CDF has been generated for a grid point and ensemble member, we determine the forecast amount (here, 3 mm) and its nonexceedance probability (here ∼0.895). The analyzed amount associated with

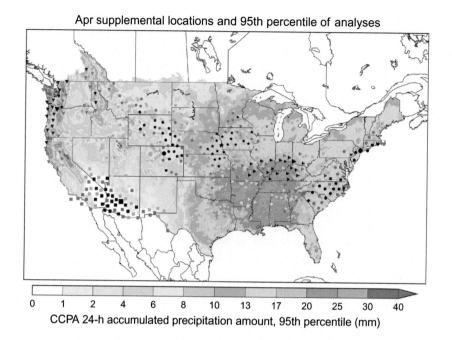

FIG. 7.8

Illustration of supplemental locations for the month of April. Larger symbols denote the locations for which supplemental locations were calculated (roughly Portland, OR; Phoenix, AZ; Boulder, CO; Omaha, NE; Cincinnati, OH; and New York City, NY). Smaller symbols indicate the supplemental locations. Darker symbols indicate a better match; lighter symbols a poorer match. The colors on the map denote the 95th percentile of the 24-h accumulated precipitation amounts for the month, determined from a climatology of 2002–15 CCPA data.

Reprinted with permission from Hamill, T. M., Engle, E., Myrick, D., Peroutka, M., Finan, C., & Scheuerer, M. (2017). The US National Blend of Models statistical post-processing of probability of precipitation and deterministic precipitation amount. Monthly Weather Review, 145, 3441–3463. (Also: online Appendix A and online Appendix B).

the same nonexceedance probability is determined (4 mm), and the forecast amount is adjusted to this value. The procedure is repeated for each output grid point and each ensemble member. This procedure permits the conditional bias to be mitigated. Should the analysis be available on a more finely spaced grid with more detail, then the algorithm is also implicitly performing a statistical downscaling.

Other procedures such as Bayesian Model Averaging (Raftery, Gneiting, Balabdaoui, & Polakowski, 2005; Sloughter, Raftery, Gneiting, & Fraley, 2007) adjust for forecast bias through regression approaches. As noted in Wilks (2006) and Hodyss, Satterfield, McLay, Hamill, and Scheuerer (2016), when regression equations are applied to ensemble members under situations where there is little predictive relationship between forecast and analysis, the ensemble members are regressed toward the mean analyzed value, resulting in an ensemble with a reduction in spread. The reason ensembles are generated in the first place is to provide situational estimates of the forecast uncertainty, and raw ensembles are commonly under-spread; regression of each member can make the underdispersion problem worse. It is for this reason that quantile mapping was preferred over regression; the ensemble spread is less affected, and the subsequent dressing step discussed below has less "work" to do.

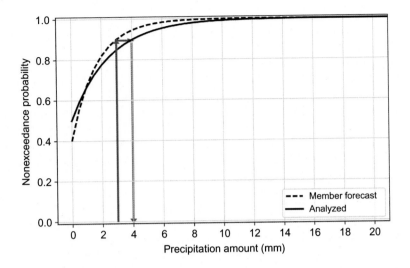

FIG. 7.9

Illustration of the deterministic quantile mapping procedure applied to ensemble members. Forecast and analyzed distributions adjust the raw forecast to the analyzed value associated with the same cumulative probability. Gray arrows denote the mapping process.

An additional feature of the POP12 quantile-mapping procedure is briefly described. Should probabilities now be determined from the 42 ensemble members, one would expect some unreliability and loss of skill in part from the relatively modest size of the ensemble (Richardson, 2001). To ameliorate this and to deal with overconfidence in ensemble systems in the positioning of precipitation features, the quantile mapping utilizes not just the forecast at the grid point of interest, but also quantile-maps forecasts from surrounding grid points. In particular, that grid point and eight nearby points are used as input to the quantile mapping. For each nearby point, the forecast CDF used is the one associated with that nearby grid point, while the analyzed CDF is the one associated with center grid point of interest. In this way, forecasts from surrounding locations, even if they are in mountainous terrain with different climatologies, are mapped to be consistent with the analyzed distribution at the interior point. This process then provides a nine-fold larger ensemble, minimizing errors attributable to finite ensemble size. See Hamill et al. (2017) for more rationale and figures illustrating the process, and see Scheuerer and Hamill (2015) for another application of a similar procedure.

At this stage of the procedure, a ninefold larger quantile-mapped ensemble has been produced for each output grid point, with location-dependent conditional bias reduced. There may be remaining errors, such as insufficient ensemble spread, but they are assumed to be independent of location, though likely dependent on the precipitation amount. The errors of the quantile-mapped members are also assumed at this point to be exchangeable; forecast member 1's quantile-mapped error statistics are assumed the same as forecast member 42's. Inspired by Fortin, Favre, and Saïd (2006), a best-member dressing procedure is now applied. Each quantile mapped member's value is perturbed with random, normally distributed noise with mean zero and standard deviation $0.2 + 0.3 \times$ the quantile-mapped value. Dressed values below zero precipitation are reset to zero. While this procedure is ad hoc, it was informed by other experiments (not shown) where Gamma dressing distributions were objectively fitted.

The next step of the procedure is relatively straightforward. POP12 is estimated from the ensemble relative frequency, that is, if 30% of the members have precipitation above the POP12 threshold of 0.254 mm, the probability is set to 30%.

The final step improves the visual appearance of the forecasts. There are small-scale variations in the POP12 field that are attributable to finite sample size and the application of random dressing noise. However, not all small-scale variations are noise. Over mountainous regions, small-scale variations may reflect orographically enhanced precipitation. Accordingly, we do a final Savitzky-Golay (Press, Teukolsky, Vetterling, & Flannery, 1992) smoothing of the POP12, with more aggressive smoothing in flatter areas and less smoothing in areas with more variations in elevation. There are also procedures to taper the probabilities from their calibrated values to raw multimodel ensemble values beyond the borders of the U.S. See Hamill et al. (2017) for specifics.

Figs. 7.10 and 7.11 provide a case study of how +60 to +72 h POPs are changed through each stage of the postprocessing. Fig. 7.10a shows the verifying precipitation analysis, with heavy precipitation in the central U.S., from Texas north to Kansas. There was also a smaller, north-south band northward from Mississippi to Wisconsin and Michigan. Scattered lighter precipitation occurred the Rocky Mountains of Colorado and Wyoming. The raw NCEP ensemble in Fig. 7.10b was overconfident of precipitation in many regions where no precipitation occurred, including in Arkansas, North Carolina and South Carolina, and in the northwest U.S. The raw CMC ensemble in Fig. 7.10c also forecast elevated POP12 in the northwest U.S. and probabilities above 80% in a wide swath of the central U.S. As expected, the raw combined data shown in Fig. 7.10d portrays intermediate probabilities between these two.

Fig. 7.11a presents the results if only quantile mapping is applied using the center of a 3×3 array of grid points, that is, not using the surrounding data nor thereby increasing the ensemble size ninefold. The quantile mapping reduces the areal extent with low but nonzero POP12s in the western U.S., adjusting for the model tendency to over-forecast light precipitation amounts. The areal extent with very high POP12 in the central U.S. was also decreased, with many areas with 95% or greater probability reduced to ∼80%. The effects of statistical downscaling are also evident in the western U.S., where, for example, POP12 was decreased in eastern Oregon, but much less so along peaks of the Cascade Range, so they now appear as local maxima. The nonzero POP12 in North Carolina and South Carolina was reduced to near zero in many locations. When quantile mapping included the 3×3 array of surrounding points (Fig. 7.11b), there were many grid points whose probabilities were lowered further, and the probabilities east of the Rocky Mountains had a more smooth characteristic. The dressing algorithm (Fig. 7.11c) also made the forecasts less sharp in general, but they add some undesirable small-scale noise. This is largely diminished in the final product, shown in Fig. 7.11d. This final product still has deficiencies; for example, the final POP12 has a single north-south band of higher probabilities in the central-southern U.S., while the observed precipitation had two bands. Ideally, the postprocessing would have reduced probabilities to zero throughout most of the western U.S., as that region was analyzed as dry. Nonetheless, the overall product exploits the diversity in the positioning of precipitation between the two systems, and it reduced POP12 in many regions with high raw probability but no occurrence of precipitation.

The various steps of the algorithm each contribute to the improvements in reliability (i.e., calibration) and skill. Fig. 7.12 shows reliability diagrams at the +60 to +72 h lead time for dates from 1 April to 6 July 2016. Quantile mapping using only the center point improves reliability and skill substantially,

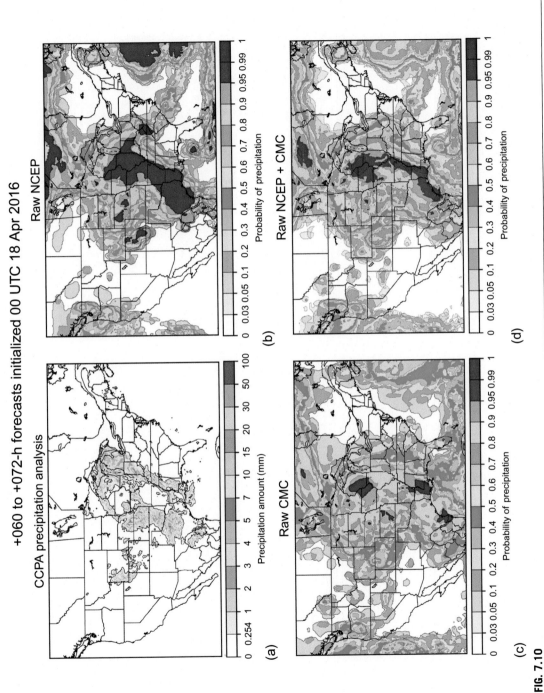

FIG. 7.10

Case study of the steps in POP12 postprocessing for a +60 to +72 h forecast initialized at 00 UTC 18 April 2016. (a) CCPA precipitation analysis. (b) Raw NCEP POP12 forecast. (c) Raw CMC POP12 forecast. (d) Raw CMC+NCEP POP12 forecast.

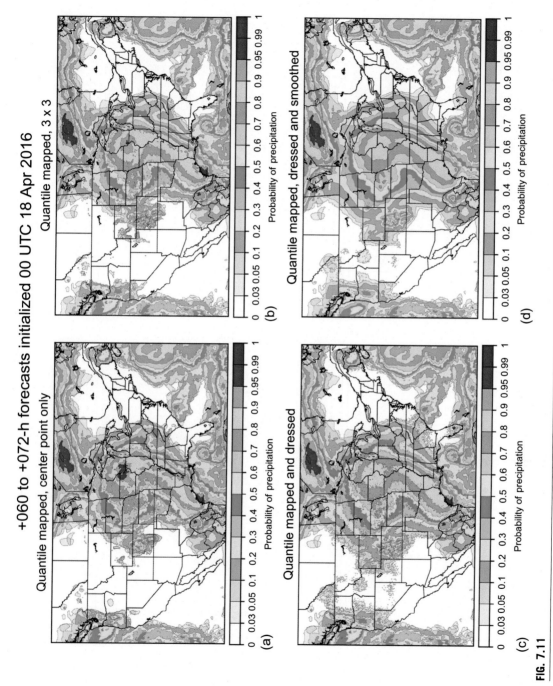

FIG. 7.11

Continuation of the case study of the steps in POP12 postprocessing for a +60 to +72 h forecast initialized at 00 UTC 18 April 2016. (a) postprocessing with quantile mapping using only the grid point in question. (b) Quantile mapping of a 3 × 3 array of grid points, centered on each point of interest. (c) Quantile-mapped and dressed POP12 forecast, and (d) the final product, with 3 × 3 quantile mapping, dressing, and smoothing.

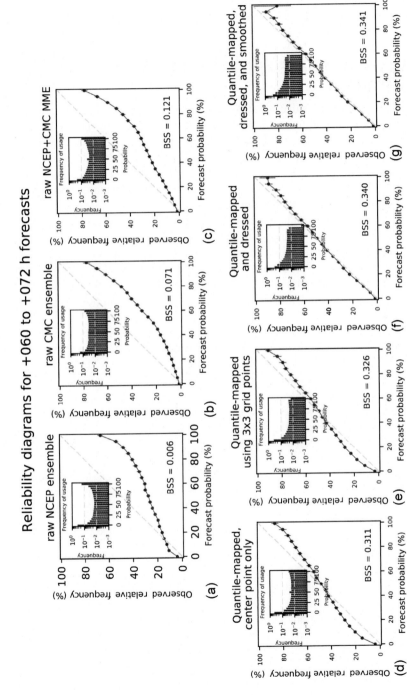

FIG. 7.12

Reliability diagrams for +60 to +72 h POP12 forecasts over the CONUS. Inset histograms show overall frequency with which forecasts are issued, and Brier Skill Scores are noted. (a) Raw NCEP ensemble forecasts, (b) raw CMC ensemble forecasts, (c) raw multimodel ensemble forecasts, (d) postprocessed guidance after quantile mapping using the center point only, (e) after quantile mapping using 3 × 3 stencil of points, (f) after dressing, and (g) after smoothing. Error bars represent the 5th and 95th percentiles of from a 1000-sample bootstrap distribution generated by sampling case days with replacement.

and the use of the 3×3 stencil improves them further. Application of the dressing algorithm improves the reliability and skill a bit more. Smoothing does little to improve the skill, despite the improvement in the visual appearance of the forecasts (Fig. 7.11c and d).

7.6 COLLABORATING ON SOFTWARE AND TEST DATA TO ACCELERATE POSTPROCESSING IMPROVEMENT

Finally, let us turn attention to possible ways that the statistical postprocessing community can work more effectively together than individually. Recognizing the complexity of developing weather prediction components, data assimilation systems and forecast models are increasingly maintained and supported as community endeavors (Skamarock et al., 2005). Users are free to download the code, compile it on the computer system of their choice, generate assimilations and forecasts, and develop and test algorithmic improvements. Commonly with such systems, there is a protocol for submitting algorithmic changes to be incorporated back into the community software. If they are coded to predefined standards and demonstrated to improve the forecasts, a change review board can accept these software modifications, which are then incorporated into future releases.

Envision a similar community infrastructure for postprocessing, including a software and test-data repository. Code for reading these data and writing postprocessed output would be available. A variety of postprocessing algorithms, verification routines, and data visualization tools would be part of the software library. Data sets for common problems of interest would be available in portable formats such as netCDF (Unidata, 2012).

With these components in place, answering research questions could become much more straightforward. A university investigator who seeks to develop a new postprocessing methodology and to test a hypothesis that the method improves upon existing methods could start with data sets and a code infrastructure in place. Their time and effort could be concentrated on the science question at hand, not wrangling with the data and supporting code.

Comparisons with existing benchmarks would be straightforward, and presuming confirmation of the hypothesis of an improved method, the resulting journal articles would be more valuable. Readers would have confidence that the new method had been sufficiently demonstrated to provide an improvement over existing standards.

How do we go about building such a community? This topic was discussed at a 2016 workshop on postprocessing, hosted by the U.S. NWS; the workshop recommendations are shown in the box below. They require some moderate amounts of resources and commitment from a few key individuals, hopefully supported by one or more weather prediction organizations. With the rapid growth of open-source software, there are many established best practices that can be followed to ensure that our new community would have a greater likelihood of flourishing. Standard software version control systems such as "git" would be used; these allow a new user to replicate (create a branch of) the community software on their local computer system, make modifications, but never lose the original. A governance procedure would be established to enable diverse groups to work together and make decisions about software and data changes. Following established best practices, a ticket-tracking system would be established to monitor suggested product improvements and their disposition. A change-control board would be instituted to manage code contributions. These code contributions would be expected to follow predefined testing, documentation, and metadata standards. Documentation would be centralized and consolidated into a few core documents. Ideally, assistance would be provided to help collaborators.

RECOMMENDATIONS FROM A WORKSHOP ON STATISTICAL POSTPROCESSING

Here are some general recommendations from the February 2016 workshop on statistical postprocessing (http://www.dtcenter.org/events/workshops16/NWPprocessing/). Many focus on the actions that government weather prediction centers should take to support the postprocessing community. Recommendations focus on science, community infrastructure, and data.

Science

1. Entrain professional statisticians to assist meteorologists with the development of improved statistical postprocessing methodologies.
2. Perform more intercomparisons of existing algorithms to determine which are the most skillful and reliable.
3. Standard-setting algorithms should be coded so that they are efficient and easily usable by the broader community.
4. Given the challenges with developing and storing high-quality training data sets, further research and development is particularly needed on methodologies that permit high-quality results to be developed with minimal training data.
5. Algorithms developed in the future should be validated against relevant standards of comparison such as those developed in (3).

Community infrastructure

The postprocessing community should collaborate to build high-quality shared code and data repositories and should maintain these. Ideally, a community repository would have characteristics such as:

- Tracking of software changes using a community-standard version control system such as git.
- Support for tiers in the repository, from tightly controlled (operational prediction code) to more loosely controlled (community scientists).
- An established process to manage incorporation of code contributions into the inner tiers of the repository, that is, a change control board.
- A ticket-tracking system to monitor requested code changes and their disposition.
- Established standards for metadata, tests, and documentation.
- Use of an agreed-upon common vocabulary.
- A centralized location for documentation and data access, with focus on a few core documents.
- Use of two or three modern common data formats (e.g., netCDF, HDF, geoJSON) that will satisfy operational, research, collaboration, and archival purposes.

Data

1. Prediction centers, if possible, should regularly generate high-quality reanalysis and reforecast data.
2. Prediction centers should ensure that future high-performance computing and disk procurement reflect the computational and storage needs of reforecasts and reanalyses.
3. Prediction centers should also generate high-quality, high-resolution analysis data for training and validation.

Continued

RECOMMENDATIONS FROM A WORKSHOP ON STATISTICAL POSTPROCESSING—CONT'D

4. Prediction centers should postprocess and make readily available commonly used "foundational data" (e.g., temperature, precipitation) for use inside that center and across the broader enterprise.
5. Prediction centers should make training data easily accessible.
6. Given the challenges with transmission of voluminous training data, prediction centers are encouraged to either set aside computational resources for external collaborators working on postprocessing, or to permit read access to storage systems with training data.
7. Survey postprocessing product developers to ensure prediction centers are saving the relevant predictor information.

How might a diverse suite of software be maintained such that it serves the joint needs of a more free-wheeling academic community and the more controlled needs of weather services? How do we build a community that fosters intellectual diversity while containing software entropy, an uncontrolled growth of code size and diversity? Following a suggestion by Tom Auligné (personal communication, 2016), a software repository might have several tiers. Users could contribute their modified software branch back to an outer tier of the repository, which would be home for a wide diversity of software branches that could be shared between investigators working on a common problem. Software in this tier would not be rigorously vetted. Should a community user wish to see the software become incorporated into the community "trunk," then the software would be reviewed by a change-control board that evaluates the software for coding clarity, adherence to documentation and test standards, and results. Presuming acceptance, the software would then become part of a more limited suite of broadly supported community algorithms. A final inner tier would be for a particular agency like the U.S. NWS. The software that is run on operational supercomputers would be subject to further refinement and quality control, per standards established by that agency. Conceivably, one might envision multiple agencies having similar or slightly different inner tiers, but sharing community contributions migrating in from the middle tier.

7.7 RECOMMENDATIONS AND CONCLUSIONS

This chapter discussed many of the practical aspects related to statistical postprocessing. To achieve the highest quality result, the statistician must attend to the practical realities of the data to be used in addition to the algorithmic design of the postprocessing software. Here are some recommendations on how we can make more rapid progress:

(a) *Postprocessing scientists should engage with prediction system developers about the data needs.* There can be tension between the desires of our model-development colleagues to improve the prediction system as quickly as possible and the desire of statisticians for data that are homogeneous in their error characteristics as well as high in quality. Finding the balance is challenging. The earlier the data requirements are communicated with the prediction system developers, the easier it will be for them to accommodate postprocessing needs in their plans.

For example, the prediction system developers may have a choice between two possible upgrade paths, one that minimizes RMS error at the expense of some bias, or one that minimizes forecast bias at the expense of slightly higher RMS error. Changes in bias from one model version to the next are generally more challenging to address in postprocessing. Hence, if we agree that the end goal is high-quality postprocessed guidance rather than the lowest-error raw guidance, then the latter upgrade path (minimizing forecast bias) may be a more sensible path forward.

(b) *Challenge ourselves to use the existing training data more efficiently.* Given the expense and work required to generate lengthy retrospective data, our algorithms should extract the most information possible from the limited data at hand.

(c) *Work together to build a postprocessing community, sharing data and software.* By building algorithms in isolation, we are unsure as to whether our design represents an improvement over existing methods. If standard data sets are available to researchers, and if we share code for data input, output, and verification, then it becomes much simpler to test our methodology against other reference standards. Rome wasn't built in a day; this is an ambitious goal, but we can start with simple and productive steps. After we finish a project, we can make our data freely available and share our algorithms in public portals such as github. Here is my personal example, a reforecast precipitation data set and associated analog method software (https://github.com/ThomasMoreHamill/analog).

(d) More generally, *prediction centers should share their data.* TIGGE (Bougeault et al., 2009; Swinbank et al., 2016) was an international research project that archived global ensemble data for research purposes. Many scientists have used data from the TIGGE archive to demonstrate the improvements possible from the postprocessing of multicenter ensemble data. We all have much to gain and little to lose by sharing more data in real time, leveraging each others' investments in research and computing. In particular, such data is of exceptional importance to developing countries that cannot yet afford to develop their own prediction systems.

(e) *Collaborate with professional statisticians.* Chances are, you, the reader, are trained as an atmospheric scientist or hydrologist. You have knowledge about data characteristics and potential predictors that a statistician will not have. However, a statistician is likely to have a more thorough grounding in such areas as Bayesian methods, spatial statistics, and machine learning. Together you may be able to produce higher-quality products than you could working individually.

REFERENCES

Allen, R. L., & Erickson, M. C. (2001). *AVN-based MOS precipitation type guidance for the United States* (pp. 9). NWS technical procedures bulletin no. 476, NOAA, U.S. Dept. of Commerce.

Anthes, R., Ector, D., Hunt, D., Kuo, Y., Rocken, C., Schreiner, W., et al. (2008). The COSMIC/FORMOSAT-3 mission: Early results. *Bulletin of the American Meteorological Society*, *89*, 313–333.

Auligné, T., McNally, A. P., & Dee, D. P. (2007). Adaptive bias correction for satellite data in a numerical weather prediction system. *Quarterly Journal of the Royal Meteorological Society*, *133*, 631–642.

Benjamin, S. G., Jamison, B. D., Moninger, W. R., Sahm, S. R., Schwartz, B. E., & Schlatter, T. W. (2010). Relative short-range forecast impact from aircraft, profiler, radiosonde, VAD, GPS-PW, METAR, and mesonet observations via the RUC hourly assimilation cycle. *Monthly Weather Review*, *138*, 1319–1343.

Bi, L., Jung, J. A., Morgan, M. C., & Le Marshall, J. F. (2011). Assessment of assimilating ASCAT surface wind retrievals in the NCEP global data assimilation system. *Monthly Weather Review*, *139*, 3405–3421.

Boisserie, M., Decharme, B., Descamps, L., & Arbogast, P. (2016). Land surface initialization strategy for a global reforecast dataset. *Quarterly Journal of the Royal Meteorological Society, 142,* 880–888.

Bougeault, P., et al. (2009). TheTHORPEX Interactive Grand Global Ensemble (TIGGE). *Bulletin of the American Meteorological Society,* **91,** 1059–1072.

Buehner, M., Morneau, J., & Charette, C. (2013). Four-dimensional ensemble-variational data assimilation for global deterministic weather prediction. *Nonlinear Processes in Geophysics, 20,* 669–682.

Candille, G. (2009). The multiensemble approach: The NAEFS example. *Monthly Weather Review, 137,* 1655–1665.

Collard, A. D., & McNally, A. P. (2009). The assimilation of infrared atmospheric sounding interferometer radiances at ECMWF. *Quarterly Journal of the Royal Meteorological Society, 135,* 1044–1058.

Courtier, P., Thépaut, J.-N., & Hollingsworth, A. (1994). A strategy for operational implementation of 4D-Var, using an incremental approach. *Quarterly Journal of the Royal Meteorological Society, 120,* 1367–1387.

Cui, B., Toth, Z., Zhu, Y., & Hou, D. (2012). Bias correction for global ensemble forecast. *Weather and Forecasting, 27,* 396–410.

Dabernig, M., Mayr, G. J., Messner, J. W., & Zeileis, A. (2017). Spatial ensemble post-processing with standardized anomalies. *Quarterly Journal of the Royal Meteorological Society, 143,* 909–916.

Daley, R. (1991). *Atmospheric Data Analysis* (pp. 457). Cambridge: Cambridge University Press.

Dee, D. P. (2005). Bias and data assimilation. *Quarterly Journal of the Royal Meteorological Society, 131,* 3323–3343.

Dee, D. P., et al. (2011). The ERA-Interim reanalysis: configuration and performance of the data assimilation system. *Quarterly Journal of the Royal Meteorological Society,* **137,** 553–597.

Dee, D. P., Balmaseda, M., Balsamo, G., Engelen, R., Simmons, J., & Thépaut, J. -N. (2014). Towards a consistent reanalysis of the climate system. *Bulletin of the American Meteorological Society, 95,* 1236–1248.

De Pondeca, M., Manikin, G., DiMego, G., Benjamin, S., Parrish, D., Purser, R., et al. (2011). The real-time mesoscale analysis at NOAA's National Centers for Environmental Prediction: Current status and development. *Weather and Forecasting, 26,* 593–612.

Durran, D. R. (2010). *Numerical Methods for Fluid Dynamics* (2nd ed., 516 pp.). New York: Springer.

Evensen, G. (2014). *Data Assimilation, the Ensemble Kalman Filter* (2nd ed., 307 pp.). Dordrecht: Springer.

Fortin, V., Favre, A. -c., & Saïd, M. (2006). Probabilistic forecasting from ensemble prediction systems: Improving upon the best-member method by using a different weight and dressing kernel for each member. *Quarterly Journal of the Royal Meteorological Society, 132,* 1349–1369.

Gagnon, N., Deng, X., Houtekamer, P. L., Beauregard, S., Erfani, A., Charron, M., et al. (2014). *Improvements to the Global Ensemble Prediction System (GEPS) from version 3.1.0 to version 4.0.0.* Environment Canada Tech Note. Available from: http://collaboration.cmc.ec.gc.ca/cmc/cmoi/product_guide/docs/lib/technote_geps-400_20141118_e.pdf.

Gagnon, N., Deng, X., Houtekamer, P. L., Erfani, A., Charron, M., Beauregard, S., et al. (2015). *Improvements to the Global Ensemble Prediction System from version 4.0.1 to version 4.1.1.* Environment Canada Tech Note. Available from: http://collaboration.cmc.ec.gc.ca/cmc/cmoi/product_guide/docs/lib/technote_geps-411_20151215_e.pdf.

Glahn, B., Gilbert, K., Cosgrove, R., Ruth, D., & Sheets, K. (2009). The Gridding of MOS. *Weather and Forecasting, 24,* 520–529.

Hagedorn, R., Buizza, R., Hamill, T. M., Leutbecher, M., & Palmer, T. N. (2012). Comparing TIGGE multi-model forecasts with reforecast-calibrated ECMWF ensemble forecasts. *Quarterly Journal of the Royal Meteorological Society, 138,* 1814–1827.

Hamill, T. M. (2006). Ensemble-based atmospheric data assimilation. In *Chapter 6 of Predictability of Weather and Climate* (pp. 124–156). Cambridge: Cambridge Press.

Hamill, T. M. (2017). Changes in the systematic errors of global reforecasts due to an evolving data assimilation system. *Monthly Weather Review, 145,* 2479–2485.

Hamill, T. M., Bates, G. T., Whitaker, J. S., Murray, D. R., Fiorino, M., Galarneau, T. J., Jr., et al. (2013). NOAA's second-generation global medium-range ensemble reforecast data set. *Bulletin of the American Meteorological Society, 94,* 1553–1565.

Hamill, T. M., Engle, E., Myrick, D., Peroutka, M., Finan, C., & Scheuerer, M. (2017). The US National Blend of Models statistical post-processing of probability of precipitation and deterministic precipitation amount. *Monthly Weather Review, 145,* 3441–3463. Also: online Appendix A and online Appendix B.

Hamill, T. M., Hagedorn, R., & Whitaker, J. S. (2008). Probabilistic forecast calibration using ECMWF and GFS ensemble reforecasts. Part II: precipitation. *Monthly Weather Review, 136,* 2620–2632.

Hamill, T. M., Scheuerer, M., & Bates, G. T. (2015). Analog probabilistic precipitation forecasts using GEFS Reforecasts and Climatology-Calibrated Precipitation Analyses. *Monthly Weather Review, 143,* 3300–3309 [also: online Appendix A and Appendix B].

Hamill, T. M., Whitaker, J. S., & Mullen, S. L. (2006). Reforecasts, an important dataset for improving weather predictions. *Bulletin of the American Meteorological Society, 87,* 33–46.

Hamill, T. M., Whitaker, J. S., & Wei, X. (2004). Ensemble re-forecasting: Improving medium-range forecast skill using retrospective forecasts. *Monthly Weather Review, 132,* 1434–1447.

Hastie, T. J., & Tibshirani, R. J. (1990). *Generalized Additive Models* (335 pp.). London: Chapman and Hall.

Hastie, T., Tibshirani, R., & Friedman, J. (2001). *The Elements of Statistical Learning* (533 pp.). New York: Springer.

Hemri, S. (2018). Applications of postprocessing for hydrological forecasts. In S. Vannitsem, D. S. Wilks, & J. W. Messner (Eds.), *Statistical Postprocessing of Ensemble Forecasts.* Elsevier.

Hodyss, D., Satterfield, E., McLay, J., Hamill, T. M., & Scheuerer, M. (2016). Inaccuracies with multimodel postprocessing methods involving weighted, regression-corrected forecasts. *Monthly Weather Review, 144,* 1649–1668.

Hopson, T. M., & Webster, P. J. (2010). A 1–10-day ensemble forecasting scheme for the major river basins of Bangladesh: Forecasting severe floods of 2003–07. *Journal of Hydrometeorology, 11,* 618–641.

Hou, D., Charles, M., Luo, Y., Toth, Z., Zhu, Y., Krzysztofowicz, R., et al. (2014). Climatology-calibrated precipitation analysis at fine scales: Statistical adjustment of stage IV toward CPC gauge-based analysis. *Journal of Hydrometeorology, 15,* 2542–2557.

Kalnay, E. (2003). *Atmospheric Modeling, Data Assimilation, and Predictability* (341 pp.). Cambridge: Cambridge University Press.

Kleiber, W., Raftery, A. E., Baars, J., Gneiting, T., Mass, C. F., & Grimit, E. (2011a). Locally calibrated probabilistic temperature forecasting using geostatistical model averaging and local Bayesian model averaging. *Monthly Weather Review, 139,* 2630–2649.

Kleiber, W., Raftery, A. E., & Gneiting, T. (2011b). Geostatistical model averaging for locally calibrated probabilistic quantitative precipitation forecasting. *Journal of the American Statistical Association, 106,* 1291–1303.

Kleist, D. T., & Ide, K. (2015). An OSSE-based evaluation of hybrid variational-ensemble data assimilation for the NCEP GFS. Part I. System description and 3D-hybrid results. *Monthly Weather Review, 143,* 433–451.

Lalaurette, F. (2003). Early detection of abnormal weather conditions using a probabilistic extreme forecast index. *Quarterly Journal of the Royal Meteorological Society, 129,* 3037–3057.

Lerch, S., & Baran, S. (2017). Similarity-based semilocal estimation of post-processing models. *Journal of the Royal Statistical Society: Series C, 66,* 29–51.

Lespinas, F., Fortin, V., Roy, G., Rasmussen, P., & Stadnyk, T. (2015). Performance evaluation of the Canadian precipitation analysis (CaPA). *Journal of Hydrometeorology, 16,* 2045–2064.

Lin, H., Gagnon, N., Beauregard, S., Muncaster, R., Markovic, M., Denis, B., et al. (2016). GEPS based monthly prediction at the Canadian Meteorological Centre. *Monthly Weather Review, 144,* 4867–4883.

Maraun, D. (2013). Bias correction, quantile mapping, and downscaling: Revisiting the inflation issue. *Journal of Climate, 26,* 2137–2143.

Mass, C. F., Baars, J., Wedam, G., Grimit, E., & Steed, R. (2008). Removal of systematic model bias on a model grid. *Weather and Forecasting, 23,* 438–459.

McNally, A. P., Watts, P. D., Smith, J. A., Engelen, R., Kelly, G. A., Thépaut, J. N., et al. (2006). The assimilation of AIRS radiance data at ECMWF. *Quarterly Journal of the Royal Meteorological Society, 132,* 935–957.

Neter, J., Wasserman, W., & Kutner, M. H. (1990). *Applied Linear Statistical Models* (3rd ed., 1181 pp.). Homewood, IL: Irwin Press.

Ou, M., Charles, M., & Collins, D. (2016). Sensitivity of calibrated week-2 probabilistic forecast skill to reforecast sampling of the NCEP global ensemble forecast system. *Weather and Forecasting, 31,* 1093–1107.

Park, Y. -Y., Buizza, R., & Leutbecher, M. (2008). TIGGE: Preliminary results on comparing and combining ensembles. *Quarterly Journal of the Royal Meteorological Society, 134,* 2029–2050.

Petroliagis, T. I., & Pinson, P. (2014). Early warnings of extreme winds using the ECMWF extreme forecast index. *Meteorological Applications, 21,* 171–185.

Press, W. H., Teukolsky, S. A., Vetterling, W. T., & Flannery, B. P. (1992). *Numerical Recipes in Fortran* (2nd ed., 963 pp.). Cambridge: Cambridge Press.

Raftery, A. E., Gneiting, T., Balabdaoui, F., & Polakowski, M. (2005). Using Bayesian model averaging to calibrate forecast ensembles. *Monthly Weather Review, 133,* 1155–1174.

Richardson, D. S. (2001). Measures of skill and value of ensemble prediction systems, their interrelationship and the effect of ensemble size. *Quarterly Journal of the Royal Meteorological Society, 127,* 2473–2489.

Saha, S., et al. (2010). The NCEP climate forecast system reanalysis. *Bulletin of the American Meteorological Society, 91,* 1015–1057.

Schaake, J. C., Hamill, T. M., Buizza, R., & Clark, M. (2007). HEPEX, the hydrological ensemble prediction experiment. *Bulletin of the American Meteorological Society, 88,* 1541–1547.

Scheuerer, M., & Büermann, L. (2014). Spatially adaptive post-processing of ensemble forecasts for temperature. *Journal of the Royal Statistical Society: Series C, 63,* 405–422.

Scheuerer, M., & Hamill, T. M. (2015). Statistical post-processing of ensemble precipitation forecasts by fitting censored, shifted gamma distributions. *Monthly Weather Review, 143,* 4578–4596.

Scheuerer, M., & König, G. (2014). Gridded, locally calibrated, probabilistic temperature forecasts based on ensemble model output statistics. *Quarterly Journal of the Royal Meteorological Society, 140,* 2582–2590.

Scheuerer, M., & Möller, D. (2015). Probabilistic wind speed forecasting on a grid based on ensemble model output statistics. *The Annals of Applied Statistics, 9,* 1328–1349.

Skamarock, W. C., J. B. Klemp, J. Dudhia, D. O. Gill, D. M. Barker, W. Wang, J. G. Powers, 2005: A description of the advanced research WRF version 2. NCAR Technical Note, NCAR/TN–468+STR (88 pp.). Available from: http://www2.mmm.ucar.edu/wrf/users/pub-doc.html.

Sloughter, J. M., Raftery, A. E., Gneiting, T., & Fraley, C. (2007). Probabilistic quantitative precipitation forecasting using Bayesian model averaging. *Monthly Weather Review, 135,* 3209–3220.

Stauffer, R., Umlauf, N., Messner, J. W., Mayr, G. J., & Zeileis, A. (2017). Ensemble postprocessing of daily precipitation sums over complex terrain using censored high-resolution standardized anomalies. *Monthly Weather Review, 145,* 955–969.

Stensrud, D. J. (2007). *Parameterization Schemes. Keys to Understanding Numerical Weather Prediction Models* (459 pp.). Cambridge: Cambridge Press.

Swinbank, R., et al. (2016). The TIGGE project and its achievements. *Bulletin of the American Meteorological Society, 97,* 49–67.

Unidata. (2012). Integrated Data Viewer (IDV) version 3.1 *[software].* Boulder, CO: UCAR/Unidata. https://www.unidata.ucar.edu/software/netcdf/.

Uppala, S. M., Kållberg, P. W., Simmons, A. J., Andrae, U., Bechtold, V. D. C., Fiorino, M., et al. (2005). The ERA-40 re-analysis. *Quarterly Journal of the Royal Meteorological Society, 131,* 2961–3012.

Vannitsem, S., & Hagedorn, R. (2011). Ensemble forecast post-processing over Belgium: Comparison of deterministic-like and ensemble regression methods. *Meteorological Applications, 18,* 94–104.

Velden, C. S., Daniels, J., Stettner, D., Santek, D., Key, J., Dunion, J., et al. (2005). Recent nnovations in deriving tropospheric winds from meteorological satellites. *Bulletin of the American Meteorological Society, 86,* 205–223.

Vitart. F., G. Balsamo, R. Buizza, L. Ferranti, S. Keeley, L. Magnusson, F. Molteni and A. Weisheimer, 2014: Sub-seasonal predictions. ECMWF Tech Memo 738. Available from: https://www.ecmwf.int/sites/default/files/elibrary/2014/12943-sub-seasonal-predictions.pdf

Voisin, N., Schaake, J. C., & Lettenmaier, D. P. (2010). Calibration and downscaling methods for quantitative ensemble precipitation forecasts. *Weather and Forecasting, 25,* 1603–1627.

Warner, T. T. (2011). *Numerical Weather and Climate Prediction* (526 pp.). Cambridge: Cambridge University Press.

Wikipedia. (2016). https://en.wikipedia.org/wiki/Bias%E2%80%93variance_tradeoff.

Wilks, D. S. (2006). Comparison of ensemble-MOS methods in the Lorenz '96 setting. *Meteorological Applications, 13,* 246–256.

Wilks, D. S. (2011). *Statistical Methods in the Atmospheric Sciences* (3rd ed., 676 pp.). Academic Press.

Zhang, J., et al. (2016). Multi-radar sensor (MRMS) quantitative precipitation estimation: Initial operating capabilities. *Bulletin of the American Meteorological Society, 97,* 621–638.

Zhou, X., Zhu, Y., Hou, D., Luo, Y., Peng, J., & Wobus, R. (2017). Performance of the new NCEP global ensemble forecast system in a parallel experiment. *Weather and Forecasting, 32,* 1989–2004.

FURTHER READING

NCAR/MMM, 2017: ARW version 3 modeling system user's guide (434 pp.). Available from: http://www2.mmm.ucar.edu/wrf/users/pub-doc.html.

Schättler, U., G. Doms, and C. Schraff, 2016: A description of the nonhydrostatic regional COSMO-model. Part VII. Users guide (221 pp.). Available from: http://www2.cosmo-model.org/content/model/documentation/core/.

APPLICATIONS OF POSTPROCESSING FOR HYDROLOGICAL FORECASTS

8

Stephan Hemri[*,†]

Heidelberg Institute for Theoretical Studies, Heidelberg, Germany[*]
Presently at: Federal Office of Meteorology and Climatology MeteoSwiss, Zurich-Airport, Switzerland[†]

CHAPTER OUTLINE

8.1 INTRODUCTION

Hydrological forecasts, that is, forecasts of river runoff or gauge level, are crucial for flood prevention, water supply management, navigation, the operation of hydropower plants, and further surface water related activities. For medium range forecasts of about 2–15 days, hydrological models are driven by dynamical weather prediction models (Cloke & Pappenberger, 2009). As, for instance, stated by Ajami, Duan, and Sorooshian (2007), the predictive uncertainty of hydrological forecasts, that is, the uncertainty conditional on the information and knowledge available at forecast issue time (Krzysztofowicz, 1999; Todini, 2008), emerges from different sources of uncertainty. These are related to the parameters of the hydrological model, deficiencies in both model structure and meteorological input (models), unreliable measurements of input data as well as misspecified initial and boundary conditions of the hydrological model. Typically, the meteorological input is the largest cause of hydrological uncertainty. Hence, a large amount thereof can be accounted for by running a hydrological model several times in parallel, each run being driven by a member of the meteorological input ensemble. This approach results in a hydrological output ensemble forecast.

The increasing importance of hydrological ensemble forecasts is reflected by the Hydrological Ensemble Prediction Experiment (HEPEX; Schaake, Hamill, Buizza, & Clark, 2007, see also https://hepex.irstea.fr/about-hepex/). Established in 2004, HEPEX is an open, bottom-up process designed by hydrologists, meteorologists, and users. It aims to foster the advancement of hydrological ensemble postprocessing as well as its usage for rational risk-based decision making. Its main activities are the organization of international scientific exchange, promoting the exchange between researchers, forecast producers, and users, as well as planning and coordination of experiments and testbeds (Wood, Wetterhall, & Ramos, 2015). Testbeds facilitate shared experiments and the comparison of different ensemble prediction methods. An example of a HEPEX intercomparison experiment is described in Van Andel, Weerts, Schaake, and Bogner (2013). Its goal was to enable the systematic assessment of different postprocessing techniques in varying settings. More specifically, this experiment was designed to test if postprocessing atmospheric input ensembles enhances the meteorological inputs with regard to hydrological ensemble forecasting.

Prominent examples of hydrological ensemble prediction systems are the Hydrologic Ensemble Forecast Service (HEFS; Demargne et al., 2014, see also http://www.nws.noaa.gov/oh/hrl/general/indexdoc.htm#hefs), the European Flood Awareness System (EFAS; Bartholmes, Thielen, Ramos, & Gentilini, 2009; Thielen, Bartholmes, Ramos, & De Roo, 2009, see also https://www.efas.eu/about-efas.html), and the Global Flood Awareness System (GloFAS; Alfieri et al., 2013, see also http://globalfloods.jrc.ec.europa.eu/). HEFS is an end-to-end ensemble forecasting platform implemented by the National Oceanic and Atmospheric Administration (NOAA)'s National Weather Service (NWS). Being designed for many different end users, HEFS provides a wide range of forecasts from flash flood predictions to large area water supply forecasts at forecast horizons greater than a year. The generation of hydrologic ensemble forecasts based on HEFS can be divided into three parts. First, input ensemble forecasts of temperature and precipitation are processed using the meteorological ensemble forecast processor (MEFP; Schaake et al., 2007; Schaake, Demargne, Hartman, Mullusky, Welles, Wu, et al., 2007; Wu et al., 2011). Second, the processed meteorological ensemble is used to drive several parallel runs of a hydrological model. Third, the hydrological forecast ensemble is postprocessed using the EnsPost approach (Seo, Herr, & Schaake, 2006) in order to assess the hydrologic uncertainty and to apply bias correction.

EFAS is an initiative by the European Commission that aims to reduce the impact of transnational floods in Europe. Since 2005, EFAS provides twice daily forecasts of river flow and early flood warnings up to 10 days before a flood event (Pappenberger et al., 2015) based on ensemble runs of the hydrological model LISFLOOD (De Roo, 1999; Van der Knijff, Younis, & De Roo, 2010). The meteorological input models comprise two deterministic and two ensemble models. The former are the high-resolution forecast of the European Center for Medium-Range Weather Forecasts (ECMWF-HRES; Molteni, Buizza, Palmer, & Petroliagis, 1996) and the icosahedral nonhydrostatic (ICON; Wan et al., 2013; Dipankar et al., 2015; Zängl, Reinert, Rípodas, & Baldauf, 2015) model of the German Meteorological Service (DWD). The latter are composed of the 51 member ensemble forecast of the ECMWF (ECMWF-ENS; Molteni et al., 1996; Persson, 2015) and the 16 member COnsortium for Small-scale MOdelling Limited-area Ensemble Prediction System (COSMO-LEPS; Montani, Cesari, Marsigli, & Paccagnella, 2011). EFAS disseminates warnings to the corresponding national hydrological authorities if specific flood alert thresholds are persistently exceeded. In addition to early flood forecasting, EFAS is also designed for monitoring the hydrological conditions, analyzing climate trends, and exchanging flood forecasting experiences (Thielen et al., 2009). Since 2010 the EFAS framework has included an operational postprocessing routine that reduces forecast biases by applying a combination of wavelet transformation and time series modeling (Bogner & Kalas, 2008; Bogner & Pappenberger, 2011; Bogner, Pappenberger, & Cloke, 2012).

Encouraged by the good experiences with EFAS, GloFAS has been set up jointly by the Joint Research Centre (JRC) of the European Commission and the ECMWF. GloFAS has been operational since 2011 and provides daily global runoff forecasts over a forecast horizon of up to 45 days on a grid of 0.1 degree resolution, which is about 10 km in mid-latitudes. While forecasts up to lead day 15 are (hydro-) meteorologically driven, those beyond 15 days are based on catchment depletion processes and river routing. Catchment depletion processes refer to runoff generation from catchment-specific water storages such as snow cover, soil moisture, and groundwater drainage. River routing refers to predicting runoff and water levels along a river system by following the water as it flows along its course from smaller tributaries to the main stream and then along the main stream to the estuary. The hydrological model LISFLOOD is run on a global grid with the 51 ECMWF-ENS members as meteorological input. Being designed for the world's major catchments, GloFAS performs well for locations with upstream areas greater 10,000 km^2. Data scarcity is overcome by using ERA-Interim analysis-based LISFLOOD runs to construct initial conditions and flood warning thresholds (Alfieri et al., 2013).

Despite being the state of the art in probabilistic hydrological forecasting, biases and dispersion errors of the meteorological ensembles (Buizza, 2018, Chapter 2 of this book) cascade down to the hydrological model. Furthermore, hydrological errors are neglected by this approach. In other words, just like meteorological ensembles, hydrological ensembles need to be postprocessed. Fig. 8.1 depicts the strong underdispersion of a 19-member runoff forecast ensemble at three gauges within the river Rhine basin in Western Europe over the period from November 2008 to January 2011. The ensemble consists of 19 parallel runs of the hydrological model HBV-96 (Bergström, 1995; Lindström,

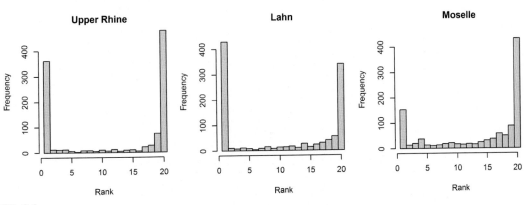

FIG. 8.1

Rank histograms of the observations relative to the ensemble forecasts with a lead time of 48 h for the gauges Maxau (Upper Rhine), Kalkofen (Lahn), and Trier (Moselle).

Johansson, Persson, Gardelin, & Bergström, 1997) driven with meteorological input from the 16-member COSMO-LEPS ensemble and three deterministic models. They comprise the DWD-GME model (Majewski et al., 2002; Majewski, Liermann, & Ritter, 2012), which is the precursor of ICON, ECMWF-HRES, and a mixture of two models, here named DWD-MER, that is, COSMO-EU (Schulz & Schättler, 2011; Steppeler, Doms, & Adrian, 2002) up to a lead time of 78 h and DWD-GME thereafter. The locations and areas of the three sub-catchments are displayed in Fig. 8.2. For more details, the reader is referred to Hemri, Lisniak, and Klein (2015). Note also that Reggiani, Renner, Weerts, and Van Gelder (2009) have used gauge Lobith at the Dutch-German border as a test gauge for the development of a Bayesian postprocessing approach (see Section 8.2.2).

The hydrological variables runoff and water level have some characteristics that affect postprocessing. First, they usually follow a skewed distribution. Runoff (and/or water level) is mostly low to medium, but during occasional flood events it may be drastically higher. Second, hydrological forecast and observation time series, which are called hydrographs, are strongly autocorrelated. And third, hydrographs at different gauges within the same river system are cross-correlated. Auto- and

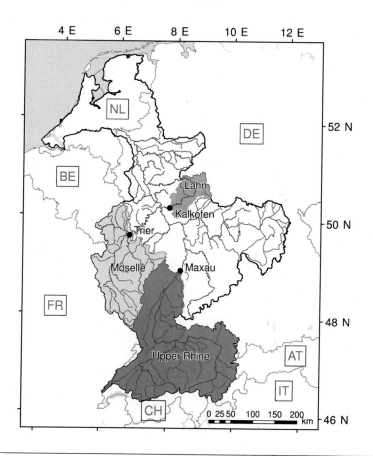

FIG. 8.2

Selection of sub-catchments within the basin of river Rhine. *Reproduced from Hemri, S., Lisniak, D., & Klein, B. (2015). Multivariate post-processing techniques for probabilistic hydrological forecasting. Water Resources Research, 51, 7436–7451.*

cross-correlations of the errors of the mean forecast of the COSMO-LEPS driven hydrological ensemble for the three example gauges mentioned herein are shown in Fig. 8.3. At all three gauges, the autocorrelations are very strong, while cross-correlations between the gauges are weaker, but still substantial at longer lead times. Note also that cross-correlations are typically not maximal on the diagonal, as shown by the cross-correlations between Moselle and Lahn.

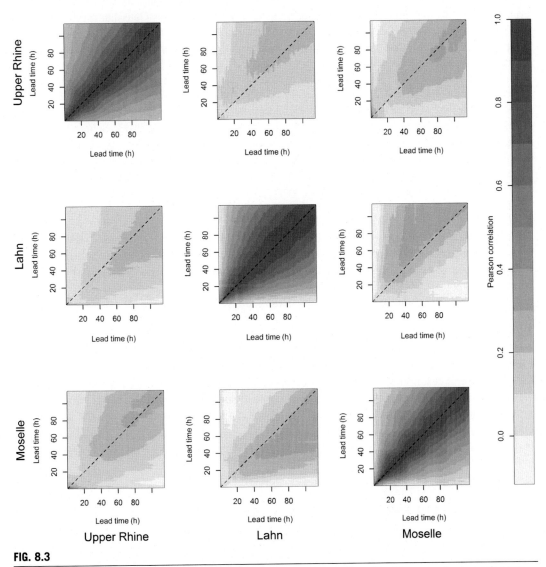

FIG. 8.3

Correlation matrices of mean COSMO-LEPS forecast errors between different lead times and sub-catchments. The panels on the diagonal depict autocorrelations, and the off-diagonal panels refer to cross-correlations between different sub-catchments of river Rhine. Within the panels the diagonals are indicated by the dashed lines.

Prior to presenting different methods to postprocess hydrological ensembles, an illustration of a combination of univariate and multivariate postprocessing approaches in hydrology is given now. The example shown in Fig. 8.4 is taken from Hemri et al. (2015). The goal here is to calibrate ensemble runoff forecasts for gauge Trier in such a way that realistic forecast scenarios over the entire forecast horizon can be generated. First, forecast and observation data are Box-Cox (Box & Cox, 1964; Wilks, 2018, Chapter 3 of this book) transformed to achieve approximate Gaussianity. This step can be omitted, if one applies a nonparametric method or a parametric method that can represent skewness appropriately. Then, we postprocess, for instance, by using nonhomogeneous Gaussian regression (NGR; Gneiting, Raftery, Westveld, & Goldman, 2005; Thorarinsdottir & Gneiting, 2010; Wilks, 2018, Chapter 3 of this book), the forecasts at each lead time (and gauge) separately to obtain univariate predictive distributions (see panels (a) and (b) of Fig. 8.4). Panels (c) and (d) show the trajectories of the raw ensemble and the quantiles of the postprocessed univariate predictive distributions over the entire forecast horizon, respectively. In order to obtain realistic forecast scenarios suitable quantiles from the NGR distributions need to be extracted using appropriate multivariate approaches. In this example, ensemble copula coupling (ECC; Schefzik, Thorarinsdottir, & Gneiting, 2013; Schefzik & Möller, 2018, Chapter 4 of this book) and the Gaussian copula approach (GCA; Pinson & Girard, 2012) have been applied to obtain the forecast trajectories shown in panels (e) and (f), respectively. Owing to the different ways to generate the forecast scenarios, which will be presented in Section 8.3.1, the GCA trajectories look less realistic than the ECC ones.

The univariate postprocessing methods described in Wilks (2018, Chapter 3 of this book) and the multivariate methods presented in Schefzik & Möller (2018, Chapter 4 of this book), can generally be used for hydrological ensembles. For instance, examples of Bayesian model averaging (BMA; Raftery, Gneiting, Balabdaoui, & Polakowski, 2005; Wilks, 2018, Chapter 3 of this book) in hydrology can be found, for example, in Ajami et al. (2007), Duan, Ajami, Gao, and Sorooshian (2007), Parrish, Moradkhani, and DeChant (2012), Rings, Vrugt, Schoups, Husman, and Vereecken (2012), Vrugt and Robinson (2007), and hydrological applications of NGR in, for example, Hemri, Lisniak, and Klein (2014); Hemri et al. (2015) and Skøien, Bogner, Salamon, Smith, and Pappenberger (2016). In Section 8.2, we first discuss methods to transform hydrological data such that approximate Gaussianity can be assumed. Then, we present methods for univariate postprocessing of hydrological forecasts that have been developed particularly for hydrological ensembles, and are hence not covered by the more general description of postprocessing approaches in Chapter 3. In the last part of Section 8.2 we give some insights into the question of whether the meteorological input ensemble should be postprocessed prior to running the hydrological model compared to just postprocessing the hydrological output ensemble. In Section 8.3, we summarize methods to include temporal, spatial, and spatio-temporal dependencies into the postprocessing of hydrological ensembles. This is followed by a short outlook on future research topics in Section 8.4.

8.2 UNIVARIATE HYDROLOGICAL POSTPROCESSING

8.2.1 SKEWNESS AND THE ASSUMPTION OF GAUSSIANITY

Typically, runoff and water level follow heavily skewed distributions. In contrast, most of the ensemble postprocessing approaches applied to hydrological forecasts are based on an assumption of Gaussianity. The usual work-around to this problem is to transform both forecast and verification data such that they are approximately normal, or at least the forecast errors are approximately normal. This implies

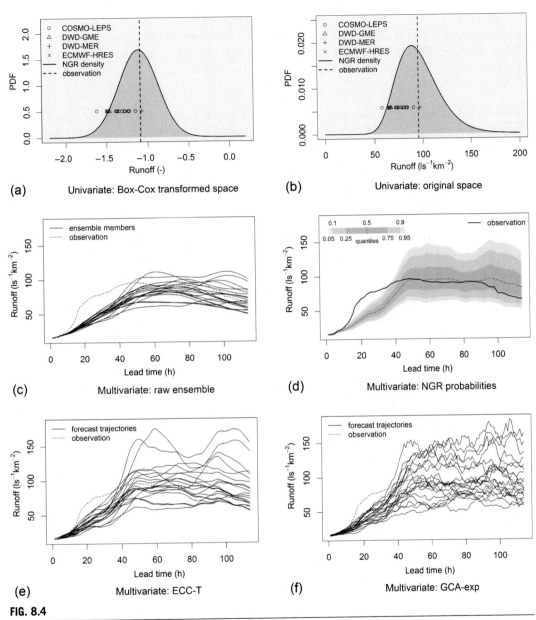

FIG. 8.4

Illustration of raw ensemble runoff forecasts and the corresponding postprocessed NGR forecasts for river Moselle at Trier issued on 6 January 2011 at 6 GMT. Panels (a) and (b) depict the univariate raw ensemble and the NGR probability density functions at lead time 48 h, on the Box-Cox transformed and on the original space, respectively. Each symbol represents an ensemble member. The observed value is represented by the dashed, vertical line. The raw ensemble trajectories and the observed trajectory over the entire forecast horizon are shown in panel (c). Panel (d) shows quantiles of the NGR forecast distributions, while multivariate NGR forecast scenarios with correlation structure obtained by ECC-T or GCA-exp are depicted in (e) and (f), respectively. A similar figure can be found in Hemri et al. (2015).

that the skewness of the conditional error distribution is removed and its variance is stabilized. The safest way to achieve Gaussianity is the normal quantile transform (NQT), as applied by for example, Seo et al. (2006), Reggiani et al. (2009), Yuan and Wood (2012), Bogner, Liechti, and Zappa (2016), and Zhao, Duan, Schaake, Ye, and Xia (2011). It can be written as

$$z = \Phi^{-1}[F(y)] \tag{8.1}$$

where y denotes the observed (or forecast) value on the original space. The empirical cumulative distribution function (CDF) of the observed (or forecast) variable is denoted by F, and Φ^{-1} refers to the quantile function, that is, the inverse of the CDF, of the standard normal distribution. While this approach leads to transformed values that are Gaussian by definition, its main drawback is revealed when values outside of the range of the collection of past observations, that is, the climatology, should be back transformed to the original space. In practice, values outside that range need to be covered by parametric approximations.

The Box-Cox (Box & Cox, 1964) transformation, which has already been introduced in Wilks (2018, Chapter 3 of this book), has also been used in the field of hydrological postprocessing (see, e.g., Duan et al., 2007; Engeland & Steinsland, 2014; Hemri et al., 2015). The log transformation, which is a special case of the Box-Cox transformation, has also been applied in hydrological postprocessing studies (see e.g., Van den Bergh & Roulin, 2016; Wood & Lettenmaier, 2008; Zalachori, Ramos, Garçon, Mathevet, & Gailhard, 2012). Wang, Shrestha, Robertson, and Pokhrel (2012) proposed the log-sinh transformation as an alternative to the Box-Cox transformation. The log-sinh transformation, which has been developed especially for hydrological applications, is given by

$$z = \frac{1}{b} \log[\sinh(a + by)] \tag{8.2}$$

where y, and z, denote again the variable of interest in the original and the transformed space, respectively. The parameters a and b need to be estimated based on the data at hand. According to Wang et al. (2012), it is particularly well suited for variables with error spread that first increases rapidly as its value increases, but for larger values this increase in error spread decelerates, and for very high values it is eventually constant. Fig. 8.5 depicts the non-Gaussianity of untransformed forecast errors typically

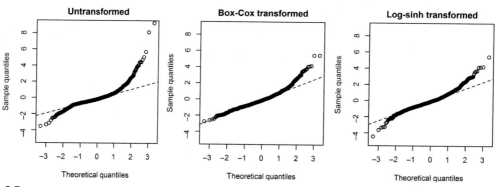

FIG. 8.5

QQ-plots of normalized mean HBV COSMO-LEPS forecasts compared with a standard normal distribution. From left to right: untransformed, Box-Cox transformed, log-sinh transformed. The forecast, on which the QQ-plots are based, covers the period from November 2008 to October 2011 for gauge Maxau at river Rhine.

encountered in hydrological forecasting. More importantly, it reveals also that Box-Cox and log-sinh transformations are not able to achieve normality. Likewise, the quality of the transformation in terms of variance stabilization depends on the dataset at hand. Though such transformations are frequently used in hydrological practice, one may argue from a technical point of view that both normality and variance stabilization are not necessary for methods like NGR, because NGR can capture dispersion variations, and, depending on the distribution family used, can also model non normal forecast variables. Hence, for any hydrological postprocessing study the benefit from data transformation needs to be assessed individually.

8.2.2 UNIVARIATE HYDROLOGICAL ENSEMBLE POSTPROCESSING

Most postprocessing methods applied in the field of hydrology are parametric. However, runoff and water level are the result of many different atmospheric and hydrological processes most likely not leading to outcomes that follow a parametric distribution. As discussed herein, transformation approaches like NQT, Box-Cox, or log-sinh transformation usually enable the use of standard parametric postprocessing methods, which are summarized in Wilks (2018, Chapter 3 of this book). But such transformations are typically imperfect. Parameter estimation on the transformed space is suboptimal (Brown & Seo, 2010) and susceptible to artifacts caused by (back-) transforming the forecasts. In the following, we describe postprocessing approaches that have been developed particularly for hydrological applications. They all allow for flexible postprocessing of ensemble forecasts of heavily skewed variables like runoff or gauge level.

Krzysztofowicz (1999) developed the Bayesian forecasting system (BFS) as a general framework to quantify the different sources of uncertainty in hydrological forecasting. The BFS, originally developed to assess the uncertainty of deterministic forecasts, contains a Bayesian uncertainty processor (BUP) that provides a posterior probability density forecast of runoff or water level. Based on the BUP, Reggiani et al. (2009) developed the Bayesian ensemble uncertainty processor (BEUP), which constructs posterior forecast densities based on prior probabilities and hydrological ensemble forecasts. In a first step, Reggiani et al. (2009) transform the forecasts and the observations using NQT to ensure Gaussianity, and hence linear dependencies. As in BMA (Raftery et al., 2005; Wilks, 2018, Chapter 3 of this book), the BEUP forecast density is given by a mixture of component densities

$$f_{\text{BEUP}}(y_{t,1:L}) = \sum_{k=1}^{m} w_k f_k(y_{t,1:L}) \tag{8.3}$$

where t denotes forecast occasion, w_k are the weights assigned to the kernel densities f_k of the members $x_{t,k,1:L}$ with $k = 1, \dots, m$, and $1{:}L$ is the vector of lead times covered by the forecast. The densities f_k are obtained in a Bayesian way. They are proportional to the product of the prior density of the observed values $y_{t,1:L}$ given a series of historical observations and the likelihood of the ensemble member vectors $x_{t,k,1:L}$ given the historical observations and the observations to be forecast. Normalizing this product by dividing it by the density of the kth forecast member given all information available at the time of issuing the forecast yields the actual kernel densities f_k. Note that both prior density of the observed values and likelihood of the ensemble members are estimated by linear regression. If the ensemble consists only of exchangeable members as in Reggiani et al. (2009), the weights w_k are uniform.

As stated previously, runoff follows a strongly skewed distribution. In order to avoid issues with data transformation, Madadgar and Moradkhani (2014) propose a copula based BMA (Cop-BMA) approach that is tailored to hydrological forecasts, does not need any transformation, and can be implemented in a straightforward manner. The Cop-BMA mixture forecast density is given by

$$f_{\text{Cop-BMA}}(y_t) = \sum_{k=1}^{m} w_k c[F(y_t), F(x_{t,k})]f(y) \tag{8.4}$$

where $c[F(y_t), F(x_{t,k})]$ denotes the copula density function between the observations and the forecast member $x_{t,k}$, and $f(y)$ is the prior density of the observations. The subscript t, for instance in $F(y_t)$, indicates time dependence. In contrast, the prior density $f(y)$ is time invariant. Refer to Schefzik and Möller (2018, Chapter 4 of this book) for Sklar's theorem that introduces the notion of a copula. For hydrological applications it makes sense to use seasonal climatologies as the prior. In contrast to the standard BMA approach by Raftery et al. (2005), Cop-BMA kernels are obtained directly from the copula density and the climatological prior. The expectation-maximization algorithm (EM; Dempster, Laird, & Rubin, 1977; McLachlan & Krishnan, 1997) is only needed to estimate the weights w_k. Details on the implementation of the EM algorithm can be found in Madadgar and Moradkhani (2014). Furthermore, this copula approach implicitly removes systematic biases. Hence, the additional bias correction step applied in standard BMA is unnecessary in Cop-BMA. In the study of Madadgar and Moradkhani (2014) the Gumbel copula proved to perform best among the five tested copula models (elliptical copulae: Gaussian and T copula; Archimedean copulae: Gumbel, Clayton, and Frank copula; an overview over these copulae can be found in Madadgar, Moradkhani, & Garen, 2014).

The conditional bias penalized indicator cokriging method (CBP-ICK; Brown & Seo, 2010, 2013) is a nonparametric postprocessing approach well suited for strongly non-Gaussian variables such as river runoff. CBP-ICK links the observation threshold exceedance probability to threshold exceedance probabilities of the ensemble members. To this end, we consider the deviations of observed threshold exceedance probabilities from climatological exceedance probabilities and similar exceedance threshold deviations of the forecast members $x_{t,k}$ with regard to their climatologies. The threshold exceedance probability of a single member $x_{t,k}$ can only be zero or one, whereas the climatological exceedance probabilities can take any value in the unit interval $[0, 1]$. For each observation threshold of interest, an estimate of the threshold exceedance probability can now be obtained by regressing the associated deviation on the corresponding deviations of the different forecast members $x_{t,k}$ at many different forecast thresholds. The regression coefficients are obtained by minimizing an objective function that considers both the conditional bias of the observations given the forecasts and vice versa the conditional bias of the forecasts given the observations.

Klein, Meißner, Kobialka, and Reggiani (2016) developed a pair-copula-based approach to estimate the predictive uncertainty of hydrological ensemble forecasts, which is implicitly also a way to postprocess ensemble forecasts. Using an appropriate copula function, the $m+1$ dimensional joint distribution of the forecast ensemble members and the observation can be represented by a combination of the marginal distributions and an $m+1$ dimensional copula. Though there are many different pair-copulae, that is, copulae over two dimensions, which are very flexible; only a few copula functions are available for higher dimensions. As presented by Aas, Czado, Frigessi, and Bakken (2009), sophisticated multivariate dependence structures can be modeled by combining pair-copulae. This allows for very flexible copula modeling also in dimensions >2. Klein et al. (2016) used the concept of C(anonical)-Vine copulae (Aas et al., 2009) to build a 4-dimensional copula (three forecasts and one observation variable) using pair-copulae only. An $m+1$ variate copula can be represented by a graph

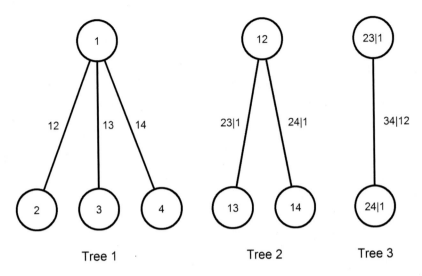

FIG. 8.6

Graph showing the three trees of a 4-dimensional C-Vine copula with dimension 1 as root node. Each edge represents a pair-copula (Aas et al., 2009; Klein et al., 2016).

consisting of m trees, of which each edge stands for a pair-copula density. C-Vine copulae exhibit the nice property of having a key dimension that can be placed at the root of the graph. In a typical hydrological setting it makes sense to use observed runoff as the root variable. An example of a graphical representation of a 4-dimensional C-Vine copula is shown in Fig. 8.6. Klein and Meißner (2016) used the R-package CDVine by Brechmann and Schepsmeier (2013) to select appropriate pair-copulae and to estimate their parameters in the C-Vine copula setting.

8.2.3 POSTPROCESSING OF HYDROLOGICAL FORECASTS VERSUS POSTPROCESSING OF METEOROLOGICAL INPUT

As already mentioned, hydrological forecasts suffer from modeling errors, incorrect initial conditions, and erroneous forecasts of meteorological forcings such as precipitation and temperature. This raises the question of which error source should be addressed preferably by statistical postprocessing. Though many studies have been conducted that apply statistical postprocessing either to meteorological input ensembles or to hydrological output ensembles, there are only a few studies that compare the two approaches.

Using the ECMWF-ENS ensemble as meteorological input to generate a hydrological forecast ensemble, Zalachori et al. (2012) assessed four different scenarios of daily runoff forecasts for 11 test catchments in France ranging from 220 to 3600 km^2. The first scenario relates to the raw ensemble, the second scenario to a hydrological ensemble generated using postprocessed precipitation and temperature inputs, the third scenario to a postprocessed hydrological ensemble forecast, which has been primarily generated based on raw precipitation and temperature inputs, and the fourth scenario is a combination of scenarios two and three, that is, both meteorological input and hydrological output are postprocessed. Evaluating the relative predictive performance of the different scenarios, Zalachori et al. (2012) concluded that postprocessing only the meteorological inputs is not sufficient to obtain well calibrated hydrological forecasts. On one hand, the corrections made to the meteorological inputs do not entirely propagate through the hydrological model, and on the other hand, the hydrological modeling uncertainties need to be accounted for as well.

Roulin and Vannitsem (2015) provide a thorough comparison based on a medium-sized test basin of 317 km^2 in Belgium. Using a 5-member ECMWF reforecast meteorological forcing, they compare one procedure to postprocess the meteorological input with two approaches that postprocess the output ensemble of the hydrological model. For postprocessing of the meteorological forcing ensemble they rely on extended logistic regression (Wilks, 2009, 2018, Chapter 3 of this book). The first method they use to postprocess the hydrological ensemble is an adapted version of variance inflation (INFL; Wood & Schaake, 2008; Johnson & Bowler, 2009). This method is a special case of the member-by-member postprocessing approach described in Van Schaeybroeck & Vannitsem (2015) and Wilks (2018, Chapter 3 of this book). However, as the method of Roulin and Vannitsem (2015) is an adaptation to hydrological forecasts, we briefly discuss it here. Omitting any spatial indices, the kth INFL corrected ensemble member forecast is given by

$$\hat{y}_{t,k} = \mu_y + \alpha(\bar{x}_t - \mu_x) + \beta e_{t,k} \tag{8.5}$$

where μ_y and μ_x denote the observed and the forecast mean over the training set, $\bar{x}_t$ is the current forecast mean, and e_k the difference between $\bar{x}_t$ and the kth forecast, i.e., $x_{t,k} = \bar{x}_t + e_{t,k}$. Based on an appropriate training set, the parameters α and β can be obtained in a straightforward manner. The other method they used to postprocess the hydrological ensemble is error-in-variable model output statistics (EVMOS; Vannitsem, 2009). It is a parsimonious method that can be applied to both deterministic and ensemble forecasts. Both INFL and EVMOS rely on the assumption of normally distributed forecast errors. Accordingly, Roulin and Vannitsem (2015) applied a logarithmic transformation to the runoff data prior to INFL and EVMOS in order to ensure approximate Gaussianity. In their setting, only postprocessing the precipitation input could not improve forecast skill as much as postprocessing the hydrological output ensemble. Hence, they concluded that there is a need for hydrological reforecasts.

In the context of monthly forecasting of the inflow to large dams in South Korea, Kang, Kim, and Hong (2010) assessed the differences in skill between statistical postprocessing of the meteorological inputs and postprocessing the hydrological outputs. In line with the findings by Zalachori et al. (2012) and Roulin and Vannitsem (2015), they concluded that the latter approach was more successful in reducing forecast uncertainty than the former, in particular for the dry season.

8.3 MULTIVARIATE HYDROLOGICAL POSTPROCESSING

Alhough adequate modeling of the spatio-temporal dependencies is crucial in order to obtain realistic scenarios of the development of future runoff values and gauge levels in space and time, very little work has been done on spatio-temporal postprocessing of hydrological ensemble forecasts, and there have been very few papers devoted to this topic. While Hemri, Fundel, and Zappa (2013) and Hemri et al. (2015) consider only temporal dependencies and Skøien et al. (2016) consider only spatial dependencies, Engeland and Steinsland (2014) performed the only study on spatio-temporally coherent postprocessing.

8.3.1 TEMPORAL DEPENDENCIES

Temporal dependence structures can be reintroduced to postprocessed hydrological ensemble forecasts basically in the same way as for meteorological forecasts. Following Berrocal, Raftery, and Gneiting (2007), who did a similar analysis in a spatial meteorological setting, Hemri et al. (2013) applied the geostatistical output perturbation method (GOP; Gel, Raftery, & Gneiting, 2004) to BMA postprocessed hydrological ensemble forecasts. A related parametric method to introduce a correlation structure is the Gaussian copula approach (GCA; Pinson & Girard, 2012), while ensemble copula coupling (ECC; Schefzik et al., 2013; Schefzik & Möller, 2018, Chapter 4 of this book) comprises a nonparametric method. Both GCA and ECC have been used in Hemri et al. (2015) to reintroduce a correlation structure to NGR postprocessed hydrological ensemble forecasts.

The BMA-GOP model leads to a mixture of m multivariate normal distributions. Conditional on member k performing best, the BMA multivariate normal kernel follows

$$y_{t,1:L}|x_{t,k,1:L} \sim \mathcal{N}\left(\mu_{t,k,1:L}, \Sigma_{t,k}\right) \tag{8.6}$$

where $x_{t,k,1:L}$ is the, possibly Box-Cox transformed, forecast vector of member k covering lead times $1, \ldots, L$. The vector $\mu_{t,k,1:L}$ denotes the forecast mean, which is typically a bias-corrected version of the transformed raw ensemble member k. The covariance matrix $\Sigma_{t,k}$ needs to be estimated from training data that may differ for different instances of t. First, the empirical variogram $\hat{\gamma}_{t,k}(d)$ is obtained by computing the mean squared differences in the forecast errors

$$\hat{\gamma}_{t,k}(d) = \frac{1}{2T(L-d)} \sum_{z \in TR} \sum_{\ell=1}^{L-d} [(x_{z,k,\ell} - y_{z,\ell}) - (x_{z,k,\ell+d} - y_{z,\ell+d})]^2 \tag{8.7}$$

where ℓ and d are lead time and the time lag, respectively. The training period is denoted by TR and its length by T. The subscript z identifies the different forecast initialization dates within the training period. Then, a theoretical variogram is fitted to the empirical variogram $\hat{\gamma}_{t,k}(d)$. Following Gel et al. (2004) and Berrocal et al. (2007), Hemri et al. (2013) used the exponential variogram

$$\gamma_{t,k}(d) = \rho_{t,k}^2 + \tau_{t,k}^2 \left(1 - e^{\frac{-d}{r_{t,k}}}\right) \tag{8.8}$$

where the coefficients $\rho_{t,k}^2$, $\tau_{t,k}^2$, and $r_{t,k}$ could be interpreted as nugget, scale, and range parameters in a geostatistical setting (Chilès & Delfiner, 1999; Cressie, 1993). From this, the elements $\delta_{t,k,i,j}$ of a covariance matrix $\Sigma_{t,k}^0$ can be obtained by

$$\delta_{t,k,i,j} = \rho_{t,k}^2 I_{[i=j]} + \tau_{t,k}^2 e^{\frac{|i-j|}{r_{t,k}}} \tag{8.9}$$

where I denotes the indicator function and i, j are the different lead times. For each ensemble forecast member the optimal variogram parameters can be estimated by minimizing

$$U\left(\rho_{t,k}^2, \tau_{t,k}^2, r_{t,k}\right) = \sum_{d=1}^{L-1} (L-d) \left\{ \frac{\hat{\gamma}_{t,k}(d) - \left[\rho_{t,k}^2 + \tau_{t,k}^2 \left(1 - e^{\frac{d}{r_{t,k}}}\right)\right]}{\rho_{t,k}^2 + \tau_{t,k}^2 \left(1 - e^{\frac{d}{r_{t,k}}}\right)} \right\}^2 \tag{8.10}$$

While the univariate BMA estimate $\hat{\sigma}_{t,k}^2$ of the variance is conditional on member k being best, $\Sigma_{t,k}^0$ is unconditional. Hence, the covariance matrix $\Sigma_{t,k}^0$ should be deflated by the factor

$$\alpha_{t,k} = \frac{\hat{\sigma}_{t,k}^2}{\rho_{t,k}^2 + \tau_{t,k}^2}, \text{ i.e., } \Sigma_{t,k} = \alpha_{t,k}\Sigma_{t,k}^0$$

The assumption of equal conditional marginal variances $\sigma_{t,k}^2$ over the entire range of lead times may not be fulfilled. In this case, it may make sense to partition the forecast horizon into several blocks of lead time ranges. As any multivariate distribution $f(y_1, y_2, \ldots, y_L)$ can be represented as

$$f(y_1, y_2, \ldots, y_L) = f(y_1)f(y_2 \mid y_1)f(y_3 \mid y_1, y_2) \cdots f(y_L \mid y_1, \ldots, y_{L-1}) \tag{8.11}$$

partitions of arbitrary size may be used. For multivariate normal models, these conditional distributions can be obtained in a straightforward manner. Hence, trajectories of forecast scenarios are obtained by first sampling a forecast value for lead time 1, then a value for lead time 2 conditional on lead time 1, and so on. Details on this procedure are given in Hemri et al. (2013).

As already mentioned, Hemri et al. (2015) employed ECC and GCA to model the temporal dependence structure of runoff forecasts. For these methods too, ensemble forecasts and observations are first, for example, Box-Cox, transformed to achieve approximate normality. Then, univariate left-truncated normal NGR models (Thorarinsdottir & Gneiting, 2010) are estimated in order to obtain postprocessed univariate predictive distributions $F_{t,\ell}$ for each lead time $\ell = 1, \ldots, L$. In order to avoid discontinuities in the marginal distributions from lead time ℓ to lead time $\ell+1$, the parameters of the forecast distributions $F_{t,\ell}$ for the different lead times are smoothed. Finally, the temporal correlation structures are re-established using ECC and GCA.

As ECC has already been introduced by Schefzik & Möller (2018, Chapter 4 of this book), only a short description of the application of ECC to hydrological forecasts is provided here. Given that the raw ensemble variance is considerably larger than zero, it turned out that among the different ECC variants described in Schefzik et al. (2013), ECC-T led to the most realistic forecast scenarios. Its application can be summarized as follows:

1. Unique ordered indices $k = 1, \ldots, m$ are assigned to the possibly Box-Cox transformed members of the raw ensemble. With this, the raw ensemble at lead time ℓ rewrites as $(x_1^\ell, \ldots, x_m^\ell)$. The index k assigned to a particular ensemble member does not change with lead time.
2. At each lead time ℓ, fit a parametric distribution to the possibly Box-Cox transformed raw ensemble. Its CDF is denoted by $\hat{S}_{t,\ell}$. Usually, it makes sense to use the same distribution family as for the NGR model. In Hemri et al. (2015) a left-truncated normal distribution is used for $\hat{S}_{t,\ell}$. Usually maximum likelihood estimation of $\hat{S}_{t,\ell}$ leads to good results, but in cases of very low raw ensemble variance, one has to be careful in order to avoid unrealistically high forecast quantiles. This has to be avoided by applying ad-hoc methods tailored to the problem at hand.
3. With this, the ECC-T reordered NGR forecast scenario members $\hat{y}_{t,k}^\ell$ can be obtained as

$$\hat{y}_{t,k}^\ell = F_{t,\ell}^{-1}\left[\hat{S}_{t,\ell}(x_{t,k}^\ell)\right] \tag{8.12}$$

where $F_{t,\ell}^{-1}$ denotes the inverse of the marginal NGR CDF at lead time ℓ.

GCA is an alternative to ECC that models the dependence structure independently from the raw ensemble. The application of GCA includes the following steps:

1. The observations from the training periods are used to obtain an empirical correlogram among lead times 1 to L.

2. As the empirical correlogram can be interpreted as the empirical variogram of a process with variance 1, we can fit a theoretical variogram to the empirical correlogram. There are many different variogram models. Hemri et al. (2015) tested, for Box-Cox transformed hydrological data, the exponential, the Matérn, and the Cauchy correlation model, though other models may perform just as well. Note that the forecast trajectories shown in Fig. 8.4f have been obtained using the exponential correlation model (GCA-exp). Schlather (1999) provides a comprehensive review of correlation functions.

3. Sample n times from a standard L-variate Gaussian distribution $\mathcal{N}(\mathbf{0}, \Sigma)$ with diagonal elements $\Sigma_{\ell,\ell} = 1$ and correlation structure from the fitted variogram model, that is $\Sigma_{\ell,\ell'} = e^{-\frac{|\ell-\ell'|}{r}}$ in case of an exponential model with free parameter r. The n resamples $(x_i^1, x_i^2, \ldots, x_i^L)$ with $i = 1, \ldots, n$ define n forecast trajectories in the standard normal space.

4. Similarly as for ECC-T, the GCA NGR forecast scenario members $\hat{y}_{t,i}^\ell$ can be obtained as

$$\hat{y}_{t,i}^\ell = F_{t,\ell}^{-1}\left[\Phi\left(x_i^\ell\right)\right] \tag{8.13}$$

where Φ denotes the CDF of the standard normal distribution and n can be arbitrarily large. If a comparison with the raw ensemble or ECC is desired, it makes sense to set $n = m$.

8.3.2 SPATIAL DEPENDENCIES

Skøien et al. (2016) present a method for spatially coherent postprocessing of hydrological ensemble forecasts. Based on NGR and top-kriging (Skøien et al., 2014; Skøien, Merz, & Blöschl, 2006), they generate probabilistic forecasts that consider spatial correlations in a sound way and allow for interpolation at ungauged sites. Top-kriging is very similar to ordinary kriging. The difference lies only in the estimation of the variogram model. Top-kriging is well suited for runoff related variables as it considers both the catchment area and nesting of sub-catchments. As shown by Skøien et al. (2016) it can be used to add the spatial correlation structure to univariate NGR fits in a straightforward manner. It allows spatial scenarios of runoff over large, heavily interconnected river systems to be obtained through simulation. Assuming Gaussian conditional predictive errors, the univariate NGR model can be written as

$$\hat{y}_{t,i} \sim \mathcal{N}\left(\mu_{t,i}, \sigma_{t,i}^2\right) \tag{8.14}$$

where the index i denotes now location and $\mu_{t,i} = a_{t,i,0} + a_{t,i,1} x_{t,i,1} + a_{t,i,2} x_{t,i,2} + \ldots + a_{t,i,m} x_{t,i,m}$ and $\sigma_{t,i}^2 = c_{t,i} + d_{t,i} S_{t,i}^2$. The bias correction coefficients are denoted by $a_{t,i,0}, a_{t,i,1}, \ldots, a_{t,i,m}$, and the spread correction coefficients by $c_{t,i}$ and $d_{t,i}$. The raw ensemble of size m at location i is given by $x_{t,i,1}, \ldots, x_{t,i,m}$ and $S_{t,i}^2$ denotes its variance. Top-kriging is then used to interpolate the bias and the spread coefficients. Closely following Skøien et al. (2016), it can be summarized as follows:

1. Estimate a sample variogram from the observations at the different gauges. For variables such as runoff that are spatial aggregates over an entire catchment area, spatial averages have to be taken. The variogram distances are calculated based not on the locations of the gauges, but on the centers of the contributing catchments. Binning of the variogram values is done not only based on distances, but also based on catchment size.

2. Fit a theoretical variogram function to the empirical variograms obtained in step 1.
3. Considering catchment size and location, construct covariance matrices from the expected semivariances. Covariance matrices among gauged locations are obtained as well as covariance matrices between gauged and ungauged locations.
4. With these covariance matrices runoff at ungauged sites can be interpolated as in standard kriging.

This approach is quite appealing in that it considers the topology of the river network in addition to distance. Hence, the modeled dependence between locations that are connected through the river network is stronger than between equally distant locations that are not hydrologically interconnected.

In cases of highly correlated predictors $x_{t,i,1}, \ldots, x_{t,i,m}$ the optimal set of bias correction coefficients is not unique. That is, there may be several different combinations of coefficients that lead to almost equal postprocessed forecasts. Obviously, in such cases naive interpolation of weights between gauges may lead to problematic artifacts. More specifically, even among neighboring and strongly correlated locations the weights may be uncorrelated. Hence, the weights need to be forced to exhibit reasonable correlation. The penalty function at location $i=0$ is given by

$$S_{pen,t} = P_c \sum_{i=1}^{n} \frac{1}{\gamma_{t,0i}} \sum_{k=0}^{m} \frac{2|a_{t,0,k} - a_{t,i,k}|}{|a_{t,0,k}| + |a_{t,i,k}|} \tag{8.15}$$

where $i=1, \ldots, n$ refers here to the n locations with the highest correlations in runoff with the location of interest, with n being set manually. From the already estimated variograms, the expected semivariance $\gamma_{t,0,i}$ for mean runoff can easily be obtained. Here, the joint subscript $0i$ indicates that the expected semivariance is calculated between the location of interest $i=0$ and another location $i \neq 0$. The scaling parameter P_c is used to set the relative weight of the spatial penalty compared with the CRPS in the estimation process. Hence, the objective function for model estimation at an individual location is

$$U_t = \text{CRPS}_t + S_{pen,t} \tag{8.16}$$

where CRPS_t refers to the mean CRPS value over the training period of the univariate NGR model at the location of interest. The actual parameter estimation is then performed iteratively:

1. In the first iteration the spatial penalty is omitted. That is, for each gauge the univariate NGR models are estimated by minimizing the CRPS over the training period, with $S_{pen,t} = \text{const.}$
2. Then, in a second iteration, the univariate NGR models are updated by minimizing U_t over the training period for each gauge separately. The scaling parameter P_c is set to 1.
3. For the third iteration, we set P_c to two times the ratio between the CRPS component of U_t and the spatial penalty from the last iteration. With this, the univariate NGR models are re-estimated for each gauge.

Note that in each iteration step the gauges are visited in a random order. As soon as the coefficients for a location are updated, the updated values are used for the estimation of the coefficients at the not yet updated neighboring gauges.

8.3.3 SPATIO-TEMPORAL DEPENDENCIES

Engeland and Steinsland (2014) propose the following spatio-temporal Gaussian model after Box-Cox transformation

$$\hat{y}_{t,1:L,1:n} \sim \mathcal{N}\left(\mu_{t,1:L,1:n}, \Sigma_t\right) \tag{8.17}$$

where t denotes issue time, the vector $1:L$ denotes the lead times, and $1:n$ the different gauges $i=1\dots n$. The mean $\mu_{t,\ell,i}$ is given in an NGR like manner as

$$\mu_{t,\ell,i} = a_{t,\ell,i}^0 + a_{t,\ell,i}^1 x_{t+\ell,i}^1 + \dots + a_{t,\ell,i}^m x_{t+\ell,i}^m \tag{8.18}$$

where $x_{t+\ell,i}^1, \dots, x_{t+\ell,i}^m$ are the members of the raw ensemble forecast issued at time t for gauge i with lead time ℓ. From now on, the index of issue time t is omitted for the sake of clarity. As already mentioned in Section 8.3.1, any multivariate distribution $f(y_1, y_2, \dots, y_L)$ can be rewritten as

$$f(y_1, y_2, \dots, y_L) = f(y_1)f(y_2|y_1)f(y_3|y_1, y_2)\cdots f(y_L|y_1, \dots, y_{L-1}) \tag{8.19}$$

If—as in Engeland and Steinsland (2014)—a Markov property is appropriate, i.e., $y_{\ell+1}$ is conditionally independent from $y_{1:(\ell-1)}$ given y_ℓ is known, this simplifies to

$$f(y_1, y_2, \dots, y_L) = f(y_1)f(y_2|y_1)f(y_3|y_2)\cdots f(y_L|y_{L-1}) \tag{8.20}$$

Assuming that $\hat{y}_{t,1:L,1:n}$ follows a multivariate normal distribution, its joint distribution at lead time 1 only, i.e., marginal in time, is given by

$$\hat{y}_{1:n} \sim \mathcal{N}\left(\mu_{1:n}, \Sigma_{1,1}\right) \tag{8.21}$$

where $\Sigma_{1,1}$ denotes the $n \times n$ covariance matrix between the different locations at lead time 1. In the approach by Engeland and Steinsland (2014) the Markovian simplification is not only applied for consecutive lead times at the same gauge, but also for consecutive lead times at different gauges. If runoff at gauge i at lead time $\ell-1$ is known, the runoff at the other gauges at lead time $\ell-1$ do not provide any additional information on the runoff at gauge i at lead time ℓ.

These simplifications on one hand reduce the number of parameters needed to estimate the dependence model, and on the other hand, they allow estimation of the parameters using a cascade of regressions. This cascade starts at lead time 1 and iterates through all subsequent lead times. As these regression equations do not share parameters, they can be estimated one after the other. The different gauges are linked through the correlations in the residuals. This method is also called seemingly unrelated regression (Zellner, 1962). The regression coefficients and the covariance matrices can be estimated using the R-package systemfit (Henningsen & Hamann, 2007). For further details on the actual model estimation procedure refer to Engeland and Steinsland (2014).

8.4 OUTLOOK

Having summarized the state of the art of postprocessing hydrological ensemble forecasts, let us now have a brief look at potential future developments and work still to be done. Currently, there is research devoted to the development of improved methods for spatio-temporal postprocessing of both the meteorological inputs and the hydrological output ensembles. For instance, a very recent paper by Scheuerer, Hamill, Whitin, He, and Henkel (2017) proposes a novel variant of the Schaake shuffle (Clark, Gangopadhyay, Rajagalopalan, & Wilby, 2004; Schefzik & Möller, 2018, Chapter 4 of this book) that selects historical templates of spatio-temporal and intervariable (precipitation and temperature) dependence structures based on forecast analogs (see Hamill, Whitaker, & Mullen, 2006, for details on the concept of forecast analogs). In a different context, Skøien et al. (2016) mention the need for more detailed analyses on how to take account of non-Gaussianity in spatio-temporal hydrological postprocessing. Furthermore, an intercomparison experiment of the different spatio-temporal postprocessing methods, for instance within the framework of HEPEX, may help to get further insights into the advantages and disadvantages of the different multivariate approaches.

As stated by Demargne et al. (2014), uncertainty modeling of rare events and the generation of spatio-temporally consistent seamless predictions from short range flash flood forecasts to long range or even seasonal forecasts continues to represent a challenge for the hydrometeorological forecasting community. Though the methods summarized in this chapter provide powerful approaches, the synthesis over the entire forecast range is still incomplete. Last but not least, additional studies in the vein of Roulin and Vannitsem (2015) and Van den Bergh and Roulin (2016) on the benefits of using hydrological reforecasts may help to further improve the skill of postprocessed hydrological predictions.

ACKNOWLEDGMENTS

This work was funded by the European Union Seventh Framework Programme under grant agreement 290976. Stephan Hemri furthermore gratefully acknowledges the support of the Klaus Tschira Foundation and thanks Tilmann Gneiting for valuable comments and suggestions. Likewise, the author is grateful to the external reviewer and the editors for their helpful comments. The dataset used for illustration of the methods presented in this chapter as well as Fig. 8.2 have been provided by the German Federal Institute of Hydrology (BfG).

REFERENCES

Aas, K., Czado, C., Frigessi, A., & Bakken, H. (2009). Pair-copula constructions of multiple dependence. *Insurance: Mathematics and Economics*, *44*, 182–198.

Ajami, N. K., Duan, Q., & Sorooshian, S. (2007). An integrated hydrologic Bayesian multimodel combination framework: Confronting input, parameter, and model structural uncertainty in hydrologic prediction. *Water Resources Research*, *43*, 1–19.

Alfieri, L., Burek, P., Dutra, E., Krzeminski, B., Muraro, D., Thielen, J., et al. (2013). GloFAS—Global ensemble streamflow forecasting and flood early warning. *Hydrology and Earth System Sciences*, *17*, 1161–1175.

Bartholmes, J. C., Thielen, J., Ramos, M. H., & Gentilini, S. (2009). The European flood alert system EFAS. Part 2. Statistical skill assessment of probabilistic and deterministic operational forecasts. *Hydrology and Earth System Sciences*, *13*, 141–153.

Bergström, S. (1995). The HBV model. In V. Singh (Ed.), *Computer Models of Watershed Hydrology* (pp. 443–476). Highlands Ranch, CO: Water Resources Publications.

Berrocal, V. J., Raftery, A. E., & Gneiting, T. (2007). Combining spatial statistical and ensemble information in probabilistic weather forecasts. *Monthly Weather Review*, *135*, 1386–1402.

Bogner, K., & Kalas, M. (2008). Error-correction methods and evaluation of an ensemble based hydrological forecasting system for the upper Danube catchment. *Atmospheric Science Letters*, *9*, 95–102.

Bogner, K., Liechti, K., & Zappa, M. (2016). Post-processing of stream flows in Switzerland with an emphasis on low flows and floods. *Water*, *8*, 20.

Bogner, K., & Pappenberger, F. (2011). Multiscale error analysis, correction, and predictive uncertainty estimation in a flood forecasting system. *Water Resources Research*, *47*, W07524.

Bogner, K., Pappenberger, F., & Cloke, H. L. (2012). Technical note: The normal quantile transformation and its application in a flood forecasting system. *Hydrology and Earth System Sciences*, *16*, 1085–1094.

Box, G., & Cox, D. (1964). An analysis of transformations. *Journal of the Royal Statistical Society, Series B, 26*, 211–252.

Brechmann, E. C., & Schepsmeier, U. (2013). Modeling dependence with C- and D-vine copulas: The R-package CDvine. *Journal of Statistical Software, 52*, 1–27.

Brown, J. D., & Seo, D.-J. (2010). A non-parametric post-processor for bias-correction of hydrometeorological and hydrologic ensemble forecasts. *Journal of Hydrometeorology, 11*, 642–665.

Brown, J. D., & Seo, D.-J. (2013). Evaluation of a nonparametric post-processor for bias correction and uncertainty estimation of hydrologic predictions. *Hydrological Processes, 27*, 83–105.

Buizza, R. (2018). Ensemble forecasting and the need for calibration. In S. Vannitsem, D. S. Wilks, & J. W. Messner (Eds.), *Statistical Postprocessing of Ensemble Forecasts*. Elsevier.

Chilès, J.-P., & Delfiner, P. (1999). *Geostatistics: Modeling Spatial Uncertainty*. New York: Wiley.

Clark, M., Gangopadhyay, S., Rajagalopalan, L., & Wilby, R. (2004). The Schaake shuffle: A method for reconstructing space-time variability in forecasted precipitation and temperature fields. *Journal of Hydrometeorology, 5*, 243–262.

Cloke, H. L., & Pappenberger, F. (2009). Ensemble flood forecasting: A review. *Journal of Hydrology, 375*, 613–626.

Cressie, N. A. C. (1993). In Rev (Ed.), *Statistics for Spatial Data*. New York: Wiley.

De Roo, A. P. J. (1999). LISFLOOD: A rainfall-runoff model for large river basins to assess the influence of land use changes on flood risk. In P. Balabanis (Ed.), *Ribamod: River Basin Modelling, Management and Flood Mitigation: Concerted Action* (pp. 349–357). European Commission. EUR 18287 EN.

Demargne, J., Wu, L., Regonda, S. K., Brown, J. D., Lee, H., He, M., et al. (2014). The science of NOAA's operational hydrologic ensemble forecast system. *Bulletin of the American Meteorological Society, 95*, 79–98.

Dempster, A. P., Laird, N. M., & Rubin, D. B. (1977). Maximum likelihood from incomplete data via the EM algorithm. *Journal of the Royal Statistical Society, Series B, 39*, 1–39.

Dipankar, A., Stevens, B., Heinze, R., Moseley, C., Zängl, G., Giorgetta, M., et al. (2015). Large eddy simulation using the general circulation model ICON. *Journal of Advances in Modeling Earth Systems, 7*, 963–986.

Duan, Q., Ajami, N. K., Gao, X., & Sorooshian, S. (2007). Multi-model ensemble hydrologic prediction using Bayesian model averaging. *Advances in Water Resources, 30*, 1371–1386.

Engeland, K., & Steinsland, I. (2014). Probabilistic postprocessing models for flow forecasts for a system of catchments and several lead times. *Water Resources Research, 50*, 182–197.

Gel, Y., Raftery, A. E., & Gneiting, T. (2004). Calibrated probabilistic mesoscale weather field forecasting: The geostatistical output perturbation method. *Journal of the American Statistical Association, 99*, 575–583.

Gneiting, T., Raftery, A. E., Westveld, A. H., & Goldman, T. (2005). Calibrated probabilistic forecasting using ensemble model output statistics and minimum CRPS estimation. *Monthly Weather Review, 133*, 1098–1118.

Hamill, T. M., Whitaker, J. S., & Mullen, S. L. (2006). Reforecasts: An important dataset for improving weather predictions. *Bulletin of the American Meteorological Society, 87*, 33–46.

Hemri, S., Fundel, F., & Zappa, M. (2013). Simultaneous calibration of ensemble river flow predictions over an entire range of lead-times. *Water Resources Research, 49*, 6744–6755.

Hemri, S., Lisniak, D., & Klein, B. (2014). Ascertainment of probabilistic runoff forecasts considering censored data (in German). *Hydrologie und Wasserbewirtschaftung, 58*, 84–94.

Hemri, S., Lisniak, D., & Klein, B. (2015). Multivariate post-processing techniques for probabilistic hydrological forecasting. *Water Resources Research, 51*, 7436–7451.

Henningsen, A., & Hamann, J. D. (2007). Systemfit: A package for estimating systems of simultaneous equations in R. *Journal of Statistical Software, 23*, 1–40.

Johnson, C., & Bowler, N. (2009). On the reliability and calibration of ensemble forecasts. *Monthly Weather Review, 137*, 1717–1720.

Kang, T. -H., Kim, Y. -O., & Hong, I. -P. (2010). Comparison of pre- and post-processors for ensemble streamflow prediction. *Atmospheric Science Letters*, *11*, 153–159.

Klein, B., & Meißner, D. (2016). *Vulnerability of inland waterway transport and waterway management on hydrometeorological extremes*. IMPREX Project Report.

Klein, B., Meißner, D., Kobialka, H. -U., & Reggiani, P. (2016). Predictive uncertainty estimation of hydrological multi-model ensembles using pair-copula construction. *Water*, *8*, 125.

Krzysztofowicz, R. (1999). Bayesian theory of probabilistic forecasting via deterministic hydrologic model. *Water Resources Research*, *35*, 2739–2750.

Lindström, G., Johansson, B., Persson, M., Gardelin, M., & Bergström, S. (1997). Development and test of the distributed HBV-96 hydrological model. *Journal of Hydrology*, *201*, 272–288.

Madadgar, S., & Moradkhani, H. (2014). Improved Bayesian multimodeling: Integration of copulas and Bayesian model averaging. *Water Resources Research*, *50*, 9586–9603.

Madadgar, S., Moradkhani, H., & Garen, D. (2014). Towards improved post-processing of hydrologic forecast ensembles. *Hydrological Processes*, *28*, 104–122.

Majewski, D., Liermann, D., Prohl, P., Ritter, B., Buchhold, M., Hanisch, T., et al. (2002). The operational global icosahedral-hexagonal gridpoint model GME: Description and high-resolution tests. *Monthly Weather Review*, *130*, 319–338.

Majewski, D., Liermann, D., & Ritter, B. (2012). *Kurze Beschreibung des Globalmodells GME (20 km/L60) und seiner Datenbanken auf dem Datenserver des DWD (in German)*. Technical report Offenbach, Germany: Deutscher Wetterdienst (DWD).

McLachlan, G. J., & Krishnan, T. (1997). *The EM Algorithm and Extensions*. New York: Wiley.

Molteni, F., Buizza, R., Palmer, T., & Petroliagis, T. (1996). The ECMWF ensemble prediction system: Methodology and validation. *Quarterly Journal of the Royal Meteorological Society*, *122*, 73–119.

Montani, A., Cesari, D., Marsigli, C., & Paccagnella, T. (2011). Seven years of activity in the field of mesoscale ensemble forecasting by the COSMO-LEPS system: Main achievements and open challenges. *Tellus A*, *63*, 605–624.

Pappenberger, F., Coke, H. L., Parker, D. J., Wetterhall, F., Richardson, D. S., & Thielen, J. (2015). The monetary benefit of early flood warnings in Europe. *Environmental Science & Policy*, *51*, 278–291.

Parrish, M. A., Moradkhani, H., & DeChant, C. M. (2012). Toward reduction of model uncertainty: Integration of Bayesian model averaging and data assimilation. *Water Resources Research*, *48*, 1–18.

Persson, A. (2015). *User guide to ECMWF forecast products. Version 1.2*. Reading, UK: ECMWF. http://www.ecmwf.int/files/user-guide-ecmwf-forecast-products. Accessed 10 August 2017.

Pinson, P., & Girard, R. (2012). Evaluating the quality of scenarios of short-term wind power generation. *Applied Energy*, *96*, 12–20.

Raftery, A. E., Gneiting, T., Balabdaoui, F., & Polakowski, M. (2005). Using Bayesian model averaging to calibrate forecast ensembles. *Monthly Weather Review*, *133*, 1155–1174.

Reggiani, P., Renner, M., Weerts, A. H., & Van Gelder, P. A. H. J. M. (2009). Uncertainty assessment via Bayesian revision of ensemble streamflow predictions in the operational river Rhine forecasting system. *Water Resources Research*, *45*, W02428.

Rings, J., Vrugt, J. A., Schoups, G., Husman, J. A., & Vereecken, H. (2012). Bayesian model averaging using particle filtering and Gaussian mixture modeling: Theory, concepts, and simulation experiment. *Water Resources Research*, *48*, 1–12.

Roulin, E., & Vannitsem, S. (2015). Post-processing of medium-range probabilistic hydrological forecasting: Impact of forcing, initial conditions and model errors. *Hydrological Processes*, *29*, 1434–1449.

Schaake, J. C., Demargne, J., Hartman, R., Mullusky, M., Welles, E., Wu, L., et al. (2007). Precipitation and temperature ensemble forecasts from single-value forecasts. *Hydrology and Earth System Sciences Discussions*, *4*, 655–717.

Schaake, J. C., Hamill, T. M., Buizza, R., & Clark, M. (2007). HEPEX: The hydrological ensemble prediction experiment. *Bulletin of the American Meteorological Society, 88*, 1541–1547.

Schefzik, R., & Möller, A. (2018). Multivariate ensemble postprocessing. In S. Vannitsem, D. S. Wilks, & J. W. Messner (Eds.), *Statistical Postprocessing of Ensemble Forecasts*. Elsevier.

Schefzik, R., Thorarinsdottir, T. L., & Gneiting, T. (2013). Uncertainty quantification in complex simulation models using ensemble copula coupling. *Statistical Science, 28*, 616–640.

Scheuerer, M., Hamill, T. M., Whitin, B., He, M., & Henkel, A. (2017). A method for preferential selection of dates in the Schaake shuffle approach to constructing spatiotemporal forecast fields of temperature and precipitation. *Water Resources Research, 53*, 3029–3046.

Schlather, M. (1999). *An introduction to positive definite functions and to unconditional simulation of random fields*. Technical report ST 99-10 Lancaster, UK: Dept. of Mathematics and Statistics, Lancaster University.

Schulz, J.-P., & Schättler, U. (2011). *Kurze Beschreibung des Lokal-Modells Europa COSMO-EU (LME) und seiner Datenbanken auf dem Datenserver des DWD (in German)*. Technical report Offenbach, Germany: Deutscher Wetterdienst (DWD).

Seo, D. -J., Herr, H. D., & Schaake, J. C. (2006). A statistical post-processor for accounting of hydrologic uncertainty in short-range ensemble streamflow prediction. *Hydrology and Earth System Sciences Discussions, 3*, 1987–2035.

Skøien, J. O., Blöschl, G., Laaha, G., Pebesma, E., Parajka, J., & Viglione, A. (2014). rtop: An R package for interpolation of data with a variable spatial support, with an example from river networks. *Computational Geosciences, 67*, 180–190.

Skøien, J. O., Bogner, K., Salamon, P., Smith, P., & Pappenberger, F. (2016). Regionalization of post-processed ensemble runoff forecasts. *Proceedings of the International Association of Hydrological Sciences, 373*, 109–114.

Skøien, J. O., Merz, R., & Blöschl, G. (2006). Top-kriging-geostatistics on stream networks. *Hydrological Earth System Sciences, 10*, 277–287.

Steppeler, J., Doms, G., & Adrian, G. (2002). Das Lokal-Modell LM (in German). *Prometheus, 27*, 123–128.

Thielen, J., Bartholmes, J. C., Ramos, M. H., & De Roo, A. P. J. (2009). The European flood alert system. Part 1. Concept and development. *Hydrology and Earth System Sciences, 13*, 125–140.

Thorarinsdottir, T. L., & Gneiting, T. (2010). Probabilistic forecasts of wind speed: Ensemble model output statistics by using heteroscedastic censored regression. *Journal of the Royal Statistical Society, Series A, 173*, 371–388.

Todini, E. (2008). A model conditional processor to assess predictive uncertainty in flood forecasting. *International Journal of River Basin Management, 6*, 123–137.

Van Andel, S. J., Weerts, A., Schaake, J., & Bogner, K. (2013). Post-processing hydrological ensemble predictions intercomparison experiment. *Hydrological Processes, 27*, 158–161.

Van den Bergh, J., & Roulin, E. (2016). Postprocessing of medium range hydrological ensemble forecasts making use of reforecasts. *Hydrology, 3*, 21.

Van der Knijff, J. M., Younis, J., & De Roo, A. P. J. (2010). LISFLOOD: A GIS-based distributed model for river basin scale water balance and flood simulation. *International Journal of Geographical Information Science, 24*, 189–212.

Van Schaeybroeck, B., & Vannitsem, S. (2015). Ensemble post-processing using member-by-member approaches: Theoretical aspects. *Quarterly Journal of the Royal Meteorological Society, 141*, 807–818.

Vannitsem, S. (2009). A unified linear model output statistics scheme for both deterministic and ensemble forecasts. *Quarterly Journal of the Royal Meteorological Society, 135*, 1801–1815.

Vrugt, J. A., & Robinson, B. A. (2007). Treatment of uncertainty using ensemble methods: Comparison of sequential data assimilation and Bayesian model averaging. *Water Resources Research, 43*, 1–18.

Wan, H., Giorgetta, M. A., Zängl, G., Restelli, M., Majewski, D., Bonaventura, L., et al. (2013). The ICON-1.2 hydrostatic atmospheric dynamical core on triangular grids. Part I. Formulation and performance of the baseline version. *Geoscientific Model Development, 6*, 735–763.

Wang, Q. J., Shrestha, D. L., Robertson, D. E., & Pokhrel, P. (2012). A log-sinh transformation for data normalization and variance stabilization. *Water Resources Research, 48*, WR010973.

Wilks, D. S. (2009). Extending logistic regression to provide full-probability-distribution MOS forecasts. *Meteorological Applications, 16*, 361–368.

Wilks, D. S. (2018). Univariate ensemble postprocessing. In S. Vannitsem, D. S. Wilks, & J. W. Messner (Eds.), *Statistical Postprocessing of Ensemble Forecasts*. Elsevier.

Wood, A. W., & Lettenmaier, D. P. (2008). An ensemble approach for attribution of hydrologic prediction uncertainty. *Geophysical Research Letters, 35*, L14401.

Wood, A. W., & Schaake, J. C. (2008). Correcting errors in streamflow forecast ensemble mean and spread. *Journal of Hydrometeolorogy, 9*, 132–148.

Wood, A. W., Wetterhall, F., & Ramos, M.-H. (2015). In *The hydrologic ensemble prediction experiment (HEPEX). EGU general assembly conference abstracts 17*.

Wu, L., Seo, D.-J., Demargne, J., Brown, J. D., Cong, S., & Schaake, J. C. (2011). Generation of ensemble precipitation forecast from single-valued quantitative precipitation forecast for hydrologic ensemble prediction. *Journal of Hydrology, 399*, 281–298.

Yuan, X., & Wood, E. F. (2012). Downscaling precipitation or bias-correcting streamflow? Some implications for coupled general circulation model (CGCM)-based ensemble seasonal hydrologic forecast. *Water Resources Research, 48*, W12519.

Zalachori, I., Ramos, M. H., Garçon, R., Mathevet, T., & Gailhard, J. (2012). Statistical processing of forecasts for hydrological ensemble prediction: A comparative study of different bias correction strategies. *Advances in Science and Research, 8*, 135–141.

Zängl, G., Reinert, D., Rípodas, P., & Baldauf, M. (2015). The ICON (icosahedral non-hydrostatic) modelling framework of DWD and MPI-M: Description of the non-hydrostatic dynamical core. *Quarterly Journal of the Royal Meteorological Society, 141*, 563–579.

Zellner, A. (1962). An efficient method of estimating seemingly unrelated regressions and tests for aggregation bias. *Journal of the American Statistical Association, 57*, 348–368.

Zhao, L., Duan, Q., Schaake, J., Ye, A., & Xia, J. (2011). A hydrologic post-processor for ensemble streamflow predictions. *Advances in Geosciences, 29*, 51–59.

APPLICATION OF POSTPROCESSING FOR RENEWABLE ENERGY

Pierre Pinson, Jakob W. Messner

Technical University of Denmark, Kongens Lyngby, Denmark

CHAPTER OUTLINE

9.1 INTRODUCTION

The energy system is witnessing a rapid transition toward more renewable energy generation capacity and new ways to exchange and consume electric energy. Besides hydro power, which is a type of renewable energy that has been used for a long time already, the new types of renewable energy generation that are mainly considered recently include wind, solar, tidal, and wave energy. For an overview of the status as of 2016, the reader is referred to Renewable Energy Policy Network for the 21st Century (2017) and International Renewable Energy Agency (2017). At the time of writing (mid-2017), the overall installed wind power capacity worldwide is approximately 500 GW, while for solar power it is approximately 300 GW. Wave energy generation has not yet reached the stage of extensive operational deployment.

Renewable energy generation is directly linked to the related meteorological variables, that is, wind power is generated when the wind blows and solar power when the sun shines. The conversion from these meteorological variables to power production is nonlinear and bounded, and possibly also non-stationary due to the physical nature of the conversion process (e.g., dirt on wind turbine blades, thermal effects on solar panel performance, etc.).

It is of crucial importance to be able to predict the generation from wind farms, wave energy farms, and solar power plants, at lead times between a few minutes and several days ahead. Such forecasts link to operational decisions for the management of the energy system, for example, switching on and off conventional power generation units, offering excess energy production through electricity markets, and so forth. Forecasting is today seen as a must-have input to the vast majority of decision problems related to the operation and maintenance of power systems with a nonnegligible share of renewable energy generation. For an overview of models and applications of renewable energy forecasting, see Kariniotakis (2017).

Forecasting of renewable power generation is a very active field of research with a wealth of industrial and commercial applications. While most forecasts issued and used today are single-valued and interpreted as deterministic forecasts, there is a strong push toward probabilistic forecasting. Examples of recent reviews are Bessa et al. (2017) and Zhang, Wang, and Wang (2014). This probabilistic approach is motivated by the nonnegligible and dynamic uncertainty in renewable energy forecasts, rendering it difficult to make optimal decisions with uncertainty estimates. This uncertainty and its complex characteristics originate from weather dynamics and predictability. On top of that, the meteorological variables of core interest for predicting renewable power generation, for example, wind speed and solar irradiance, are variables that were conventionally of less focus for weather forecasters. A general overview of how weather and energy are linked, especially for the case of renewable energy generation, can be found in Troccoli, Dubus, and Haupt (2014).

A number of approaches have been proposed to issue probabilistic forecasts of renewable power generation. The basics of these methodologies are similar for all types of renewable power generation (e.g., wind, solar, wave). As wind energy is the most extensively used renewable generation type, and wind power forecasting is the most mature among these applied forecasting problems, we have chosen to concentrate on it to introduce and illustrate the concepts. It is well known that for lead times beyond approximately 6 hours, dynamical weather predictions provide essential input for the forecast models. Probabilistic forecasts can be based on deterministic forecasts where the forecast uncertainty is estimated statistically. However, such an approach usually ignores flow-dependent forecast uncertainty. Therefore, ensemble forecasts, which provide such information, have gained popularity as input to probabilistic forecasts of renewable power generation, especially for the wind power case. It is our aim here to cover "nearly" all aspects of postprocessing ensemble forecasts for relevant meteorological variables to obtain various types of probabilistic forecasts for wind power generation. We say "nearly" because relevant generic aspects, such as verification, are covered in other parts of this book.

Through this chapter, our aim is to introduce some of the basic and key aspects of renewable energy generation forecasting by postprocessing ensemble forecasts for relevant meteorological variables. It should be noted that, even though there is a vast literature on renewable energy forecasting, the literature on postprocessing of ensemble forecasts for this purpose is fairly limited. Notable examples include the work of Messner, Zeileis, Bröcker, and Mayr (2013), Pinson and Madsen (2009), and Taylor, McSharry, and Buizza (2009), to be discussed in more detail in the following. We will first introduce relevant forecast products and notation, based on current operational practice in renewable energy forecasting, accounting for the needs of forecast users. Postprocessing of ensemble forecasts to generate renewable energy forecasts is generally based on two equally important tasks that also form the two core sections of this chapter: (i) conversion of meteorological variable forecasts into power, and (ii) calibration of ensembles or related predictive densities. We will end the chapter by gathering a set of conclusions and underlying current challenges, and discussing a number of perspectives for the development of further methods linking meteorological ensemble forecasting and renewable energy forecasting.

9.2 PRELIMINARIES: RELEVANT FORECASTING PRODUCTS AND NOTATION

We first introduce the main forecast products and notations we will cover. Let us write u as wind speed, which with the time index t comprises wind speed u_t at time t. Wind speed is used here for the sake of example, because it is the most relevant variable for wind power generation. Additional variables potentially impacting generation levels could be considered, for example wind direction or temperature, as well as pressure and humidity (because air density affects the power curve of wind turbines). The most relevant meteorological variables obviously depend upon the type of renewable energy generation considered: for solar power, these would be solar irradiance and temperature, while for wave energy, one would care about wave height and period. In parallel, y_t denotes wind power production measured at time t. For simplicity, only cases for single locations will be considered in the following, though a single location may refer to either a single wind farm or a region. Therefore, no index specifying spatial location is used. There is today increasing interest in spatiotemporal modeling for renewable energy forecasting (see e.g., Dowell & Pinson, 2016; Tastu, Pinson, Trombe, & Madsen, 2011) though not for the case of postprocessing meteorological ensemble forecasts.

When aiming to generate forecasts for wind power generation for lead times further than a few hours ahead, it is commonly agreed that weather forecasts should be used as input. We generically write $\hat{u}_{t+k|t}$ and $\hat{y}_{t+k|t}$ to represent single-valued forecasts of wind speed and power generation, respectively. The latter are obtained by conversion of the wind forecasts, as will be described in Section 9.3. In practice, such single-valued forecasts correspond to the conditional expectation of y_{t+k}, that is, the expected power generation for time $t + k$, conditional on the information available at time t. This is consistent with the aim of minimizing a quadratic loss function and thus minimizing the Root Mean Square Error verification criterion. Note that it is assumed that power generation at time $t + k$ is only explained by weather forecasts for that lead time. In general, this does not need to be the case, and forecasts for previous lead times $(t + k - 1, ...)$ and further lead times $(t + k + 1, ...)$ could be employed. Similarly for location, neighboring grid point information from the weather forecasting model can be used, for example, Andrade and Bessa (2017) and Cutler, Outhred, MacGill, Kay, and Kepert (2009). Here again, additional forecasts for other relevant meteorological variables could be accommodated in a more general context.

In contrast with such single-valued forecasts, ensemble forecasts of wind speed $\hat{u}_{t+k|t}^{(j)}, j = 1, ..., J$, with J the number of ensemble members, comprise alternative scenarios for future wind conditions that may be postprocessed to obtain scenarios for future wind power generation $\hat{y}_{t+k|t}^{(j)}, j = 1, ..., J$. It is then highly likely that forecast users will still ask to be provided with single-valued forecasts based on these ensemble forecasts. In line with their definition as conditional expectations aiming to minimize a quadratic loss function, such single-valued forecasts may be readily obtained as ensemble mean,

$$\hat{y}_{t+k|t} = \frac{1}{J}\sum_{j=1}^{J}\hat{y}_{t+k|t}^{(j)}, \quad \forall t, k \tag{9.1}$$

Here we somewhat abusively use the same notation for single-valued forecasts derived from ensembles as for other single-valued forecasts that would not be based on ensembles, which may not be the case elsewhere in the literature.

Although it is well known that an ensemble mean forecast often provides better skill than a single deterministic forecast run, the real value of ensemble forecasts is that their spread provides information on the forecast uncertainty. This spread can be measured, for example, by the ensemble variances

$$s_{t+k|t}^2 = \frac{1}{J}\sum_{j=1}^{J}\left(\hat{u}_{t+k|t}^{(j)} - \hat{u}_{t+k|t}\right)^2, \quad \forall t,k \tag{9.2}$$

or

$$r_{t+k|t}^2 = \frac{1}{J}\sum_{j=1}^{J}\left(\hat{y}_{t+k|t}^{(j)} - \hat{y}_{t+k|t}\right)^2, \quad \forall t,k \tag{9.3}$$

but for making optimal decisions under uncertainty, probabilistic forecasts are required that give more complete information on the distribution of y_{t+k}, given relevant information, model, and parameters at time t.

Predictive densities and predictive cumulative distribution functions are denoted by $\hat{f}_{t+k|t}(y)$ and $\hat{F}_{t+k|t}(y)$, respectively. These may be parametric, if making an assumption about the shape of predictive densities, for example, Gaussian, censored Gaussian, Beta, or logit-Normal. Alternatively, these may be nonparametric if not relying on such assumptions. Typical nonparametric methods include kernel dressing and quantile regression. The communication of these probabilistic forecasts may only involve specific prediction intervals, quantiles, or the complete densities. Optimal communication of probabilistic forecasts, not only for renewable energy forecasting applications, is a challenging problem in itself, as, for instance, already discussed by Hyndman (1996) more than 20 years ago.

For an extensive overview of the various types of renewable energy forecasting products, their specifics, as well as a relevant verification toolbox, the reader is referred to Morales, Conejo, Madsen, Pinson, and Zugno (2014).

9.3 CONVERSION OF METEOROLOGICAL VARIABLES TO POWER

The first aspect of postprocessing meteorological ensemble forecasts is to find a way to convert them into forecasts of renewable power generation. For all types of renewable energies, this conversion will be nonlinear, bounded, and nonstationary, as will be illustrated for the case of wind power generation. This conversion may be modeled in a flexible manner based on local polynomial regression. However, the fitting of such models may have to go further than classical least-squares approaches, as will be described and discussed as follows.

Throughout this section, we only consider the conversion of single-valued wind speed forecasts. This conversion is then used as the basis for the conversion and calibration methods of ensemble forecasts that are described in the next section. Extensions of the example conversion models to inclusion of variables such as wind direction or air density are possible, but are omitted here for simplicity.

9.3.1 DATA AND EMPIRICAL FEATURES

In order to introduce and discuss the specific features of the conversion from wind to power, we use a toy model example described in Section 9.6, which allows illustration of the various features on that conversion model while also having access to the true conversion model. This consequently permits comparison of the various regression approaches we will describe in the following.

For renewable energy generation like wind, wave, and solar power, the conversion from meteorological variables to power has similar basic characteristics: it is nonlinear, bounded, and time-varying. While the first two may seem intuitive, because power generation needs to be greater than 0 and necessarily has an upper limit, the time-varying aspect may be less obvious to grasp. Nonstationarity of the conversion functions may be due to changes in the vicinity of the power generation installment (e.g., trees changing wind characteristics and having varying shadowing effects), aging of the hardware, dust and dirt, etc. These effects therefore motivated renewable energy forecasters to accept that power conversion models should be seen as time-varying, while investing in online learning approaches to track them through time.

Our toy model, which is described in detail in Section 9.6, considers all these features of real power curves. Furthermore, it also represents errors in the wind speed forecasts and power conversion depending on the wind speed level. Thus, it provides reasonably realistic wind power data where the true relationship between wind speed and power generation is known. Fig. 9.1 shows the time variations of the

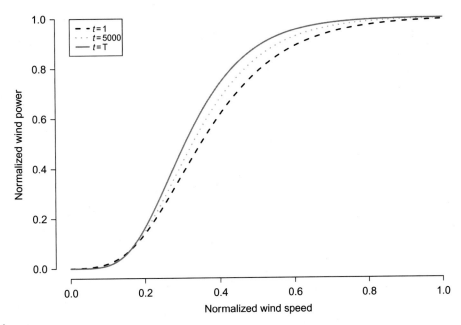

FIG. 9.1

Time-varying power curve for the conversion of wind speed to power. It is modeled as a double-exponential function with parameters varying from $\tau_1^\top = [8, 7]$ to $\tau_T^\top = [11, 9]$. The power curve at time $t = 5000$ through linear interpolation of the double-exponential model parameters.

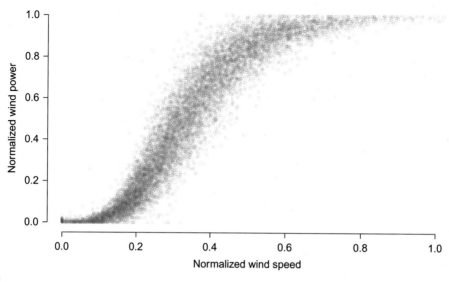

FIG. 9.2

Scatterplot of wind speed forecasts and power observations based on the previous time-varying power curve for the conversion of wind speed to power and added noise sequences.

assumed power curve model and Fig. 9.2 shows a scatterplot of the complete 10,000-point simulated wind speed and power dataset.

It should be noted that our toy model omits the fact that wind turbines have to be shut down above a certain cut-out wind speed to avoid structural damages.

9.3.2 LOCAL POLYNOMIAL REGRESSION AS A BASIS

Modeling the conversion from wind to power is a nonlinear regression problem, possibly with time-varying parameters. Consequently, many have focused on proposing alternative approaches for such nonlinear regression problems, either using modern statistical approaches such as local polynomial regression described in the following, or alternatively, machine learning approaches. If considering local polynomial regression, the most interesting and extensive description of relevance for the wind power application may be that in Nielsen, Nielsen, Joensen, Madsen, and Holst (2000), which has served as a basis also for application in, for example, solar power forecasting (Bacher, Madsen, & Nielsen, 2009). It introduces local polynomial regression while also describing a recursive weighted least-squares estimation approach with exponential forgetting, permitting adaptive estimation of the parameters of the nonlinear regression model assuming that those vary slowly.

As the true model for the conversion from predicted meteorological variables (i.e., wind speed forecast $\hat{u}_{t+k|t}$) to power generation (y_{t+k}) is not known (e.g., the double exponential choice made in our toy model), a flexible approach is to avoid assuming any specific function for that conversion model,

$$y_{t+k} = \theta(\hat{u}_{t+k|t}) + \epsilon_{t+k}, \quad t = 1, \ldots, T \tag{9.4}$$

ϵ_t is independent and identically distributed (i.i.d.) with unknown distribution, though assumed to have zero mean and finite variance. In practice, it is preferred to normalize all variables involved (i.e., wind speed and power generation). We then reformulate this general nonlinear model as a set of linear models (of polynomial transformations) to be estimated locally at a number of fitting points, u_i, $i = 1, ..., I$. As an example, one could fit a model for every ms^{-1} of wind speed. If having normalized wind speed by a maximum value of 20, u_1 would be located a 0, u_2 at 0.05 (for 1 ms^{-1}), etc. until reaching $u_{21} = 1$ (for an actual wind speed value of 20 ms^{-1}).

In the one-dimensional case that we consider here, these linear models consist of polynomials, that is,

$$y_{t+k} = \sum_{l=1}^{L} \alpha_{i,l} \left(\hat{u}_{t+k|t} - u_i \right)^l + \epsilon_{t+k}, \quad t = 1, ..., T \tag{9.5}$$

to be estimated locally in the neighborhood of each and every fitting point u_i, hence justifying the i-index for the α coefficients. L is the degree of the polynomials and usually it is sufficient to use $L = 1$ or $L = 2$. If higher dimensions are to be considered (e.g., by considering wind direction), and generalizing to conditional parametric models, the reader is referred to Cleveland and Develin (1988), Härdle (1990), and Hastie and Tibshirani (1993) for an extensive coverage. In a more compact form this can be rewritten as

$$y_{t+k} = \alpha_i^\top \mathbf{x}_{t+k} + \epsilon_{t+k}, \quad t = 1, ..., T \tag{9.6}$$

with $\mathbf{x}_{t+k} = \left[1, \hat{u}_{t+k|t} - u_i, ..., \left(\hat{u}_{t+k|t} - u_i \right)^L \right]^\top$ the vector of explanatory variables and $\alpha_i = [\alpha_{i,0}, \alpha_{i,1}, ..., \alpha_{i,L}]^\top$ the vector of model coefficients. Note that, because we have been clear that models should be set up and coefficients estimated for all lead times of interests, we will drop the k index in the following, except for the case of explanatory and response variables where time indices need to account for that lead time.

Let us now focus on a single fitting point u_i only. For the local polynomial approximation formulated in the preceding equation, model fitting translates to estimating the set of coefficients α_i. This linear model can be fitted locally based on weighted least squares,

$$\hat{\alpha}_i = \underset{\alpha_i}{\arg\min} \sum_{t=1}^{T} w_{t,i} \rho \left(y_{t+k} - \alpha_i^\top \mathbf{x}_{t+k} \right) \tag{9.7}$$

where ρ is a quadratic loss function, $\rho(\epsilon) = \epsilon^2/2$, while the weights $w_{t,i}$ are assigned by a Kernel function, for instance of the form

$$w_{t,i} = K_T \left(\frac{|\hat{u}_{t+k|t} - u_i|}{h_i} \right) \tag{9.8}$$

where h_i ($h_i > 0$) is a local bandwidth parameter controlling the size of the neighborhood over which the local model should be fitted. A classical example for K_T is the tricube function, which reads

$$K_T(v) = \begin{cases} (1 - v^3)^3, & v \in [0, 1] \\ 0, & v > 1 \end{cases} \tag{9.9}$$

as found in Cleveland and Develin (1988). Fig. 9.3 shows the local linear regression fit schematically.

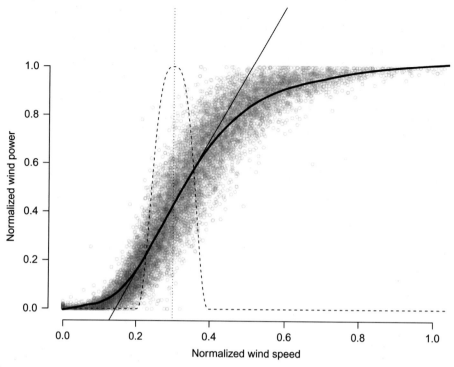

FIG. 9.3

Schematic plot of local linear regression. The *dashed line* shows a tricube kernel function with a bandwidth of 0.2 that is centered at a normalized wind speed of 0.3, as indicated by the *dotted vertical line*. The *thin black line* shows the corresponding weighted linear regression fit and the *thick black curve* shows the local linear regression fit, which combines the intersection of the *dotted vertical* and the *thin line* with similarly constructed points over the range of normalized wind speed.

As for any weighted least squares estimation problem, by assembling the appropriate vectors and matrices, estimated model coefficients are readily given as

$$\hat{\boldsymbol{\alpha}}_i = \left(\mathbf{X}^\top \mathbf{W}_i \mathbf{X}\right)^{-1} \mathbf{X}^\top \mathbf{W}_i \mathbf{y} \qquad (9.10)$$

where $\mathbf{X}$ is a design matrix containing all values of explanatory variables,

$$\mathbf{X} = \begin{bmatrix} \mathbf{x}_{1+k}^\top \\ \mathbf{x}_{2+k}^\top \\ \vdots \end{bmatrix} \qquad (9.11)$$

$\mathbf{W}_i$ is a diagonal matrix gathering the weights, which are different for each and every fitting point,

$$\mathbf{W}_i = \begin{bmatrix} w_{1+k,i} & 0 & 0 \\ 0 & w_{2+k,i} & 0 \\ 0 & 0 & \ddots \end{bmatrix} \qquad (9.12)$$

and finally where $\mathbf{y}$ is the vector of observed power values,

$$
\mathbf{y} = \begin{bmatrix} y_{1+k} \\ y_{2+k} \\ \vdots \end{bmatrix} \tag{9.13}
$$

Having fitted these local polynomial models at all the I training-data points, values for the original nonlinear regression model may be obtained by interpolation through the values at those fitting points. Such interpolation may be linear, based on splines, etc.

Local polynomial regression is a very common technique to model nonlinear relationships such as for wind speed and wind power. Therefore, implementations of this model can be found in various software packages, for example, the `loess()` function in R.

To illustrate the outcome of fitting local polynomial regression models for the conversion of wind to power generation, we give here an example for the data generated herein. We use 41 fitting points uniformly distributed over the normalized wind speed interval [0, 1] and a bandwidth of 0.1. The estimates are obtained by using all $T = 10{,}000$ instances in the dataset. The results of model fitting are depicted in Fig. 9.4. Even though the regression curve obtained is qualitatively similar to the true power curve at the end of the dataset, the fact that all data were used without differentiation makes this model fit represent an average power curve for the whole dataset and not the most recent power curve at time $t = T$. This may then encourage the use of an approach that allows for time-adaptivity, as described in the following.

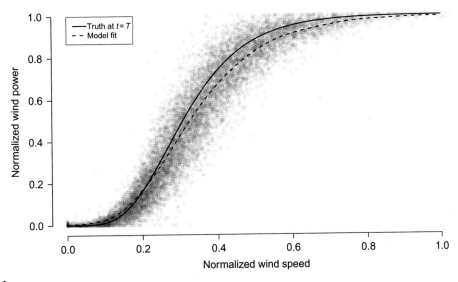

FIG. 9.4

Fitting of local polynomial regression to empirical data for the wind power conversion process, in a least squares estimation framework.

9.3.3 TIME-VARYING ESTIMATION TO ACCOMMODATE NONSTATIONARITY

The estimation in Eq. (9.7) allows local coefficients to be obtained for a given fitting point, but for the whole time period. For time-varying power curves such as that depicted in Fig. 9.1, a batch estimation over the T time steps would result in an estimated power curve resembling that for $t = T/2 = 5000$. However, in practice at a given time one would like to have estimated the most recent power curve model possible. This actually can be done fairly easily through generalization of the preceding weighted least squares approach to recursive and adaptive estimation.

With that objective in mind, placing ourselves at a given time t, the estimation problem in Eq. (9.7) is reformulated as

$$\hat{\boldsymbol{\alpha}}_{t,i} = \underset{\boldsymbol{\alpha}_i}{\arg\min} \sum_{t'=1}^{t} \beta_{t,i}(t') w_{t',i} \rho\left(y_{t'+k} - \boldsymbol{\alpha}_i^{\mathsf{T}} \mathbf{x}_{t'+k}\right) \tag{9.14}$$

where the only difference with the previous formulation consists in using an additional weighting function β to discard older information based on an exponential forgetting scheme, that is,

$$\beta_{t,i}(t') = \begin{cases} \lambda_{t',i}^{\text{eff}} \beta_{t-1,i}(t'-1), & 1 \leq t' \leq t-1 \\ 1, & t = t' \end{cases} \tag{9.15}$$

$\lambda_{t',i}^{\text{eff}}$ denotes a so-called effective forgetting factor (as coined by Nielsen et al., 2000) for the fitting point u_i of interest, being a function of the local weight $w_{t',i}$, that is,

$$\lambda_{t',i}^{\text{eff}} = 1 - (1 - \lambda) w_{t',i} \tag{9.16}$$

This effective forgetting factor ensures that old observations are downweighted only when new information is available. λ is the traditional forgetting factor used in adaptive estimation, with λ slightly less than 1.

This type of adaptive estimation formulation for time-varying models related to conversion from meteorological variables to power generation has been widely used for wind and solar power applications, mostly for the case of single-valued forecasting. Based on this formulation, and after a little algebra, one can derive the following three-step updating procedure:

$$\epsilon_{t,i} = y_{t+k} - \hat{\boldsymbol{\alpha}}_{t,i}^{\mathsf{T}} \mathbf{x}_{t+k} \tag{9.17}$$

$$\hat{\boldsymbol{\alpha}}_{t,i} = \hat{\boldsymbol{\alpha}}_{t-1,i} + \epsilon_{t,i} w_{t,i} \left(\mathbf{R}_{t,i}\right)^{-1} \mathbf{x}_{t+k} \tag{9.18}$$

$$\mathbf{R}_{t,i} = \lambda_{t,i}^{\text{eff}} \mathbf{R}_{t-1,i} + w_{t,i} \mathbf{x}_{t+k} \mathbf{x}_{t+k}^{\mathsf{T}} \tag{9.19}$$

where $\mathbf{R}_{t,i}$ is the time and distance weighted covariance matrix of $\mathbf{x}_{t+k}$ and is initialized with a diagonal matrix with arbitrary small values on the diagonal. The model coefficients $\hat{\boldsymbol{\alpha}}_{t,i}$ are initialized with arbitrary values, for example, a vector of zeros. In the preceding, in the case where $w_{t,i} = 0$, this means that related explanatory variables at that time are not in the neighborhood of the fitting point of interest, and hence this case should not be used for updating the model coefficients. In practice, this will yield $\hat{\boldsymbol{\alpha}}_{t,i} = \hat{\boldsymbol{\alpha}}_{t-1,i}$ and $\mathbf{R}_{t,i} = \mathbf{R}_{t-1,i}$.

Fig. 9.5 shows that this time adaptive model is in general able to better follow the changes in the power curve. However, the fitted curve is still slightly too flat, which could be caused by a delay in the time adaptation, but could also stem from noise along the wind speed axis, which is not considered. An extension to also consider this noise will be presented in the following.

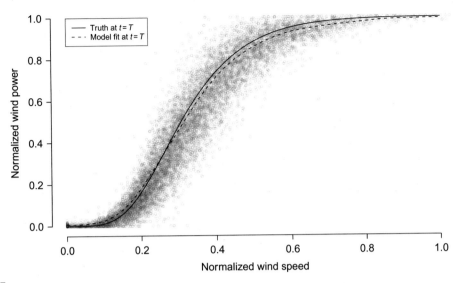

FIG. 9.5

Fitting of local polynomial regression to empirical data for the wind power conversion process, with an adaptive and recursive least squares estimation framework.

9.3.4 FROM LEAST SQUARES ESTIMATION TO FITTING OF PRINCIPAL CURVES

The previous models were fitted in a least squares framework, hence only considering noise along the power axis. However, as illustrated in Section 9.3.1 and 9.6, explanatory variables consist of forecasts that necessarily have an error component that should be accounted for in order to approximate the true conversion model from wind to power. Ideally, this could be done within an error-in-variables modeling framework, requiring estimation and tracking of the local covariance of errors in both explanatory and response variables at all fitting points. In a more data-driven framework, the approach presented before may be generalized to estimate so-called principal curves, as originally introduced and discussed by Hastie and Stuetzle (1989) and Tibshirani (1992). In terms of estimation, this can be done by using a total least squares criterion instead of the more conventional least squares. For a complete introduction to total least squares, we refer to Golub and Van Loan (1980), as well as to de Groen (1996) for the case of model fitting. The generalization to total least squares we describe here is largely based on Pinson, Nielsen, Madsen, and Nielsen (2008) for the case of the wind power application.

Principal curves and total least squares model fitting aim at obtaining the curve that minimizes the distance between pairs of explanatory/response variable values and the regression line, but perpendicular to that line. Consequently, instead of considering residuals in the form of $y_{t+k} - \boldsymbol{\alpha}_{t,i}^\top \mathbf{x}_{t+k}$, hence along the y-axis only, these are replaced by a distance perpendicular to the regression line. A schematic plot, showing the difference between least squares and orthogonal fitting is provided by Fig. 9.6.

Let us limit ourselves to first-order polynomials for the local approximation of the conversion, that is, with $\boldsymbol{\alpha}_{t,i} = [\alpha_{t,i,1}, \alpha_{t,i,2}]^\top$ and $\mathbf{x}_{t+k} = [1, \hat{u}_{t+k|t}]^\top$. We write $l_{t,i}$ the line parametrized by $\boldsymbol{\alpha}_{t,i}$, p_{t+k} the

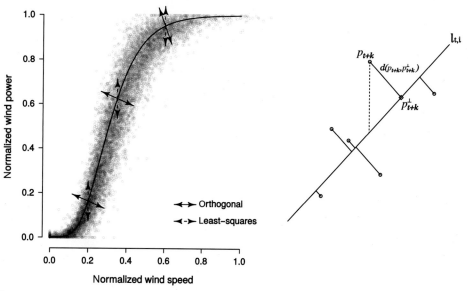

FIG. 9.6

Schematic plots showing the difference between classical least-squares and orthogonal fitting. The *left figure* shows the different directions of the residuals and the *right figure* shows the orthogonal residuals (*thin solid lines*) and the ordinary least squares residual (*dashed line*) for a specific linear line segment $l_{t,i}$.

pair of explanatory/response variable values for time $t + k$ and $p_{t+k}^{\perp}$ the projection of p_{t+k} on $l_{t,i}$, which may be readily obtained based on geometrical considerations. The model residual is then defined as

$$\epsilon_{t+k}^{\perp} = d(p_{t+k}, p_{t+k}^{\perp}) \tag{9.20}$$

where d is the Euclidean distance.

Subsequently, in line with our previous developments, the total least squares estimates at a given fitting point u_i are obtained as

$$\hat{\alpha}_{t,i}^{\perp} = \underset{\alpha}{\mathrm{argmin}} \sum_{t'=1}^{t} \beta_{t,i}(t') w_{t',i}^{\perp} \rho\left(\epsilon_{t+k}^{\perp}\right) \tag{9.21}$$

While the residual is defined based on a distance perpendicular to the regression line, the weight $w_{t',i}^{\perp}$ is to be defined as a function of a distance along that regression line. It should be noted that the estimation problem in Eq. (9.21) is hard to solve directly, although it is more easily feasible in a recursive estimation framework.

For that, let us place ourselves at time t, hence having access to the previous estimates $\hat{\alpha}_{t-1,i}$ at the fitting point u_i and then also to the regression line $l_{t-1,i}$. The model residual $\epsilon_{t+k}^{\perp}$ can be easily obtained as the distance between p_{t+k} and its projection on that regression line. In parallel, if writing

$$\tilde{p}_i = [u_i \ \hat{\alpha}_{t-1,i}^{\top} x_{t+k}] \tag{9.22}$$

the value of the regression line $l_{t-1,i}$ at the fitting point, the distance along the regression line is denoted by $d_l(\widetilde{p}_i, p_{t+k})$. This can also be calculated using geometrical considerations. Eventually, the weight to be assigned to that observation is given by

$$w_{t,i}^{\perp} = K_T\left(\frac{d_l(\widetilde{p}_i, p_{t+k})}{h_i}\right) \tag{9.23}$$

As a generalization of the design matrix $\mathbf{X}$ introduced in Eq. (9.11), the augmented design matrix is given as

$$\mathbf{X}^+ = \mathbf{W}_i^{1/2}[\mathbf{X} \quad \mathbf{y}] \tag{9.24}$$

where $\mathbf{W}_i$ and $\mathbf{y}$ are defined as in Eqs. (9.12), (9.13). That is, $\mathbf{X}^+$ is the original design matrix, to which is added an extra column consisting in the vector of observed power values, then weighted by the square root of the weights given to the observations as a function of their distance to the fitting point of interest.

The idea of Pinson et al. (2008) for recursive estimation of the total least squares estimates $\hat{\alpha}_{t,i}^{\perp}$ in Eq. (9.21) is to track the singular vector (for an introduction to singular value decomposition, see Golub & Van Loan, 1996) related to the smallest singular value of $\mathbf{X}^+$. This eventually leads to tracking the eigenvector related to largest eigenvalue of $\mathbf{P}^+$, the augmented covariance matrix for the design matrix $\mathbf{X}^+$. This eigenvector is commonly referred to as the left-most eigenvector.

We write $\mathbf{x}_{t+k}^+$ the augmented vector of observations at time $t + k$,

$$\mathbf{x}_{t+k}^+ = [\mathbf{x}_{t+k} \quad y_{t+k}] \tag{9.25}$$

Consequently, if having $\mathbf{P}_{t-1}^+$ the augmented covariance matrix based on data available up to time $t-1$, an update of that matrix at time t can be obtained through application of the matrix inversion lemma,

$$\mathbf{P}_t^+ = \frac{1}{\lambda_t^{\text{eff}}}\left[\mathbf{P}_{t-1}^+ - \gamma_t \frac{\mathbf{P}_{t-1}^+ \mathbf{x}_{t+k}^+ \mathbf{x}_{t+k}^{+\top} \mathbf{P}_{t-1}^+}{1 + \gamma_t \mathbf{x}_{t+k}^{+\top} \mathbf{P}_{t-1}^+ \mathbf{x}_{t+k}^+}\right] \tag{9.26}$$

with $\gamma_t = w_t^{\perp}/\lambda_t^{\text{eff}}$.

After having updated the augmented covariance matrix $\mathbf{P}_t^+$ at time t, the power method described by Golub and Van Loan (1996) can be readily used to obtain the largest eigenvalue and corresponding left-most eigenvector. By first initializing $\mathbf{v}_t^{(0)}$ as a unit vector of dimension 3 and then iteratively computing

$$\mathbf{v}_t^{(j)} = \frac{\mathbf{P}_t^+ \mathbf{v}_t^{(j-1)}}{\|\mathbf{P}_t^+ \mathbf{v}_t^{(j-1)}\|_2} \tag{9.27}$$

$\mathbf{v}_t^{(j)}$ eventually converges to the left-most eigenvector $\mathbf{v}_t$ of $\mathbf{P}_t^+$. In addition, because the coefficients we are to recursively estimate are smoothly varying in time, one may initialize that iterative sequence with $\mathbf{v}_t^{(0)} = \mathbf{v}_{t-1}$ to speed up convergence. The left-most eigenvector of $\mathbf{P}_t^+$ is of dimension 3, with $\mathbf{v}_t = [v_{t,1}, v_{t,2}, v_{t,3}]$.

Finally, the estimate $\hat{\alpha}_{t,i}^{\perp}$ of the model coefficients is obtained as

$$\hat{\alpha}_{t,i}^{\perp} = \frac{1}{v_{t,3}}[v_{t,1} \quad v_{t,2}] \tag{9.28}$$

For initializing this recursive procedure, principles similar to those described for the least squares case may be employed. Because we are working with covariance matrices instead of inverse covariance

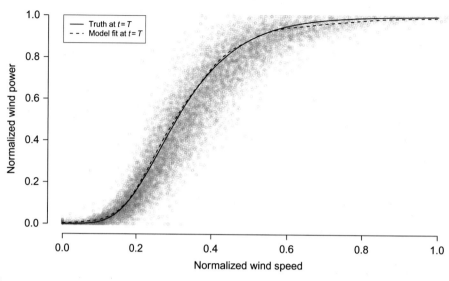

FIG. 9.7

Fitting of local polynomial regression to empirical data for the wind power conversion process, with an adaptive and recursive total least squares estimation framework.

matrices, the initialization of the augmented covariance matrices $\mathbf{P}_t^+$ consists of defining it as a diagonal matrix with very large values on its diagonal (say, 10^5). The model coefficients for each and every fitting point can be set to a vector of zeros.

Fig. 9.7 shows that the time-adaptive power curve fitted with orthogonal least squares provides a very good estimate of the true power curve.

9.4 CALIBRATED PREDICTIVE DENSITIES OF POWER GENERATION

The previous section presented methods to convert wind speed to wind power or wind speed predictions to wind power predictions. These methods concentrate on the single-valued expected power generation for a given wind speed, and so do not consider any uncertainty of the resulting wind power forecasts. In principle, probabilistic forecasts could be derived by using a power curve model to individually transform the members of a dynamical ensemble prediction. However, the resulting power generation ensembles would clearly lack calibration just as much as the meteorological ensembles themselves so that statistical postprocessing should be applied. In addition to calibration, statistical postprocessing is also used to derive full predictive distributions or specific predictive quantiles, which are often required to solve decision-making problems.

This section discusses different approaches to deriving predictive distributions or quantiles of wind power. These include approaches that calibrate the wind speed ensemble predictions prior to conversion, approaches that calibrate the ensemble after it has been transformed into wind power, and approaches that do the calibration and conversion in one step.

9.4.1 CALIBRATION PRIOR TO CONVERSION

A straightforward approach to generate calibrated predictive distributions of power generation is to first calibrate the meteorological ensemble and then use a power curve model to convert samples or quantiles of the predictive wind distribution to power generation.

Power curve functions for this approach are usually estimated on a data set with measured meteorological variables and corresponding power outputs, which does not have to be the same as the training data set for the statistical postprocessing model. For example, for single turbines, the power curves of the turbine manufacturers may be used.

Even though virtually any postprocessing method that is appropriate for the respective variable (wind speed, direction, wind vector, etc.) can be used, we only present two example postprocessing approaches that were explicitly proposed for wind power forecasting. Further potential uni- or multivariate models are reviewed in Wilks (2018, Chapter 3 of this book) and Schefzik and Möller (2018, Chapter 4 of this book).

Kernel dressing of wind speed

Taylor et al. (2009) proposed a combination of member-by-member calibration and ensemble dressing (Sections 3.3 and 3.5 in Wilks, 2018, Chapter 3 of this book) to postprocess and calibrate wind speed ensembles. First, the level and the spread of the ensemble are rescaled similar to the member-by-member approach of Eckel, Allen, and Sittel (2012), that is,

$$\hat{u}_{t+k|t}^{(j)*} = \hat{u}_{t+k|t} - a + b\left(\hat{u}_{t+k|t}^{(j)} - \hat{u}_{t+k|t}\right) \tag{9.29}$$

where $\hat{u}_{t+k|t}^{(j)*}, j = 1, \ldots, J$ are the corrected ensemble members, $\hat{u}_{t+k|t}$ is the ensemble mean, $\hat{u}_{t+k|t}^{(j)}$ are the original ensemble member forecasts, and a and b are calibration parameters. Subsequently, a continuous predictive density $\hat{f}_{t+k|t}(u)$ is constructed with kernel density smoothing

$$\hat{f}_{t+k|t}(u) = \frac{1}{J}\sum_{i=1}^{J} K\left(u, \hat{u}_{t+k|t}^{(j)*}, \sigma\right) \tag{9.30}$$

where $K\left(u, \hat{u}_{j,t+k|t}^{*}, \sigma\right)$ denotes the kernel function, which is a zero-truncated Gaussian kernel with mean $\hat{u}_{t+k|t}^{(j)*}$ and bandwidth σ. The zero-truncated normal distribution is a normal distribution that has zero density for negative values and so is well suited to model nonnegative wind speed (e.g., Baran, 2014; Thorarinsdottir & Gneiting, 2010). It is defined as

$$K\left(u, \hat{u}_{t+k|t}^{(j)*}, \sigma\right) = \frac{\phi\left(\dfrac{u - \hat{u}_{t+k|t}^{(j)*}}{\sigma}\right)}{\sigma\Phi\left(\dfrac{\hat{u}_{t+k|t}^{(j)*}}{\sigma}\right)} \tag{9.31}$$

where $\phi()$ and $\Phi()$ are the probability density function and the cumulative distribution function of the standard normal distribution, respectively.

The parameters $\tau = [a, b, \sigma]^{\top}$ can be estimated with maximum likelihood estimation

$$\hat{\tau} = \underset{\tau}{\mathrm{argmax}} \sum_{t=1}^{T} \hat{f}_{t|t-k}(u_t) \tag{9.32}$$

which can be derived using a nonlinear optimization algorithm (e.g., Nelder & Mead, 1965). A power curve model can be used to convert large random samples from the wind speed predictive distribution to estimate the full predictive distribution of power generation.

Inverse power curve transformation

One problem of calibrating wind speed prior to conversion is that a data set of wind speed measurements has to be available to train the statistical postprocessing model. Although wind speed is measured at most turbine nacelles, these measurements are often unreliable because they are made downwind of the turbine blades. To circumvent this issue, Messner et al. (2013) propose an inverse power curve transformation to generate proxy wind speed measurements $\tilde{u}_t$ from wind power data y_t or in other words, the wind speed is measured directly by the turbine, which can be seen as a huge anemometer.

$$\tilde{u}_t = \theta^{-1}(y_t) \tag{9.33}$$

Here θ is a power curve model, which can be derived from data, such as in the previous section, or taken from the turbine manufacturer. Because below a certain cut in wind speed c_1 no power is produced and above a certain nominal wind speed c_2 power production is constant, the inverse power curve function is not unique for all values. More specifically $\tilde{u}_t$ is a censored version of the true wind speed u_t (see also Wilks, 2018, Chapter 3 of this book)

$$\tilde{u}_t = \begin{cases} c_1 & \text{if } u_t \leq c_1 \\ u_t & \text{if } c_1 < u_t < c_2 \\ c_2 & \text{if } u_t \geq c_2 \end{cases} \tag{9.34}$$

To consider this censoring in the calibration step, Messner et al. (2013) applied a censored nonhomogeneous regression model where they assumed a normal distribution $N(\mu_{t+k|t}, \sigma^2_{t+k|t})$ for u_{t+k} with mean $\mu_{t+k|t}$ and variance $\sigma^2_{t+k|t}$ conditional on regressor variables $x_{t+k|t}$ and $z_{t+k|t}$.

$$u_{t+k} \sim N(\mu_{t+k|t}, \sigma^2_{t+k|t}) \tag{9.35}$$

$$\mu_{t+k|t} = \beta^\top x_{t+k|t} \tag{9.36}$$

$$\log(\sigma_{t+k|t}) = \gamma^\top z_{t+k|t} \tag{9.37}$$

A natural choice for the regressor variables are $x_{t+k|t} = [1, \hat{u}_{t+k|t}]^\top$ and $z_{t+k|t} = [1, s_{t+k|t}]^\top$ with the wind speed ensemble mean $\hat{u}_{t+k|t}$, and standard deviation $s_{t+k|t}$ but also other variables might be included such as transformations of the ensemble mean. The coefficients $\beta = [\beta_0, \beta_1]^\top$ and $\gamma = [\gamma_0, \gamma_1]^\top$ are estimated by maximum likelihood estimation where the censoring is considered with

$$\left[\hat{\beta}^\top, \hat{\gamma}^\top\right]^\top = \underset{[\beta^\top, \gamma^\top]^\top}{\text{argmax}} \sum_{t=1}^{T} \left\{ I(c_1 < \tilde{u}_t < c_2) \log\left[\frac{1}{\sigma_{t|t-k}} \phi\left(\frac{\tilde{u}_t - \mu_{t|t-k}}{\sigma_t}\right)\right] \right.$$

$$\left. + I(\tilde{u}_t = c_1) \log\left[\Phi\left(\frac{c_1 - \mu_{t|t-k}}{\sigma_{t|t-k}}\right)\right] + I(\tilde{u}_t = c_2) \log\left[1 - \Phi\left(\frac{c_1 - \mu_{t|t-k}}{\sigma_{t|t-k}}\right)\right] \right\} \tag{9.38}$$

where $I()$ is the indicator function that is 1 if the condition in brackets is true and 0 otherwise. With this model, predictive normal distributions for u_t can easily be derived as

$$\hat{f}_{t+k|t}(u) = \frac{1}{\sigma_{t+k|t}} \phi\left(\frac{u - \mu_{t+k|t}}{\sigma_{t+k|t}}\right) \tag{9.39}$$

from which random samples or quantiles can easily be derived and be transformed back to power output. Due to the back and forth transformation with the same power curve, one of the main advantages of this inverse power curve approach is that it is not very sensitive to the exact form of the power curve.

9.4.2 CALIBRATION AFTER CONVERSION

Alternatively to calibrating the meteorological ensemble prior to conversion, the calibration can also take place after transforming the raw ensemble to power generation. Therefore, a power curve model such as those presented in Section 9.3 is used to convert each member of the ensemble individually. Although potential biases in the raw ensemble are usually removed in the conversion step (if power curves are derived as a relationship between measured power generation and dynamical ensemble forecasts), the power generation ensembles are just as uncalibrated as the input meteorological ensemble, which are often underdispersive. Similar methods, as in the previous section or in Wilks (2018, Chapter 3 of this book), can be used to calibrate these power output ensemble forecasts, and in the following we present potential parametric and nonparametric approaches that consider the specific properties of wind power.

Nonhomogeneous regression of wind power

Although nonhomogeneous regression similar to Eqs. (9.35)–(9.37) has not explicitly been applied to power generation, parametric approaches of its kind are easily applicable. The only difficulty is to find an appropriate response distribution that should take into account the bounded range of wind power, the high number of zero power generation, and the heteroscedasticity of wind power (i.e., low variance for low and high values of power and high variance for intermediate values). In the context of time series models, Pinson (2012) proposed using a generalized log-normal distribution for normalized power generation

$$\hat{f}_{t+k|t}(y) = \frac{1}{\sigma\sqrt{2\pi}} \frac{\nu}{y(1-y^\nu)} \exp\left[-\frac{1}{2}\left\{\frac{\gamma(y,\nu)-\mu_{t+k|t}}{\sigma_{t+k|t}}\right\}^2\right], \quad y \in (0,1) \tag{9.40}$$

with distribution parameters $\mu_{t+k|t}$, $\sigma_{t+k|t}$, and $\nu > 0$ and

$$\gamma(y,\nu) = \log\left(\frac{y^\nu}{1-y^\nu}\right) \tag{9.41}$$

Fig. 9.8 shows this generalized logit-normal probability density function with $\nu = 1$ and $\sigma_{t+k|t} = 0.65$ for different values of $\mu_{t+k|t}$. This generalized logit-normal distribution is bounded between 0 and 1, and although the scale parameter $\sigma_{t+k|t}$ is equal, a higher variance can be seen for intermediate power values than for low and high values, which reflects well the heteroscedasicity of wind power data.

Unfortunately, the generalized logit-normal distribution is only defined on (0, 1) although there are usually a number of observations that are exactly 0 or 1. To account for these power values, Pinson (2012) proposed using a censored version of the generalized logit-normal distribution

$$y_t = \begin{cases} \xi & \text{if } y_t \leq \xi \\ y_t & \text{if } \xi < y_t < 1-\xi \\ 1-\xi & \text{if } y_t \geq 1-\xi \end{cases} \tag{9.42}$$

where ξ is the order of the measurement precision, say $\xi \leq 10^{-2}$.

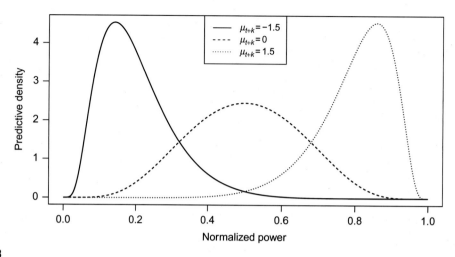

FIG. 9.8

Logit-normal distribution with $\nu = 1$ and $\sigma_{t+k} = 0.65$ for different values of μ_{t+k}.

To use this censored generalized logit-normal distribution to postprocess wind power ensembles, the distribution parameters can be related to the ensemble mean $\hat{y}_{t+k|t}$ and ensemble standard deviation $r_{t+k|t}$ similarly to Eqs. (9.35), (9.36)

$$\mu_{t+k|t} = \boldsymbol{\beta}^{\top} \boldsymbol{x}_{t+k|t} \tag{9.43}$$

$$\log(\sigma_{t+k|t}) = \boldsymbol{\gamma}^{\top} \boldsymbol{z}_{t+k|t} \tag{9.44}$$

where $\boldsymbol{x}_{t+k|t} = [1, \hat{y}_{t+k|t}]^{\top}$ and $\boldsymbol{z}_{t+k|t} = [1, r_{t+k|t}]^{\top}$ with $\hat{y}_{t+k|t}$ and $r_{t+k|t}$ being the power generation ensemble mean and standard deviation, respectively. The coefficients $\boldsymbol{\beta} = [\beta_0, \beta_1]^{\top}, \boldsymbol{\gamma} = [\gamma_0, \gamma_1]^{\top}$, and ν can be estimated by maximum likelihood estimation with

$$\left[\hat{\boldsymbol{\beta}}^{\top}, \hat{\boldsymbol{\gamma}}^{\top}, \nu\right]^{\top} = \operatorname*{argmax}_{[\boldsymbol{\beta}^{\top}, \boldsymbol{\gamma}^{\top}, \nu]^{\top}} \sum_{t=1}^{T} \left\{ I(\xi < y_t < 1 - \xi) \log\left[f_{t|t-k}(y_t,)\right] \right.$$
$$\left. + I(y_t \leq \xi) \log\left[F_{t|t-k}(\xi)\right] + I(y_t \geq 1 - \xi) \log\left[1 - F_{t|t-k}(1 - \xi)\right] \right\} \tag{9.45}$$

where $F_{t+k|t}(\xi)$ is the cumulative distribution function of the log-normal distribution with distribution parameters $\mu_{t+k|k}$, $\sigma_{t+k|k}$, and ν.

Adaptive kernel dressing

Pinson and Madsen (2009) proposed deriving predictive distributions from wind power forecast ensembles with a kernel smoothing approach, which is similar to the approaches described in Section 3.3 in Wilks (2018, Chapter 3 of this book). The predictive distribution $f_{t+k|t}(y)$ is given by

$$f_{t+k|t}(y) = \sum_{j=1}^{J} w_j \phi\left(\frac{y - \hat{y}_{t+k|t}^{(j)}}{\sigma_{t,k}^{(j)}}\right) \tag{9.46}$$

where w_j are weights for the ensemble members, $\phi()$ denotes the standard normal distribution, which is used as kernel function, and $\sigma_{t,k}^{(j)}$ is the kernel bandwidth, which can be different for each ensemble member j and time t. An important feature of wind power is that its forecast uncertainty is lower for low and high values while it is higher in the middle range of the power curve. To account for this feature, the bandwidth is modeled as a logistic function of the (normalized) ensemble forecast $\hat{y}^{(j)}_{t+k|t}$

$$\sigma^{(j)}_{t,k} = a_{t,k} + b_{t,k}\left(1 - \hat{y}^{(j)}_{t+k|t}\right)\hat{y}^{(j)}_{t+k|t} \tag{9.47}$$

Fig. 9.9 (bottom) shows an example predictive distribution with wide kernels for intermediate values and sharper kernels for lower and higher values of the ensemble forecasts. Additionally, the top row in Fig. 9.9 illustrates the kernel dressing approach on an example 48-hour 50-member ensemble forecast.

To fully specify this model, the parameters $a_{t,k}$, $b_{t,k}$, and w_j have to be estimated. For simplicity, and because input meteorological ensemble forecasts may not distinguishable anyway, it will often be appropriate that all members should be given equal weight. Consequently, w_j is set to $w_j = 1/J$. As the

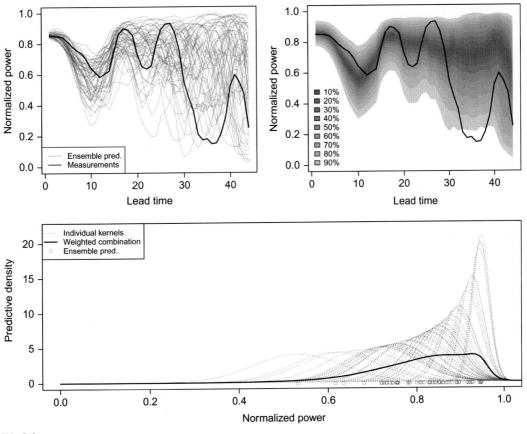

FIG. 9.9

Kernel dressing example. *Top left*: Example 48-hour 50-member ensemble predictions (*gray lines*) and corresponding measurements (*black line*). *Top right*: Predictive distribution obtained from kernel dressing. *Bottom*: Gaussian kernels and weighted combination for lead time $k = 18$.

indices already indicate, the parameters $a_{t,k}$ and $b_{t,k}$ are assumed to also vary (slowly) in time, for example, due to seasonal variations. Similar to Section 9.3.3, exponential weighting can be used to give more weight to recent data, though in a maximum likelihood estimation framework. This reads

$$[\hat{a}_{t,k}, \hat{b}_{t,k}]^\top = \underset{[a_{t,k},b_{t,k}]^\top}{\mathrm{argmin}} -\frac{1}{n_\lambda}\sum_{t=1}^{T}\lambda^{T-t}\log\left[f_{t+k|t}(y_t; a_{t,k}, b_{t,k})\right] \tag{9.48}$$

where $\lambda < 1$ is the forgetting factor, which is typically close to 1 and n_λ the equivalent window size, that is, $n_\lambda = (1-\lambda)^{-1}$.

The recursive estimation can then be performed in a manner similar to Section 9.3.3. The main steps for obtaining the recursive update formulas are given here, while complete details on the implementation are available in Pinson and Madsen (2009). Before that, it is beneficial to proceed with a change of variable, because the parameters $a_{t,k}$ and $b_{t,k}$ of the mean-variance model need to be strictly positive while it may be beneficial to keep them below a maximum value, denoted $\bar{a}_{t,k}$ and $\bar{b}_{t,k}$, respectively. A logistic transformation produces variables that may be optimized freely instead of the original variables, which would need to be restricted to a certain range. This yields

$$a_{t,k}^* = \log\left(\frac{a_{t,k}}{\bar{a}_{t,k} - a_{t,k}}\right), \quad b_{t,k}^* = \log\left(\frac{b_{t,k}}{\bar{b}_{t,k} - b_{t,k}}\right) \tag{9.49}$$

from which original parameters may be retrieved by application of the inverse transform.

Eventually, by defining the information vector

$$\mathbf{h}_{t,k} = \frac{\nabla f_{t+k|t}(y_t; a_{t,k}, b_{t,k})}{f_{t+k|t}(y_t; a_{t,k}, b_{t,k})} \tag{9.50}$$

and $\mathbf{R}_{t,k}$ the estimate of its covariance matrix, one obtains the final two-step scheme for updating estimates at time t,

$$[\hat{a}_{t,k}^*, \hat{b}_{t,k}^*]^\top = [\hat{a}_{t-1,k}^*, \hat{b}_{t-1,k}^*]^\top + \frac{1}{n_\lambda}\mathbf{R}_{t,k}^{-1}\mathbf{h}_{t,k} \tag{9.51}$$

$$\mathbf{R}_{t,k} = \lambda\mathbf{R}_{t-1,k} + \frac{1}{n_\lambda}\mathbf{h}_{t,k}\mathbf{h}_{t,k}^\top \tag{9.52}$$

Initialization of this recursive learning scheme is performed as for the recursive least square type approaches described previously.

9.4.3 DIRECT CALIBRATION OF WIND POWER

Instead of splitting the forecasting model into a calibration and a transformation step, it is also possible to combine both steps and use one model with meteorological predictions as input and calibrated predictive distributions of power generation as output. Thus, these models have to calibrate the ensemble and take into account the nonlinear relationship between wind and power generation.

To our knowledge, there is no literature on direct calibration methods that are explicitly applied to ensemble predictions. However, especially quantile regression based on deterministic predictions has been very popular to generate probabilistic wind power forecasts and can easily be extended to ensemble predictions.

Quantile regression (Koenker & Bassett, 1978) was introduced for wind power forecasting by Bremnes (2004a) and has been a popular model to generate probabilistic wind power forecasts. It can be seen as a nonparametric probabilistic extension of the power curve models of Section 9.3 where instead of the expected values of power generation, predictive quantiles are derived. As such, the formulation of the model of Bremnes (2004a) is similar to Eqs. (9.4)–(9.9) and a predictive quantile $y_{t+k}^{(q_j)}$ with exceedance probability q_j is modeled as a local polynomial function

$$y_{t+k}^{(q_j)} = \boldsymbol{\alpha}_{i,j}^{\top} \mathbf{x}_{t+k}, \quad t=1,\dots,T \tag{9.53}$$

with $\boldsymbol{\alpha}_{i,j} = [\alpha_{i,j,0} \quad \alpha_{i,j,1} \quad \dots \quad \alpha_{i,j,L}]^{\top}$ the vector of model coefficients and $\mathbf{x}_{t+k} = \left[1 \quad (\hat{u}_{t+k|t} - u_i) \quad \dots \quad (\hat{u}_{t+k|t} - u_i)^{L}\right]^{\top}$ the vector of explanatory variables to form polynomials of degree L.

Similar to the local polynomial regression model of Section 9.3.2, the model coefficients for each required quantile $y_{t+k}^{(q_j)}$ are estimated with

$$\hat{\boldsymbol{\alpha}}_{i,j} = \underset{\alpha_{i,j}}{\operatorname{argmin}} \sum_{t=1}^{T} w_{t,i} \rho_{q_j} \left(y_{t+k} - \boldsymbol{\alpha}_{i,j}^{\top} \mathbf{x}_{t+k} \right) \tag{9.54}$$

but instead of a quadratic loss the loss function

$$\rho_{q_j}(\epsilon) = \begin{cases} \epsilon q_j & \text{if } \epsilon \geq 0 \\ \epsilon(1-q_j) & \text{if } \epsilon < 0 \end{cases} \tag{9.55}$$

is used for each required quantile probability (q_j). To solve Eq. (9.54) it is usually reformulated as linear programming problem. Details on the estimation procedure can be found in Koenker and d'Orey (1994) and different implementations are provided in the R package `quantreg` (Koenker, 2017). Fig. 9.10 shows the median and the quartiles for the toy model data set of Section 9.3.

Using similar considerations as the model of Section 9.3.3, Møller, Nielsen, and Madsen (2008) proposed a recursive and time adaptive estimation method to allow the model to adapt to slow changes in the relationship between wind predictions and power generation. As an alternative to local polynomial regression, Nielsen, Madsen, and Nielsen (2006) proposed using additive quantile regression models with spline basis functions to model the nonlinear relationship between the input wind predictions and the output power generation quantiles.

Although most wind power forecasting studies have used only deterministic predictions as input, the quantile regression model can easily be extended to ensemble predictions. A straightforward approach is to replace $\hat{u}_{t+k|t}$ in $\mathbf{x}_{t+k}$ with the respective empirical quantile of the ensemble forecast $\hat{u}_{t+k|t}^{(q_j)}$

$$\mathbf{x}_{t+k} = \left[1 \quad \left(\hat{u}_{t+k|t}^{(q_j)} - u_i\right) \dots \left(\hat{u}_{t+k|t}^{(q_j)} - u_i\right)^{L}\right]^{\top} \tag{9.56}$$

Alternatively, multiple ensemble quantiles (e.g., deciles) can be combined in $\mathbf{x}_{t+k}$, similar to Bremnes (2004b) or Bentzien and Friederichs (2012). However, because these quantiles are usually highly correlated, overfitting can become a problem so that regularization methods such as lasso (Ben Bouallègue, 2017; Tibshirani, 1996) should be considered.

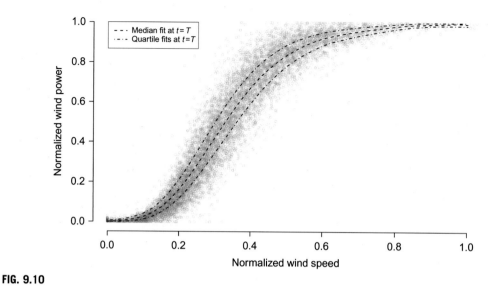

FIG. 9.10

Fitting of local polynomial quantile regression to empirical data for the wind power conversion process.

9.5 CONCLUSIONS AND PERSPECTIVES

Our aim has been to give an overview of the postprocessing of meteorological ensemble forecasts for renewable energy applications, as driven by the nonlinear, bounded, and nonstationary characteristics of the conversion from meteorological variables to power production. Even though concentrating on the wind power application, the tasks and methods to employ for other types of renewable energy generation (solar and wave energy, mainly) would be qualitatively similar. Especially, we emphasized that the models for the conversion from meteorological variables to power should not be obtained in an ordinary least squares estimation framework, because that method overlooks inherent errors in the predictor meteorological forecasts that then affect the model fitting process. In the present case, we have described a practical approach in a total least squares framework, allowing us to obtain those conversion models as principal curves.

In parallel, depending upon the type of forecast products required by users, alternative approaches to forecast calibration may be seen as most relevant. Kernel dressing is widely used, even though full parametric approaches or quantile regression may exhibit a number of advantages. Member-by-member calibration is also a good alternative when forecast users require alternative trajectories for future power generation instead of (related) predictive densities for each and every lead time. Indeed, one of the drawbacks of the calibration methods that yield predictive densities is that they may not represent the temporal dependence structure described by the original ensemble forecasts. Ensemble copula coupling (Schefzik, Thorarinsdottir, & Gneiting, 2013), the Schaake shuffle (Clark, Gangopadhyay, Hay, Rajagopalan, & Wilby, 2004), and other approaches described in Schefzik and Möller (2018) are possible approaches to retaining this dependence structure. Future work should indeed focus on better using the complete information in these meteorological ensemble forecasts, not only in terms of temporal dependence, but potentially also by using a broader range of meteorological variables as input and their interdependencies, as well as their spatial dependencies. Furthermore, extreme events can lead to very high costs or even failures of generation units or power systems and focusing on these, similar to Friederichs, Wahl, and Buschow (2018) can also be valuable.

9.6 APPENDIX: SIMULATED DATA FOR THE CONVERSION OF WIND TO POWER

Our toy model example takes the form of a semiartificial dataset, in the sense that it relies on actual wind speed measurements from a meteorological mast at a wind farm, though we then add noise and convert wind to power through a simulated power curve. Our wind measurements are assumed to be noise-free. These real wind data come from the site of a 21 MW wind farm in North Jutland known as Klim, located in the vicinity of one of the most famous windsurfing and kitesurfing sites in Northern Europe, Klitmøller.

At any time, there is a nonlinear and bounded power curve function that converts wind to power. For a given time t that power curve is

$$y_t = g_t(u_t), \quad \forall t \tag{9.57}$$

for which we consider the effect of wind speed only. Differences in air density, which also slightly influence the wind-to-power conversion are disregarded. It is also nonstationary, hence justifying the use of a time index t for the power curve model g. This nonstationary behavior translates to the assumption that the power curve is slowly varying in time due to seasonal effects and variations in the wind farm environment. The power curve for a wind turbine or a wind farm is modeled with a double exponential function as

$$g_t(u_t) = \exp\left(-\tau_{t,1} \exp\left(-\tau_{t,2} u_t\right)\right), \quad \forall t \tag{9.58}$$

making its shape fully controlled by two parameters $\tau_t^\top = [\tau_{t,1}, \ \tau_{t,2}]$ at any time t. Let us assume that over a time period of $T = 10,000$ time steps, those evolve from $\tau_1^\top = [8, 7]$ to $\tau_T^\top = [11, 9]$. The resulting power curves at both the beginning and the end of this period are represented in Fig. 9.1. As can be readily observed, this conversion is nonlinear and bounded, in addition to being nonstationary. This motivates the fact that, ideally, all those aspects should be jointly considered both in the models for the conversion of wind to power, and for the postprocessing of ensemble forecasts.

In practice, the correct conversion model from wind to power is not known and must be estimated from available data. These data often consist of wind forecasts and power observations. It is hence the relationship between wind forecasts, for each lead time k and observed wind power generation that is modeled, not the actual relationship between wind speed measurements and power generated. Consequently, instead of working directly with the model in Eq. (9.57), the power curve model to be estimated reads instead

$$y_{t+k} = g_{t,k}(\hat{u}_{t+k|t}), \quad \forall t,k \tag{9.59}$$

To first represent the fact that the power curve model is for the conversion of wind forecasts to power, and not wind measurements, we add noise to the wind speed data for the Klim wind farm to represent the discrepancy between forecasts and observations,

$$\hat{u}_{t+k|t} = u_{t+k}(1 + \eta \epsilon_{t+k}), \quad t+k \in [1,T] \tag{9.60}$$

where ϵ_{t+k} is a standard Gaussian noise, $\epsilon_{t+k} \sim \mathcal{N}(0,1)$. In the preceding equation, the noise is linearly scaled with the wind speed level (with factor η), representing the fact it may be easier to predict lower wind speeds than higher ones. This may be simplistic, though good enough for our purpose here. If aiming to consider more complex relationships, one could think of time-varying noise, several noise components, etc.

In parallel, to obtain the values for power generated at the wind farm, we add two different types of noise components to the direct conversion of wind to power $y_{t+k} = g_{t,k}(u_{t+k})$. We use here the time index $t + k$ to be in line with the lead time k of wind forecasts in the preceding. The noisy observed power $\tilde{y}_{t+k}$ is given by

$$\tilde{y}_{t+k} = g_{t,k}(u_{t+k}) + \zeta_{t+k} + \xi_{t+k}\mathcal{I}_{t+k}, \quad t+k \in [1,T] \tag{9.61}$$

The first noise component ζ_{t+k} is an additive Gaussian noise with 0 mean, and whose standard deviation σ_{t+k}^{ϵ} is a function of the power level, that is,

$$\zeta_{t+k} \sim \mathcal{N}(0, \sigma_{t+k}^{\zeta}{}^2), \quad \sigma_{t+k}^{\zeta} = \nu_0^{\zeta} + 4y_{t+k}(1 - y_{t+k})\nu_1^{\zeta}, \quad t+k \in [1,T] \tag{9.62}$$

which is for a permanent noise in the measurement process, which we assume is directly influenced by the local slope of the power curve, hence justifying the inverse U-shaped function chosen. The second component ξ_{t+k} is for an impulsive noise, based on a similar type of Gaussian noise,

$$\xi_{t+k} \sim \mathcal{N}(0, \sigma_{t+k}^{\xi}{}^2), \quad \sigma_{t+k}^{\xi} = \nu_0^{\xi} + 4y_{t+k}(1 - y_{t+k})\nu_1^{\xi}, \quad t+k \in [1,T] \tag{9.63}$$

except that it is added only at random times, depending on the binary sequence $\{\mathcal{I}_{t+k}\}$. This sequence is obtained from the successive realizations of a Bernouilli process with chance of success π_ξ. This variable can be reinterpreted as the proportion of data corrupted by this impulsive noise. It simulates the presence of outliers in the power observations. Simulated power data that are not within the range $[0, 1]$ are rounded to the closest bound.

An example of a resulting power curve, for a given set of noise parameters, is shown in Fig. 9.2. There for the noise on wind speed, we have $\eta = 0.07$, while for the noise applied to the power variable, the parameters are $\nu_0^{\zeta} = 0.01$, $\nu_1^{\zeta} = 0.09$, $\nu_0^{\xi} = 0.07$, $\nu_1^{\xi} = 0.15$, while $\pi_\xi = 0.02$.

REFERENCES

Andrade, J. R., & Bessa, R. J. (2017). Improving renewable energy forecasting with a grid of numerical weather predictions. *IEEE Transactions on Sustainable Energy, 8*, 1571–1580.

Bacher, P., Madsen, H., & Nielsen, H. A. (2009). Online short-term solar power forecasting. *Solar Energy, 83*, 1772–1783.

Baran, S. (2014). Probabilistic wind speed forecasting using Bayesian model averaging with truncated normal components. *Computational Statistics and Data Analysis, 75*, 227–238.

Ben Bouallègue, Z. (2017). Statistical postprocessing of ensemble global radiation forecasts with penalized quantile regression. *Meteorologische Zeitschrift*, *26*, 253–264.

Bentzien, S., & Friederichs, P. (2012). Generating and calibrating probabilistic quantitative precipitation forecasts from the high-resolution NWP model COSMO-DE. *Weather and Forecasting*, *27*, 988–1002.

Bessa, R., Möhrlen, C., Fundel, V., Siefert, M., Browell, J., & El Gaidi, S. H. (2017). Towards improved understanding of the applicability of uncertainty forecasts in the electric power industry. *Energies*, *10*, Article No. 1402.

Bremnes, J. B. (2004a). Probabilistic wind power forecasts using local quantile regression. *Wind Energy*, *7*, 47–54.

Bremnes, J. B. (2004b). Probabilistic forecasts of precipitation in terms of quantiles using NWP model output. *Monthly Weather Review*, *132*, 338–347.

Clark, M., Gangopadhyay, S., Hay, L., Rajagopalan, B., & Wilby, R. (2004). The Schaake shuffle: a method for reconstructing space-time variability in forecasted precipitation and temperature fields. *Journal of Hydrometeorology*, *5*, 243–262.

Cleveland, W., & Develin, S. (1988). Locally weighted regression: an approach to regression analysis by local fitting. *Journal of the American Statistical Association*, *83*, 596–610.

Cutler, N., Outhred, H., MacGill, I., Kay, M., & Kepert, J. (2009). Characterizing future large, rapid changes in aggregated wind power using numerical weather prediction spatial fields. *Wind Energy*, *12*, 542–555.

de Groen, P. (1996). An introduction to total least squares. *Nieuw Archief voor Wiskunde Vierde Series*, *14*, 237–253.

Dowell, J., & Pinson, P. (2016). Very-short-term probabilistic wind power forecasts by sparse vector autoregression. *IEEE Transactions on Smart Grid*, *7*, 480–489.

Eckel, F. A., Allen, M. S., & Sittel, M. S. (2012). Estimation of ambiguity in ensemble forecasts. *Weather and Forecasting*, *27*, 50–69.

Friederichs, P., Wahl, S., & Buschow, S. (2018). Postprocessing for extreme events. In S. Vannitsem, D. S. Wilks, & J. W. Messner (Eds.), *Statistical Postprocessing of Ensemble Forecasts*. Elsevier Chapt. 4.

Golub, G. H., & Van Loan, C. F. (1980). An analysis of the total least squares problem. *SIAM Journal of Numerical Analysis*, *17*, 883–893.

Golub, G. H., & Van Loan, C. F. (1996). *Matrix Computations*. Baltimore: John Hopkins University Press.

Härdle, W. (1990). *Applied Nonparametric Regression*. New York: Cambridge University Press.

Hastie, T., & Stuetzle, W. (1989). Principal curves. *Journal of the American Statistical Association*, *84*, 502–516.

Hastie, T., & Tibshirani, R. (1993). Varying-coefficient models. *Journal of the Royal Statistical Society B*, *55*, 757–796.

Hyndman, R. (1996). Computing and graphing highest density regions. *The American Statistician*, *50*, 120–126.

International Renewable Energy Agency. (2017). *Rethinking energy 2017*. http://www.irena.org/DocumentDownloads/Publications/IRENA_REthinking_Energy_2017.pdf (Accessed 10 October 2017).

Kariniotakis, G. (2017). *Renewable Energy Forecasting: From Models to Applications*. Duxford, United Kingdom: Elsevier.

Koenker, R. W. (2017). *quantreg: Quantile regression. R package version 5.33*. https://CRAN.R-project.org/package=quantreg (Accessed 10 October 2017).

Koenker, R., & Bassett, B. (1978). Regression quantiles. *Econometrica*, *46*, 33–49.

Koenker, R., & d'Orey. (1994). Computing regression quantiles. *Applied Statistics*, *43*, 410–414.

Messner, J. W., Zeileis, A., Bröcker, J., & Mayr, G. J. (2013). Probabilistic wind power forecasts with an inverse power curve transformation and censored regression. *Wind Energy*, *17*, 1753–1766.

Møller, J., Nielsen, H., & Madsen, H. (2008). Time-adaptive quantile regression. *Computational Statistics and Data Analysis*, *52*, 1292–1303.

Morales, J., Conejo, A., Madsen, H., Pinson, P., & Zugno, M. (2014). *Integrating Renewables in Electricity Markets: Operational Problems*. New York: Springer Verlag.

Nelder, J., & Mead, R. (1965). A simplex method for function minimization. *Computer Journal*, *7*, 308–313.

Nielsen, H. A., Madsen, H., & Nielsen, T. S. (2006). Using quantile regression to extend an existing wind power forecasting system with probabilistic forecasts. *Wind Energy*, *9*, 95–108.

Nielsen, H. A., Nielsen, T. S., Joensen, A., Madsen, H., & Holst, J. (2000). Tracking time-varying-coefficient functions. *International Journal of Adaptive Control and Signal Processing*, *14*, 813–828.

Pinson, P. (2012). Very-short-term probabilistic forecasting of wind power with generalized logit-normal distributions. *Journal of the Royal Statistical Society C*, *61*, 555–576.

Pinson, P., & Madsen, H. (2009). Ensemble-based probabilistic forecasting at Horns Rev. *Wind Energy*, *12*, 137–155.

Pinson, P., Nielsen, H. A., Madsen, H., & Nielsen, T. S. (2008). Local linear regression with adaptive orthogonal fitting for the wind power application. *Statistics and Computing*, *18*, 59–71.

Renewable Energy Policy Network for the 21st Century. *Renewables 2017 – Global status report.* http://www.ren21.net/wp-content/uploads/2017/06/GSR2017_Full-Report.pdf (Accessed 10 October 2017).

Schefzik, R., & Möller, A. (2018). Multivariate ensemble postprocessing. In S. Vannitsem, D. Wilks, & J. Messner (Eds.), *Statistical Postprocessing of Ensemble Forecasts*. Elsevier. Chapt. 4.

Schefzik, R., Thorarinsdottir, T., & Gneiting, T. (2013). Uncertainty quantification in complex simulation models using ensemble copula coupling. *Statistical Science*, *28*, 616–640.

Tastu, J., Pinson, P., Trombe, P. J., & Madsen, H. (2011). Probabilistic forecasts of wind power generation accounting for geographically dispersed information. *IEEE Transactions on Smart Grid*, *5*, 480–489.

Taylor, J., McSharry, P., & Buizza, R. (2009). Wind power density forecasting using ensemble predictions and time series models. *IEEE Transactions on Energy Conversion*, *24*, 775–782.

Thorarinsdottir, T., & Gneiting, T. (2010). Probabilistic forecasts of wind speed: ensemble model output statistics by using heteroscedastic censored regression. *Journal of the Royal Statistical Society A*, *173*, 371–388.

Tibshirani, R. (1992). Principal curves revisited. *Statistics and Computing*, *2*, 183–190.

Tibshirani, R. (1996). Regression shrinkage and selection via the lasso. *Journal of the Royal Statistical Society B*, *58*, 267–288.

Troccoli, A., Dubus, L., & Haupt, S. (2014). *Weather Matters for Energy*. Springer Verlag: New York.

Wilks, D. S. (2018). Univariate ensemble postprocessing. In S. Vannitsem, D. S. Wilks, & J. W. Messner (Eds.), *Statistical Postprocessing of Ensemble Forecasts*. Elsevier. Chapt. 3.

Zhang, Y., Wang, J., & Wang, X. (2014). Review on probabilistic forecasting of wind power generation. *Renewable and Sustainable Energy Reviews*, *32*, 255–270.

POSTPROCESSING OF LONG-RANGE FORECASTS

10

Bert Van Schaeybroeck, Stéphane Vannitsem

Royal Meteorological Institute of Belgium, Brussels, Belgium

CHAPTER OUTLINE

10.1 INTRODUCTION

Forecasts beyond the medium range of 2 weeks have attracted the attention of the climate and weather community for many decades (Royer, 1993; Shukla, 1981), triggered by a large demand for long-range forecasts from many societal sectors dealing with energy, health prevention, agriculture, or flooding and drought management (Ogallo, Bessemoulin, Ceron, Mason, & Connor, 2008). Numerous weather and climate centers have implemented operational long-range ensemble forecasting systems, and according to World Meteorological Organization (WMO) standards, long-range forecasts extend up

Statistical Postprocessing of Ensemble Forecasts. https://doi.org/10.1016/B978-0-12-812372-0.00010-8

to 2 years. In this chapter, the lead times under consideration cover monthly, seasonal, inter-annual, and decadal time scales, and all are termed "long-range".

Beyond the medium range, the unpredictable component, or weather noise, that arises from the growth of the initial uncertainties, becomes large (Royer, 1993). Therefore, predictions must be probabilistic in nature, which is made possible through the use of ensemble forecasts. There are, however, different sources of forecast skill. The added value due to good initialization of the models (as close as possible to observations) has been demonstrated for monthly and seasonal forecasts (Doblas-Reyes et al., 2013; Doblas-Reyes, Garçia-Serrano, Lienert, Biescas, & Rodrigues, 2013). The benefit of proper initialization is much less clear for decadal ranges, as initialized forecasts display important similarities with climate projections for which the predictive signal stems from the external forcing (Branstator & Teng, 2012; Meehl et al., 2014) when the internally generated variability is properly sampled (Deser, Phillips, Alexander, & Smoliak, 2014).

The source of long-range predictability within the atmosphere is usually associated with the existence of different modes of low-frequency variability (LFV). It is expected that, if models are capable of reproducing these phenomena, they may also be able to forecast them. Well-known observed low-frequency modes of internal variability include the El Niño Southern Oscillation (ENSO), monsoon rains, sudden stratospheric warmings, the Madden Julian Oscillation (MJO), the Indian Ocean dipole, the North Atlantic Oscillation (NAO), and the Pacific/North American (PNA) pattern, spanning a wide range of timescales from months to decades (Hoskins, 2013). The physical origin of LFV in the atmosphere is the subject of intensive research and usually stems from the interaction with other components of the climate system, in particular with the ocean. ENSO is the most important example of interaction between the ocean and the atmosphere in the tropical Pacific that leads to an extended skill of the tropical atmosphere up to 1 year (Hoskins, 2013). In the extratropics, the range of (potential) predictability, as revealed by the analysis of different climate models, is much lower and is usually believed to be strongly associated with teleconnections with modes of tropical variability like ENSO.

Even though considerable efforts are devoted to climate model developments, important biases remain when compared with observations (Randall et al., 2007). These biases are mainly associated with the presence of important *model errors* due to the crude representation of the climate dynamics in dynamical earth system models (ESM). The impact of model errors on the forecasts can be partly removed using postprocessing techniques (Vannitsem & Nicolis, 2008), which constitute, therefore, a crucial step in the forecasting suite of climate models.

This chapter will focus on the challenges of postprocessing of climate model output for long forecast ranges and the specific techniques that have been designed in this context to correct the model fields. We will not, however, address the specific question of statistical downscaling of climate models, in view of the thorough reviews provided recently in Benestad, Hanssen-Bauer, and Chen (2008), Maraun et al. (2010), and Maraun and Widmann (2018); although this approach is strongly related to the problem of statistical postprocessing.

Section 10.2 will be devoted to a discussion on a series of challenges when postprocessing approaches are considered for correcting long-range forecasts. The classical statistical framework in which postprocessing is usually discussed is then introduced in Section 10.3. The multimodel combination, also known as consolidation, is then presented in Section 10.4, and its direct extension to probabilistic forecasts is made in Section 10.5. Section 10.6 discusses the drift and trend correction approaches, and Section 10.7 addresses approaches of ensemble inflation, often used for long-range forecasts; and an alternative approach recently discussed for short-range forecasts in Van Schaeybroeck and Vannitsem (2015). Both approaches are then compared in Section 10.8 in the context of an idealized low-order model, Section 10.9 presents an application in a real long-range setting, and Section 10.10 concludes.

10.2 CHALLENGES OF LONG-RANGE FORECASTS

As amply discussed in this book, postprocessing techniques for ensemble forecasts rely on a sufficient set of past forecasts and observations of good quality in order to make statistical inferences. Postprocessing of long-range forecasts is faced with a set of difficulties related to that constraint. Specifically:

Small sample sizes: Reliable observational datasets only start around 1980, with the satellite era. Verification of seasonal forecasts is commonly performed for each season separately, and therefore verification datasets rarely exceed 30 samples, leading to highly uncertain verification scores (Kumar, 2009). Forecast calibration goes one step further by separating the random errors from systematic errors in an effort to correct the latter. The small sample sizes may give rise to overfitting issues, unless the applied postprocessing methods are very simple (Kharin & Zwiers, 2003; Mason, 2012).

Shortage of model integrations: In principle, a large set of independent long-range reforecasts or hindcasts with large ensembles is necessary to (i) properly sample the natural internal variability, (ii) reduce uncertainties on forecast verification scores, and (iii) determine systematic forecast errors based on which calibration can be performed. Note that these systematic errors generally depend on initialization season and lead time. However, such large training datasets are mostly lacking due to the following reasons. First, as mentioned in the previous point, only few observational realizations exist that allow initialization of a dynamical model using the same setup and quality as real-time forecasts. Second, long-range model integrations are computationally very expensive due to the use of *coupled* ESMs, that include the dynamics of the atmosphere, the ocean, sea ice, snow cover, and land surface (soil moisture), each featuring different characteristic times scales. Third, initialization schemes should be implemented for these different model components as it is known that predictability may benefit (Carrassi, Guemas, Doblas-Reyes, Volpi, & Asif, 2016; Doblas-Reyes et al., 2013; Doblas-Reyes et al., 2013; Prodhomme et al., 2016; Prodhomme, Doblas-Reyes, Bellprat, & Dutra, 2016), thereby requiring another level of computer power.

Large natural internal variability: Long-range predictions are either based on single-model or multimodel ensembles, with each ensemble member started from a distinct initial condition. The purpose is to sample the natural internal variability and model uncertainty. In this context, it is common to summarize the information content into simple forecast outputs, such as the ensemble mean or weighted multimodel ensemble mean. Categorical probabilistic forecasts are also often issued by the ensemble exceedance frequency of climatological median or tercile values (Kharin & Zwiers, 2003; Ogallo et al., 2008). These methods do not make full use of the ensemble, and it is unclear whether the ensemble spread is a good measure of uncertainty. Full-ensemble postprocessing techniques have only recently been introduced in the context of long-range forecasts (Eade et al., 2014; Johnson & Bowler, 2009; Krikken, Schmeits, Vlot, Guemas, & Hazeleger, 2016), and so far are only applied on single-model ensembles.

The need for filtering: Instantaneous and spatially local values of atmospheric fields rapidly lose their predictability due to rapidly evolving, small-scale processes (including, e.g., deep convective developments). In order to reduce their impact, spatial and/or temporal filtering is needed in addition to the ensemble averaging (Buizza & Leutbecher, 2015; Nicolis, 2016; Smith, Du, Suckling, & Niehorster, 2015; Weisheimer & Palmer, 2014). Filtering can be done through simple averaging or through specific statistical techniques such as canonical correlation analysis or empirical orthogonal functions (Benestad et al., 2008; Van den Dool, 2007). A common example of a filtered quantity is the Niño 3.4 index, which is the sea surface temperature averaged over an area in the tropical Pacific, and averaged over a monthly timescale.

Model drift and forcing sensitivity: Forecasts are often initialized as closely as possible to the observations. ESMs then drift to their own attractor at timescales typically associated with the slow component of these models (Kharin, Boer, Merryfield, Scinocca, & Lee, 2012). The drift can considerably affect the quality of the forecasts. Assuming stationarity of forecast errors for each lead time is an acceptable assumption that may hold in the range between short-range through yearly predictions. However, at decadal scales, the sensitivity of model and observations with respect to anthropogenic and natural forcing agents may be different. Such mismatch in trends complicates the drift correction (see Section 10.6).

Model forcings: The skill of weather forecasts comes mainly from initialization. The (potential) skill in long-range forecasting, on the other hand, stems predominantly from boundary forcing from other components of the climate system, the most important ones being the ocean and the cryosphere forcings, changes in greenhouse gas and aerosol concentrations, solar variability, and volcanic eruptions. Uncertainties in these forcings also considerably affect the forecasts.

The use of anomalies: Due to large model biases, anomalies are expected to convey the important information. Care must be taken when choosing the reference (period) against which anomalies are defined as results and reproducibility may critically depend on it. In general, short reference periods lead to artificial overestimation of skill (Murphy, 1990). Nonzero trends may also be present in reference periods and may lead to spurious correlations (Mason, 2012). The choice on whether and how to remove the trends in the variables adds an additional complication. Lastly, the use of different reference periods also complicates model inter-comparison.

Lack of spread-error relationship: In seasonal forecasting, ensembles are used to provide better estimates of the forecast mean, while typically the ensemble spread does not provide additional useful uncertainty information beyond the climatological variability (Kumar, Barnston, Peng, Hoerling, & Goddard, 2000; Tippett, Kleeman, & Tang, 2004).

Observational uncertainty: There are large uncertainties associated with global-scale observational datasets. More specifically, although global-scale observational or reanalysis datasets covering the whole 20th century exist, their uncertainties depend on the decade considered due to the variability in observational availability. Furthermore, the quality of observational datasets depends on the location and the variable under consideration (Massonnet, Bellprat, Guemas, & Doblas-Reyes, 2016). For instance, local precipitation estimates from global datasets differ strongly among one another, especially for earlier historical periods or for regions where ground-based measurements are sparse. The incorporation of the observational uncertainty and its spatio-temporal dependence within forecast verification (Bowler, Cullen, & Piccolo, 2015; Ebert et al., 2013) and postprocessing constitutes an important challenge.

10.3 A STATISTICAL FRAMEWORK FOR POSTPROCESSING
10.3.1 STATISTICAL HYPOTHESES

As mentioned before, long-range forecasts aim at forecasting low-frequency signals, assuming a clear separation between signal and noise for the variable under consideration. Such linear "signal-plus-noise" models were introduced, for instance, in Murphy (1990), Kharin and Zwiers (2002), Siegert et al. (2016), and references therein. More specifically, at each time t the variable y_t can be written as:

$$y_t = m_y + s_t + \varepsilon_t \tag{10.1}$$

where m_y is the average of the observable y_t, s_t is the predictable signal, and ε_t is the noise, considered as independent and temporally uncorrelated to the predictable signal, at least at the timescale considered. Eq. (10.1) represents a strong statistical hypothesis regarding the presence of a clear-cut timescale separation between the low-frequency signal and random background weather noise. This hypothesis, although useful, is rather stringent as climate fluctuations display very complicated statistical properties as a function of the time scales considered (Lovejoy, 2015). In particular, it suggests that the system is stable and cannot experience abrupt changes. The latter assumption must be relaxed in case multiple stable steady states or bifurcations are expected, for instance, in the regimes of precipitation in the Sahel region (Demarée & Nicolis, 1990), or for the thermohaline circulation or the Kuroshio current (e.g., Dijkstra & Ghil, 2005). Another important hypothesis is the fact that the noise term is independent of the predictable signal, and is violated, for instance, for the rainfall in the tropics for which the variability is known to vary with the phase of ENSO.

Within the simple framework of Eq. (10.1), averaging multiple realizations of y_t can be very powerful in extracting the signal because the noise terms are averaged out. This averaging can be approximated using averages of model forecasts. Although this scheme is very simple, it has proved to be very effective when multiple integrations from different models are available, as shown in Doblas-Reyes, Hagedorn, and Palmer (2005), and reference therein. The multimodel combination is very competitive in reducing systematic model errors and providing quite reliable seasonal forecasts in the tropical regions. These authors also indicate that linear postprocessing provides good results in extratropical regions.

The aim of postprocessing is to improve the quality of a forecast. In that sense, it is tightly related to forecast verification. The question of what defines a good forecast has been discussed in detail by Mason (2012) for long-range forecasts. A skilled forecast indicates that it is better than another one, however, providing evidence of actual predictability skill of climate forecasts is difficult. For this reason, the concept of *potential predictability* has been introduced, which measures the limit of predictability present in a perfect ESM (Boer, 2000). The approach consists of performing multiple experiments with the same model and estimation of some skill measure. The ratio of the variability of the ensemble mean to the ensemble spread is sometimes considered as an estimate of the signal-to-noise ratio.

However, even when observational uncertainties are reliably sampled initially, ensemble forecasts are generally unreliable in the sense that the observation cannot be considered a random member of the ensemble due to the presence of model error. This implies that the signal-to-noise ratio in different cases is not an upper limit nor representative of the true predictability and, as shown in Kumar, Peng, and Chen (2014), may vary from model to model. For instance, for underdispersive ensembles, the signal-to-noise ratio may lead to the conclusion that model predictability is lower than the actual predictability (Kumar et al., 2014; Siegert et al., 2016).

Eade et al. (2014) introduce the *ratio of predictable components* (RPC) that compares the actual predictable component with that of the forecast. The actual predictable component is unknown and is therefore approximated by the correlation between the observation and the ensemble mean, while the predictable component of the forecast is the correlation of the ensemble members with the ensemble mean. The ratio between these yields:

$$\text{RPC} = \frac{r_{y,\bar{x}}}{s_{\bar{x}}/s_x} \tag{10.2}$$

Here $r_{y,\bar{x}}$ is the Pearson correlation coefficient between the observation and the ensemble mean, while s_x and $s_{\bar{x}}$ are the standard deviations of all pooled forecasts (including all ensemble members) and of the ensemble mean, respectively. For perfect models and reliable initial ensemble distributions, RPC should be equal to one, apart from a correction due to finite ensemble size (Siegert et al., 2016).

Eq. (10.2) has also been used as a cost function to calibrate long-range ensemble forecasts in Eade et al. (2014), but this approach is equivalent to the method of preservation of correlation between the observation and the ensemble mean as introduced in Doblas-Reyes et al. (2005), Weigel et al. (2009) and Johnson and Bowler (2009) that will be discussed in Section 10.6. Apart from the aforementioned approximation involving the actual predictable component, there are a number of well-known issues related the use of Pearson's correlation $r_{y,\bar{x}}$ as a verification metric (Déqué, 2012; Mason, 2012), including a strong sensitivity to a few extreme values. These issues are even more important when using small data sets of forecasts and observations. Therefore care must be taken when optimizing the RPC as a calibration method without consideration of the uncertainty associated with the score itself.

10.3.2 RELIABILITY OF LONG-RANGE FORECASTS

As previously mentioned, long-range predictions are based on ensembles, and therefore, reliability becomes a central feature of the forecast. Different types of reliability can be distinguished that are of importance for long-range forecasts. First of all, *climatological reliability* is satisfied when the variability s_x^2 of all pooled forecasts (including all ensemble members) matches the observational variability s_y^2:

$$s_x^2 = s_y^2 \tag{10.3}$$

Note that in general these climatological quantities can depend on the climate forcing, an effect that can be important for decadal forecasting.

Second, two types of *ensemble reliability* can be defined. One of these is defined as the equality of the squared error of the ensemble mean and the average ensemble variance, (Johnson & Bowler, 2009; Van Schaeybroeck & Vannitsem, 2013, 2015)

$$\left\langle (\bar{x}_t - y_t)^2 \right\rangle_t = \left\langle s_t^2 \right\rangle_t \tag{10.4}$$

Here $\langle . \rangle_t$ denotes the average over all data points (realizations), $\bar{x}_t$ is the ensemble mean, y_t is the corresponding observation, and s_t^2 is the ensemble variance for forecast t. Ensemble spread and error, however, may be strongly regime dependent and one must avoid putting too much emphasis on the largest errors or spreads. Therefore, a more general notion of ensemble reliability or ensemble calibration should be introduced, referred to as *strong ensemble reliability* (SER). This condition is satisfied when the χ^2/N value, which is the standardized mean squared error (MSE) of the ensemble mean, is equal to one. The standardization is done using the ensemble variance of the corresponding forecast s_t^2. The condition for SER is therefore:

$$\chi^2/N = \left\langle \left[\frac{\bar{x}_t - y_t}{s_t}\right]^2 \right\rangle_t = 1 \tag{10.5}$$

Note that the χ^2/N value is also called the reduced centered random variable (RCRV), as introduced in Candille, Côté, Houtekamer, and Pellerin (2007). These notions of ensemble reliability provide a useful framework for the basic evaluation of ensemble forecasts. However, they do not cover all aspects of ensemble reliability. Reliability diagrams, for instance, have been used to assess seasonal (Weisheimer & Palmer, 2014) and decadal forecasts (Corti, Weisheimer, Palmer, Doblas-Reyes, & Magnusson, 2012) from a different perspective.

Whereas short- and medium-range forecast systems commonly underestimate the actual uncertainty by producing overconfident ensembles, long-range forecasts are often found to be under-confident or over-dispersive (Eade et al., 2014; Kumar et al., 2014; Siegert et al., 2016). In Ho et al. (2013), the fulfillment of ensemble reliability was studied for different time ranges for sea surface temperature, based on an ensemble of initialized predictions and historical climate simulations (uninitialized in the preindustrial period). Below 2 years of lead time, initialized ensembles were overall over-confident and hence unreliable, while beyond, all ensembles were predominantly over-dispersed. In this case, the lack of ensemble reliability was due to the lack of climatological reliability, due to inadequate underlying variability in the climate model. Given the absence of correlations at long lead times, climatological and ensemble reliability become equivalent conditions.

In Section 10.7 the use of the aforementioned reliability conditions is discussed in the context of calibration of long-range forecasts.

10.4 MULTIMODEL COMBINATION OR CONSOLIDATION

Multimodel combination, or consolidation, has been designed to improve predictions at seasonal, inter-annual, and decadal time scales. The approaches developed for that purpose have a long history (e.g., Doblas-Reyes et al., 2005; Feddersen, Navarra, & Ward, 1999; Kharin & Zwiers, 2002; Krishnamurti et al., 1999; Van den Dool & Rukhovets, 1994). The main approach consists of constructing a new "deterministic" forecast by combining multiple models. This is usually done using a linear relationship between the different model solutions, either using the ensemble mean of the individual models or all the members. Consider M models where model i forecasts x_i such that the combined forecast is:

$$x_C = a_0 + \sum_{i=1}^{M} a_i x_i \tag{10.6}$$

Here x_C is the corrected forecast, x_i the individual model solutions, a_0 and a_i a set of $M+1$ parameters that could be fixed by minimizing a cost function, typically the MSE between the corrected forecasts and the observations. Additional constraints on the weights can be imposed in order to simplify the correction scheme as in Kharin and Zwiers (2002), who showed that a very competitive approach with reduced error variance consists of equally weighting each model after correcting the bias for all models jointly. More specifically, taking $a_i = 1/M$ and $a_0 = \bar{y} - (1/M)\sum_{i=1}^{M}\bar{x}_i$ where $\bar{y}$ is the climatological mean of the observations, leads to the bias-corrected multimodel ensemble mean:

$$x_{C,EM} = \bar{y} + \frac{1}{M}\sum_{i=1}^{M}(x_i - \bar{x}_i) \tag{10.7}$$

They also found that, except for large ensemble sizes in the tropics, usually the best approach is to regress $x_{C,EM}$ against the observations such that $a_i = a = Cov(x_{C,EM}, y)/[M\ Var(x_{C,EM})]$ and $a_0 = \bar{y} - a\sum_{i=1}^{M}\bar{x}_i$ leading to

$$x_{C,A} = \bar{y} + \frac{Cov(x_{C,EM}, y)}{[M Var(x_{C,EM})]}\sum_{i=1}^{M}(x_i - \bar{x}_i) \tag{10.8}$$

where Cov and Var are the usual covariance and variance.

In a more recent paper, Casanova and Ahrens (2009) show that weighting the different members based on the skill performance of each individual member is as good or even better than the Bayesian model averaging (BMA) proposed by Raftery, Gneiting, Balabdaoui, and Polakowski (2005), and discussed in detail in Wilks (2018, Chapter 3 of this book). This, of course, implies that a set of past forecasts is available to build the skill score of each ensemble member.

The skill-weighting approach is also compared in Pena and van den Dool (2008) with the ridge regression technique already used in Van den Dool and Rukhovets (1994), and nicely generalized in a Bayesian framework by DelSole (2007). In that context, the parameters $a = (a_1, ..., a_M)$ are obtained by minimizing a cost function $\mathfrak{J}$ that also includes Lagrange multipliers multiplying some constraints that are typically of the form,

$$\mathfrak{J} = (y - Ba)^T(y - Ba) + \lambda f(a) \tag{10.9}$$

Here y is a vector containing the time series of observations, B a matrix of size $N \times P$ listing the time series of the P predictors; λ is the Lagrange multiplier, and $f(a)$ is the constraint imposed to the parameters when set to zero. DelSole (2007) discussed several specific functions $f(a)$, in particular $f(a) = (a - 1/M)^T(a - 1/M)$, which implies that the estimates should stay close to the simple multimodel average, provided λ is sufficiently large.

Pena and van den Dool (2008) compared the main approaches to building a multimodel combination, together with different strategies to stabilize the weights when only small training data sets are available. The strategies proposed are (i) to remove the bad or redundant model versions, (ii) to increase the training size by using several close grid points, and (iii) fitting the same weights to all ensemble members of each model version (and not only using the ensemble mean of each model version). One important finding of their investigation is that when the hindcast training data set is small (21 years in their example), the equally weighted approach (Eq. 10.7) or the skill-score-based approach (Eq. 10.8), are the best methods. They further advocate increasing the number of hindcasts and ensemble members in order to improve the stability of the weight estimation.

Among the different proposals for weighting the members of a multimodel ensemble that can be found in the literature, the simple multimodel averaging provides a robust approach, in particular, when the training sample is small.

10.5 THE USE OF MULTIMODELS FOR PROBABILISTIC FORECASTS

The consolidation method discussed in Section 10.4 produces a single consensus forecast. This operation, although very useful in averaging out part of the model uncertainties, does not make use of all the potential information present in the ensemble, in particular on the evaluation of the uncertainty of this consensus forecast. Approaches for constructing probabilistic forecasts from multimodel ensembles have also been developed, and provide additional information (Doblas-Reyes et al., 2005; Hagedorn, Doblas-Reyes, & Palmer, 2005; Kharin & Zwiers, 2003; Rajagopalan, Lall, & Zebiak, 2002; Smith et al., 2015; Weigel, Liniger, & Appenzeller, 2008).

The simplest method for building a probabilistic forecast based on multiple models is to pool all the ensemble members coming from the different models (Hagedorn et al., 2005). Consider M different forecasting model versions with different ensemble sizes, $m_1, ..., m_M$. Also, the probabilistic forecast is discretized in K categories (or bins) for which the frequency of occurrence can be defined, $f_{j,k} = W_{k,j}/m_j$, where $W_{k,j}$ is the number of ensemble members of model j falling in category k, for models $j = 1,...,M$ and categories $k = 1,...,K$. The method estimates the occurrence probability of an event of category k as

$$f_k^p = \frac{\sum_{j=1}^{M} W_{k,j}}{\sum_{j=1}^{M} m_j} \tag{10.10}$$

As an alternative, Rajagopalan et al. (2002) proposed a Bayesian framework for assessing the information content in categorical forecasts at the seasonal time scale. It consists of building a weighted sum of the form (see also Weigel et al., 2008),

$$f_k^w = a_0 p_k + \sum_{j=1}^{M} \frac{a_j W_{k,j}}{m_j} \tag{10.11}$$

where p_k is the climatological forecast considered as an additional potential model, and the weights a_j are determined for each model $j = 1,...,M$. These are obtained by maximizing the likelihood function

$$\mathcal{L}(a_0, ..., a_M) = \prod_{t=1}^{n} f_{k(y_t)}^w \tag{10.12}$$

where n is the number of forecast realizations and $k(y_t)$ is the category of the observation at time t. Both approaches are compared in Rajagopalan et al. (2002), Robertson, Lall, Zebiak, and Goddard (2004), and Weigel et al. (2008), suggesting that the weighted sum, Eq. (10.11), is usually better than the simple averaged frequency. Moreover, the last authors indicate that in their idealized framework, the multimodel approach is better than selecting only the best model forecasts when the probabilistic forecasts are underdispersive.

DelSole, Nattala, and Tippett (2014) propose a method to determine whether a forecast combination is significantly superior to a single forecast. This method discriminates skill improvement due to ensemble averaging versus model combination and, in their example, revealed that model combination is by far the most prominent source of skill.

10.6 DRIFT AND TREND CORRECTION TECHNIQUES

Most commonly forecast initialization is done by full-field initialization, that is, close to the actual observed state. However, in general, this state is far from the model attractor. Accordingly, the forecast will relax or drift toward this attractor. For long-term forecasts, relaxation timescales extend up to the slowest timescales of the model (i.e., those corresponding to the ocean dynamics). Therefore, classically, a simple lead-time-dependent *drift correction* is often performed. Consider the raw forecast $x_\tau(\ell)$ for lead time ℓ, initialized at time τ. The most simple drift correction is therefore:

$$x_{C,\tau}(\ell) = \alpha_d(\ell) + x_\tau(\ell) \qquad (10.13)$$

Here $\alpha_d(\ell) = \bar{y} - m_x(\ell)$ with $\bar{y}$ and $m_x(\ell)$ the observational and model climatological means, respectively. Note that this correction could be applied to an individual ensemble member as well as to the ensemble mean. This approach was applied for seasonal (Boer, 2009) and decadal (Fyfe et al., 2011; Kharin et al., 2012) global-mean temperature forecasts. These references, however, extend their methodology beyond the drift correction and also apply a *trend correction*, which is necessary due to a mismatch between model and reality in their response to anthropogenic and natural forcing.

In order to introduce the trend correction, assumptions must be made regarding the sensitivity of model and observation with respect to such forcings. The approach outlined as follows is (loosely) based on Kharin et al. (2012), who consider linear behavior as a function of time t:

$$x_t = \mu_x + t L_x + \varepsilon_{x,t} \qquad (10.14)$$

$$y_t = \mu_y + t L_y + \varepsilon_{y,t} \qquad (10.15)$$

Here L_x and L_y are the climate trends, while ε represents internal variability. The trends can be obtained by least-squares regression. Model trends, however, can only be obtained in the absence of model drift, that is, after sufficiently long lead times. It is therefore common to derive them from historical climate simulations that are initialized in the preindustrial period. The simplest trend correction applied to a forecast $x_\tau(\ell)$ for lead time ℓ and initialized at time τ is then defined as:

$$x_{C,\tau}(\ell) = G[\ell + \tau] + x_\tau(\ell) \qquad (10.16)$$

Here, the function $G[\ell + \tau]$ quantifies the distance between the climatological model mean and the observational mean at time $\ell + \tau$:

$$G[\ell + \tau] = \mu_y - \mu_x + (\ell + \tau)(L_y - L_x) \qquad (10.17)$$

This correction approach is useful in the context of long-range forecasts initialized with the so-called "anomaly initialization" where the initial state is the observational anomaly superposed on the model mean climate. In this manner, the model state remains close to its own attractor, thereby preventing shocks at short lead times.

When using full-field initialization, on the other hand, the correction approach of Eq. (10.16) does not take into account the drift correction. Simply adding the drift correction of Eq. (10.13) would amount to a deficient correction because there is an important interaction between the drift and the trend correction. More specifically, a mismatch in long-term trend implies that the "distance" between the initialized states, typically close to the observational attractor, and the model attractor depends on the date of initialization. It is exactly this distance that typically determines the drift, that is, the time and trajectory of convergence to the model attractor. In practice, this means that correcting both drift and trend requires fitting parameters that depend on both the lead time and initialization time, hence increasing the risk of overfitting. Kharin et al. (2012) reduce the number of fitted parameters by assuming an exponential decay of the model drift. The proposed correction has the form:

$$x_{C,\tau}(\ell) = G[\ell+\tau] - G[\tau]e^{-\ell/\lambda} + x_\tau \qquad (10.18)$$

Note that at initialization, the corrected and raw forecast coincide, while at long lead times, $(\ell \gg \lambda)$ the correction is equivalent to the trend correction of Eq.(10.16).

In practice, the approach of combined trend and drift correction (Eq. 10.18) requires fitting μ_y and L_y on a long observational dataset, μ_x and L_x on historical ("uninitialized") climate simulations, and finally λ on initialized forecasts. Kharin et al. (2012) apply this method to the ensemble mean, and conclude that their trend adjustment approach substantially reduces forecast errors for both initialized and uninitialized forecasts and enhances skill, particularly for the first forecast year.

10.7 ENSEMBLE POSTPROCESSING TECHNIQUES

In contrast to weather forecasts, only a small number of approaches exist for correcting long-range ensemble forecasts, including the ensemble mean and its spread. At the end of the 1990s, a popular approach for correcting long-term forecasts was proposed, known as the variance inflation method (Doblas-Reyes et al., 2005; Kharin & Zwiers, 2003; von Storch, 1999; Weigel et al., 2009) and here denoted as the "INFL" method. Consider that an m-member ensemble is built to forecast the truth y. Ensemble member k is calibrated as follows:

$$x_{C,k} = \bar{y} + \alpha\left(\bar{x} - m_x\right) + \beta\,\epsilon_k \qquad (10.19)$$

where $\bar{y}$ and m_x are the observational and model climatological means respectively; $\bar{x}$ is the ensemble mean and $\epsilon_k = x_k - \bar{x}$ the model forecast anomaly; α and β are coefficients that should be determined based on the properties of both the observations and the model forecasts. If one requires that the climatological variance of the ensemble members and of the observation be the same, $s_{x_C}^2 = s_y^2$, minimizing the MSE between the corrected ensemble mean and the observations yields

$$\alpha = \frac{r_{y,\bar{x}} s_y}{s_{\bar{x}}}$$

$$\beta = \sqrt{1 - r_{y,\bar{x}}}\left(\frac{s_{\bar{x}}}{\bar{s}}\right) \qquad (10.20)$$

where $r_{y,\bar{x}}$ is the Pearson correlation coefficient between the observation y and the uncorrected ensemble mean $\bar{x}$ while $\bar{s}^2$ is the average ensemble variance. As indicated in Johnson and Bowler (2009), minimization of the MSE between the corrected ensemble mean and the observation amounts to exactly the same result as requiring that the (Pearson) correlation between the new ensemble mean and the observation is equal to the correlation between the raw ensemble mean and the observation. This approach is also equivalent to the one used in Eade et al. (2014), obtained by requiring that the ratio of predictable components, $\text{RPC} = s_x r_{y,\bar{x}}/s_{\bar{x}}$, as introduced in Eq. (10.2), is equal to one. As we will see in the next section, this method can fail when the "weather noise" depends on the signal in Eq. (10.1).

This method falls in the category of member-by-member (MBM) approaches discussed in Wilks (2018, Chapter 3 of this book), and will be referred to as inflation (INFL) in this chapter.

State-of-the-art ensemble postprocessing techniques originally developed for weather forecasts were recently applied in the context of seasonal forecasts in Krikken et al. (2016). In this work, simple calibration methods are compared with an ensemble calibration method for the prediction of Arctic sea ice based on single-model integrations with a five-member ensemble. The postprocessing method used is heteroscedastic extended logistic regression (see Section 3.2.4, Messner, Mayr, Zeileis, & Wilks, 2014; Wilks, 2009). Due to large model biases, additional corrections are required, related to the average seasonal cycle and the strong drift toward the model attractor. Added value due to postprocessing is shown for the first 3 months, and mainly concerns the ensemble-mean correction, and to a lesser extent, the ensemble-spread correction.

Along the same line, later in this chapter an ensemble postprocessing technique developed for weather forecasts will be applied in the context of long-range forecasts. The Best$_{rel}$ approach, as briefly explained in Wilks (2018, Chapter 3 of this book), corrects individual members by maximizing a constrained likelihood (Van Schaeybroeck & Vannitsem, 2015). Both the INFL and the Best$_{rel}$ methods impose the constraint of climatological reliability (see Eq. 10.3), however, INFL uses the (simple) ensemble reliability (see Eq. 10.4) while Best$_{rel}$ imposes SER (see Eq. 10.5). Another essential difference is that for the INFL and Best$_{rel}$ method, the corrected ensemble spread is a scalar multiplication and linear function of the raw ensemble spread, respectively. The Best$_{rel}$ approach yields probabilistic scores comparable to, for instance, nonhomogeneous regression (see Section 3.1 of Chapter 3). Moreover, it was shown that this simple regression approach preserves multivariate correlations (Schefzik, 2017; Van Schaeybroeck & Vannitsem, 2015).

As explained in Section 10.2, calibration of long-range forecasts is complicated due to many reasons, most prominently by the small numbers of observations and forecasts. Overfitting is therefore a crucial problem when performing ensemble postprocessing that goes beyond a simple drift correction. It is useful to identify all uncertainties, including those associated with parameter estimation for calibration. This is done using a fully probabilistic Bayesian approach in Siegert et al. (2016) based on the linear "signal-plus-noise" model of Eq. (10.1).

10.8 APPLICATION OF POSTPROCESSING IN AN IDEALIZED MODEL SETTING

To illustrate the methods outlined herein, we will make use of a model of reduced dimensionality, developed for the understanding of coupled ocean-atmosphere dynamics (De Cruz, Demaeyer, & Vannitsem, 2016; Vannitsem, 2015; Vannitsem, Demaeyer, De Cruz, & Ghil, 2015). This model is briefly described the Appendix (Section 10.11). The atmospheric model is based on the vorticity equation defined at two levels, and the ocean dynamics are described by the reduced-gravity vorticity equation in a single homogeneous layer. The horizontal fields are developed in Fourier series and severely truncated at a low wave number. The coupled model consists of a set of 36 ordinary differential equations (ODEs) that are integrated in time using a second order explicit numerical scheme.

For certain parameter ranges, the model exhibits LFV. Two 90,000-day integrations of the model with different sets of parameters were performed, that represent the reference or true climate trajectories. One integration exhibits LFV on long time scales (20 years) when the coupling between the ocean and the atmosphere is strong. This coupling is weak, and consequently LFV is absent, for the second

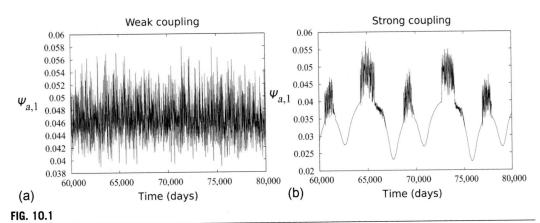

FIG. 10.1

Temporal evolution of $\psi_{a,1}$ in the two configurations of the coupled ocean–atmosphere model with the coupling parameter $d=10^{-8}\,\mathrm{s}^{-1}$ for the case of weak coupling (a) and $d=6\times10^{-8}\,\mathrm{s}^{-1}$ for the case of strong coupling (b). For more information on the meaning of these parameters, see Section 10.11 and Vannitsem et al. (2015).

integration. One of the main variables is the coefficient of the lowest wave-number Fourier mode of the atmospheric streamfunction, $\psi_{a,1}$. Its evolution in the two configurations is shown in Fig. 10.1. In panel (a) the dynamics are highly erratic in time with no visible LFV, while in panel (b) a marked LFV is present. Coming back to Eq. (10.1), the time series of panel (a) seems to conform to this simple linear scheme. The series in panel (b), exhibiting a marked modification of the "weather" variability (or noise) depending on the LFV signal, is inconsistent with Eq. (10.1). As we will see later, this difference has considerable impact on the postprocessing parameters. The two configurations will be referred as weakly (W) and strongly (S) coupled solutions, respectively.

10.8.1 EXPERIMENTAL SETUP

In order to determine the impact of postprocessing on the predictability of the system, we perform an idealized experiment along the trajectories presented above. Assume that a model is at our disposal with or without model errors and that a set of ensemble forecasts are performed. The reference model does not include model error and its output is denoted as X and used as the "truth." In order to mimic the observational error, a small random perturbation is introduced along this trajectory δX. Starting from this new initial condition, $Y=X+\delta X$, new forecasts can be performed. In order to perform ensemble forecasts, a new random initial perturbation is superimposed on the solution Y. The new initial conditions for the ensemble will be $Z(\ell=0)=Y+\delta Y$ where δY are drawn from the same distribution as the observational error, which produces a perfectly reliable initial ensemble. The perturbations are sampled from a uniform distribution between $[-5\times10^{-5},5\times10^{-5}]$ for all the variables of the model.

On top of these initial condition errors, some model error will be introduced in two key parameters: d, the coupling parameter between the ocean and the atmosphere, and k, one friction coefficient within the atmosphere (see Vannitsem et al., 2015, for more information on the parameterization of the model). The model errors for the two parameters will be denoted as δd and δk.

The number of ensemble members that are generated is fixed to 25, and the number of starting dates to 1000. The 500 first ensembles are used as training data to evaluate the postprocessing parameters, and the last 500 are used as the test set to evaluate the quality of the forecasts. The observables that are used in the present analysis are averaged over 30.4 days in order to get monthly values that can then be postprocessed for monthly, seasonal, and interannual forecasts.

In order to evaluate the forecasts, the MSE between the ensemble mean and the observation, generated with the reference model, will be computed together with the variance (SPREAD) of the ensemble forecasts. These are averaged over the 500 realizations of the test set. When the system is perfectly reliable, the MSE will be equal to the average ensemble variance or SPREAD. This will be tested in different model configurations.

10.8.2 POSTPROCESSING SINGLE-MODEL ENSEMBLES

In the following, we use different model versions separately to perform forecasts. Fig. 10.2 shows the MSE and average ensemble variance for different configurations of the model: (a) weak coupling case

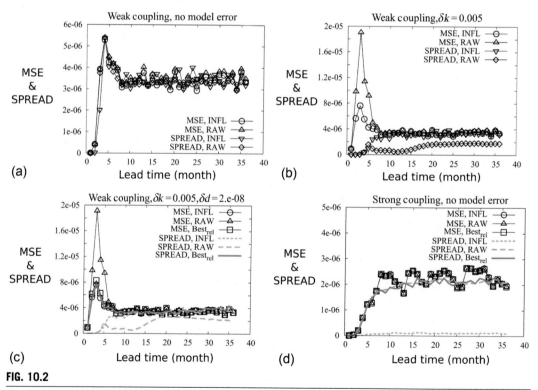

FIG. 10.2

Mean squared error (MSE) and average ensemble variance (SPREAD) for different configurations, (a) weak coupling case ($d=10^{-8}\,\mathrm{s}^{-1}$) with no model error; (b) weak coupling case with $\delta k=0.005\,\mathrm{s}^{-1}$; (c) weak coupling case with $\delta k=0.005$ and $\delta d=2\times10^{-8}\,\mathrm{s}^{-1}$; and (d) strong coupling case ($d=6\times10^{-8}\,\mathrm{s}^{-1}$) without model error. The inflation method is referred as INFL in the different panels, RAW to the raw forecasts and Best$_{\mathrm{rel}}$ as the best MBM tested in Van Schaeybroeck and Vannitsem (2015).

($d=10^{-8}$ s^{-1}), and no model error; (b) weak coupling case with $\delta k=0.005$ s^{-1}; (c) weak coupling case with $\delta k=0.005$ s^{-1} and $\delta d=2\times10^{-8}$ s^{-1}; and (d) strong coupling case ($d=6\times10^{-8}$ s^{-1}), no model error. The correction methods include the inflation method (INFL) and the Best$_{rel}$ method in the different panels, while RAW refers to the raw forecasts.

When no model error is introduced for the weak coupling case, the postprocessing based on inflation slightly reduces the average ensemble variance at short lead times. When model errors are introduced in Fig. 10.2b and c, considerable corrections to the MSE and average ensemble variance are obtained providing a more reliable forecast.

In the case of strong ocean-atmosphere coupling ($d=6\times10^{-8}$ s^{-1}), results are very different, with a correction that induces a drastic reduction of the average ensemble variance for the INFL method but not for the Best$_{rel}$ method. This implies that the INFL forecast is now strongly overconfident, that is, the truth falls mostly outside the ensemble. This feature is further confirmed in the comparison of the INFL-corrected forecasts obtained for two specific realizations for weak coupling (Fig. 10.3a) and strong coupling (Fig. 10.3b). In the latter case, the truth is often outside of the INFL-corrected set of ensemble members, indicating underdispersion.

To understand this feature, one must recall that the reference or "observational" variability in the case of strong ocean-atmosphere coupling is highly inhomogeneous in time, with periods of high chaoticity and periods of high stability (see Fig. 10.1). The use of the INFL postprocessing method is, in this case, detrimental, indicating that the parameters of the postprocessing approach depend on the LFV itself. In other words, the "weather" component ε_t in Eq. (10.1) does depend on the signal s_t. This feature also persists when model errors are introduced (not shown).

10.8.3 MULTIMODEL ENSEMBLE FORECASTS

In this section, multimodel ensemble forecasts are discussed. We will consider that the different model versions only differ through the sets of model parameters used. A grand ensemble is then built with 100 members composed of 25 members obtained with each of the 4 different model versions.

Example ensemble forecasts, obtained for both weak and strong ocean-atmosphere coupling, are given in Fig. 10.4. Fig. 10.4a shows results for weak coupling and four 25-member ensemble forecasts

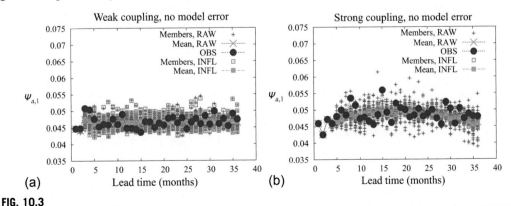

FIG. 10.3

Ensemble forecasts with 25 members for (a) weak coupling and (b) strong coupling cases without model error. Observations are represented as full blue circles. RAW and INFL refer to the raw forecasts and the corrected forecasts using inflation, respectively.

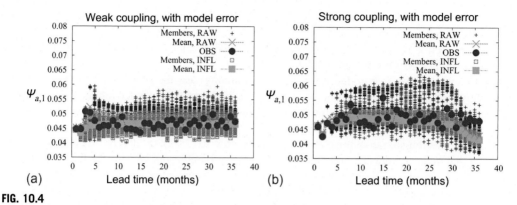

FIG. 10.4

Ensemble forecasts obtained as in Fig. 10.3 but for multimodel ensembles with 100 members, for (a) weak coupling case and (b) strong coupling case. See the text for details on the model configurations used in the multimodel ensembles.

made with the following model errors: (i) $\delta k = 0.005$ s^{-1} and $\delta d = 2 \times 10^{-8}$ s^{-1}; (ii) $\delta k = 0.001$ s^{-1} and $\delta d = 2 \times 10^{-8}$ s^{-1}; (iii) $\delta k = 0.005$ s^{-1}; and (iv) $\delta d = 10^{-8}$ s^{-1}. Fig. 10.4b applies to forecasts for the strong ocean-atmosphere coupling system, and the forecast models feature the following model errors: (i) $\delta k = 0.005$ s^{-1} and $\delta d = -5 \times 10^{-8}$ s^{-1}.; (ii) $\delta d = 2 \times 10^{-8}$ s^{-1}; (iii) $\delta d = -2 \times 10^{-8}$ s^{-1}; and (iv) $\delta d = -5 \times 10^{-8}$ s^{-1}. In both cases, the average ensemble variance is very large for the raw forecasts, while it is considerably reduced when postprocessed using the INFL method. For the case of strong coupling, the average ensemble variance is often too small for the ensembles to encompass the truth. Again, this suggests that postprocessing should be used with care when the statistical properties of the noise term depend on the signal itself.

These results are confirmed by the analysis of the MSE and average ensemble variance. Fig. 10.5a displays the MSE and ensemble spread obtained for weak coupling, where averages are taken over 500

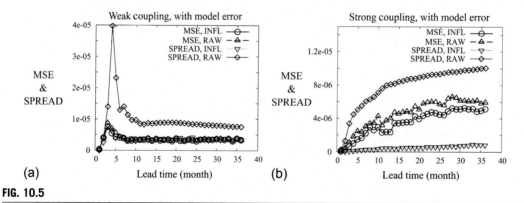

FIG. 10.5

Mean squared error (MSE) and average ensemble variance (SPREAD) as in Fig. 10.4, but for the multimodel ensembles, for (a) weak coupling case ($d = 10^{-8}$ s^{-1}) and (b) strong coupling case ($d = 6 \times 10^{-8}$ s^{-1}). See the text for details on the model configurations used in the multimodel ensembles.

realizations. Here, the average ensemble variance of the multimodel raw ensemble is very large, but is well corrected by the postprocessing based on inflation. This result contrasts with the one obtained for strong coupling, illustrated in Fig. 10.5b, where for the raw ensemble, the average ensemble variance is much larger than the MSE, indicating underconfidence of the ensemble forecasts. The inflation method allows correction of the MSE, but at the same time considerably reduces the average ensemble variance. The ensemble becomes now overconfident.

10.9 APPLICATION USING AN OPERATIONAL LONG-RANGE FORECASTING SYSTEM

An application of postprocessing based on an operational forecasting system for the prediction of the Niño 3.4 index is presented in this section. This is the most commonly-used index for ENSO and it is calculated as the mean sea surface temperature over a region of the tropical Pacific (5 N–5S, 170 W–120 W). Monthly forecasts with lead times up to 7 months are taken from the European Centre for Medium-Range Weather Forecasts (ECMWF), more specifically the IFS system 4 (Molteni et al., 2011). The hindcast dataset initiated each month during the period 1981–2010 is supplemented with the operational forecasts from the period 2010–17. While the operational runs all feature 51 ensemble members, this is only the case every 3 start months for the hindcasts and there are 15 members for all others.

Fig. 10.6a shows the mean error of the raw ensemble forecast as a function of lead month initialized for the four meteorological seasons. All biases at lead month one range between $-0.5°C$ and $0.5°C$ and, at month seven, are negative and range between $-1°C$ and $-2°C$. For intermediate lead times, on the other hand, the lead time dependence of the model drift strongly depends on the period of initialization. Therefore, the optimal manner to remove biases may be expected to be season dependent. However, that approach reduces the available training set by a factor of four, which might impact the verification scores. Such scores are shown in Fig. 10.6b and c for two postprocessing methods: a simple drift correction (green lines) and the ensemble postprocessing method Best$_{rel}$ (red lines). All results were obtained using cross validations by iteratively isolating all months of the same year for evaluation while training using the remaining data. The uncertainty intervals delineate the 95% confidence intervals.

Fig. 10.6b shows the probabilistic forecast skill in terms of CRPS with the full (green and red) lines obtained using calibration models fitted separately for each season, while the dotted lines show results when all seasons are pooled. The season-dependent calibration is clearly superior to the season-independent one while the ensemble correction method systematically improves upon the drift correction, even for lead times of 7 months, due to the correction of the ensemble spread.

In Fig. 10.6c the ensemble reliability is studied by comparing the RMSE of the ensemble mean (circles) with the ensemble standard deviation (triangles). Calibration is done for each season separately. The RMSE is strongly reduced by trend correction but, in line with the results from Fig. 10.6b, additional improvement is obtained by introducing ensemble correction. The ensemble reliability of the ensemble-corrected forecast is fairly well satisfied as seen by a near matching of RMSE with ensemble standard deviation.

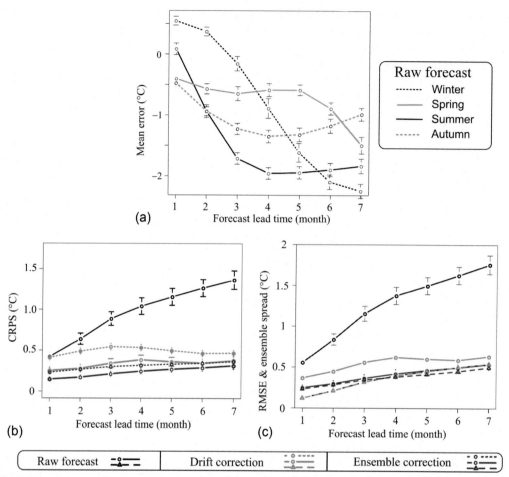

FIG. 10.6

Forecast characteristics of the ECMWF IFS raw forecasts *(black lines)*, the drift-corrected forecasts *(green lines)* and the ensemble-corrected forecasts *(red lines)* using the Best_rel approach. (a) Mean error or bias of the raw forecast against lead time for different meteorological seasons. (b) Continuous ranked probability score (CRPS) against lead time. The full green and red line are obtained by using four calibrations, one for each season, while the dotted lines are obtained using one calibration, using all available data. (c) RMSE *(circles)* and ensemble standard deviation *(triangles)* against lead time. Uncertainty intervals delineate the 95% confidence intervals assuming normal error statistics.

10.10 CONCLUSIONS

There are many factors that complicate the verification of long-range forecasts, the most prominent of which are the small numbers of observations and forecasts. These factors also impact postprocessing. On the other hand, due to large model biases, the added value of long-range forecasts can most often only be seen after forecast calibration. Whereas the performance of drift corrections are by now standard, only recently has it been shown that ensemble correction techniques are able to improve these forecasts. Still, a fully probabilistic approach to postprocessing is called for and should be elaborated upon further (Siegert et al., 2016).

A set of idealized experiments have been performed in the context of a low-order coupled ocean-atmosphere system in order to evaluate the ability of MBM approaches developed in the context of seasonal, interannual, and decadal forecasts. Two sets of experiments were performed, one with single-model ensemble forecast and the other with multimodel ensemble forecasts. Two techniques were analyzed, first the inflation method originally proposed by von Storch (1999), and further elaborated upon in Kharin and Zwiers (2002), Doblas-Reyes et al. (2005), Johnson and Bowler (2009), Weigel et al. (2009), Eade et al. (2014) and Siegert et al. (2016); and second a MBM method developed in Van Schaeybroeck and Vannitsem (2015). One difference between these two approaches is the cost function that is minimized: The former is based on minimizing the MSE between the observations and the corrected forecasts (see Johnson & Bowler, 2009) and the latter employs minimization of a *SER constraint* in which the individual spreads of the raw ensemble forecasts are taken into account. This method is referred to as $Best_{rel}$. The results indicates that the second approach performs much better in correcting the ensemble when the properties of the "weather noise," ε_t in Eq. (10.1), depend on the signal, s_t, itself.

Due to the limited number of ensemble members in both single-model and multimodel long-range forecasts, probabilistic approaches like the ones discussed in the different chapters of this book rarely apply (see Krikken et al., 2016 Siegert et al., 2016, for a few exceptions). Nowadays more and more long-range forecasts are performed, opening the possibility for applying the distribution-based approaches that could also be compared with the MBM approaches discussed in this chapter.

10.11 APPENDIX: THE IDEALIZED MODEL

For the present illustrative purpose, we will make use of a model developed for the understanding of coupled ocean-atmosphere dynamics (De Cruz et al., 2016, Vannitsem, 2015, Vannitsem et al., 2015).

The atmospheric model is based on the vorticity equation

$$\frac{\partial q}{\partial t} + (v.\nabla) q = F \tag{10.21}$$

where

$$q = \nabla^2 \psi + f + f_0^2 \frac{\partial}{\partial p} \sigma^{-1} \frac{\partial \psi}{\partial p} \tag{10.22}$$

Here ψ is the streamfunction, f_0, the dominant contribution of the Coriolis force estimated at a mid-latitudes, say φ_0, σ, the static stability parameter, $v = \left(-\frac{\partial \psi}{\partial y}, \frac{\partial \psi}{\partial x}\right)$ is the nondivergent horizontal velocity field and F contains all the dissipative and forcing terms. This vorticity equation is defined at two superimposed atmospheric levels, say 1 and 2, which constitutes a minimal representation for the development of the so-called baroclinic instabilities that are the main source of mid-latitude weather variability.

The ocean dynamics confined to a single homogeneous layer is based on a vorticity Eq. (10.21) with the potential vorticity given by,

$$q = \nabla^2 \Psi + \beta y + \frac{1}{L_d^2} \Psi \tag{10.23}$$

where $\beta = df/dy$ at a latitude φ_0, and L_d is a typical space scale involving the depth of the water layer, the Coriolis parameter and the acceleration of gravity.

Finally an advection equation for the temperature in the ocean considered as a passive scalar is incorporated in the model,

$$\frac{\partial T_o}{\partial t} + (\mathbf{v_o} \cdot \nabla) T_o = -\lambda(T_o - T_a) + E_R(t) \tag{10.24}$$

where $\mathbf{v_o}$ is the (nondivergent) ocean velocity, $E_R(t)$ the radiative input in the ocean and $= -\lambda(T_o - T_a)$ is the heat exchange between the ocean and the atmosphere. For more details on the equations of the model and the parameters, see Vannitsem et al. (2015) and Vannitsem (2015).

All equations are projected onto a set of Fourier modes compatible with the boundary conditions, after linearizing the temperature equation around a reference spatially averaged temperature. The projection is performed using the usual scalar product,

$$\langle f, g \rangle = \frac{n}{2\pi^2} \int_0^\pi dy' \int_0^{2\pi/n} dx' f g \tag{10.25}$$

for the dimensionless equations.

The ocean model then consists of eight ODEs for the dynamics within the ocean, one equation for the spatially averaged ocean temperature and six equations for the anomaly temperature field within the ocean. In addition 20 ODEs are obtained for the atmosphere: 10 for the barotropic streamfunction field $(\psi_1 + \psi_2)/2$ and 10 for the baroclinic streamfunction field $(\psi_1 - \psi_2)/2$. An additional equation for the spatially averaged temperature is also deduced. These form a set of 36 ODEs, which is fully described in the Supplement of Vannitsem (2015).

ACKNOWLEDGMENTS

This work is partly supported by the ERA4CS project, MEDSCOPE. Project MEDSCOPE is part of ERA4CS, an ERA-NET initiated by JPI Climate with co-funding of the European Union (Grant no 690462). Useful remarks from Emmanuel Roulin, an external reviewer and the editors are greatly appreciated.

REFERENCES

Benestad, R. E., Hanssen-Bauer, I., & Chen, D. (2008). *Empirical-Statistical Downscaling*. World Scientific Publishing Co Inc.

Boer, G. (2000). A study of atmosphere-ocean predictability on long time scales. *Climate Dynamics, 16*, 469–477.

Boer, G. (2009). Climate trends in a seasonal forecasting system. *Atmosphere-Ocean, 47*, 123–138.

Bowler, N., Cullen, M., & Piccolo, C. (2015). Verification against perturbed analyses and observations. *Nonlinear Processes in Geophysics, 22*, 403–411.

Branstator, G., & Teng, H. (2012). Potential impact of initialization on decadal predictions as assessed for CMIP5 models. *Geophysical Research Letters*, *39*, GL051974.

Buizza, R., & Leutbecher, M. (2015). The forecast skill horizon. *Quarterly Journal of the Royal Meteorological Society*, *141*, 3366–3382.

Candille, G., Côté, C., Houtekamer, P. L., & Pellerin, G. (2007). Verification of an ensemble prediction system against observations. *Monthly Weather Review*, *135*, 2688–2699.

Carrassi, A., Guemas, V., Doblas-Reyes, F., Volpi, D., & Asif, M. (2016). Sources of skill in near-term climate prediction: Generating initial conditions. *Climate Dynamics*, *47*, 3693–3712.

Casanova, S., & Ahrens, B. (2009). On the weighting of multi-model ensembles in seasonal and short-range weather forecasting. *Monthly Weather Review*, *137*, 3811–3822.

Corti, S., Weisheimer, A., Palmer, T., Doblas-Reyes, F., & Magnusson, L. (2012). Reliability of decadal predictions. *Geophysical Research Letters*, *39*, GL053354.

De Cruz, L., Demaeyer, J., & Vannitsem, S. (2016). The modular arbitrary-order ocean-atmosphere model: MAOOAM V1.0. *Geoscientific Model Development*, *9*, 2793–2808.

DelSole, T. (2007). A Bayesian framework for multi-model regression. *Journal of Climate*, *20*, 2810–2826.

DelSole, T., Nattala, J., & Tippett, M. K. (2014). Skill improvement from increased ensemble size and model diversity. *Geophysical Research Letters*, *41*, 7331–7342.

Demarée, G., & Nicolis, C. (1990). Onset of Sahelian drought viewed as a fluctuation-induced transition. *Quarterly Journal of the Royal Meteorological Society*, *116*, 221–238.

Déqué, M. (2012). Deterministic forecasts of continuous variables. In I. T. Jolliffe & D. S. Stephenson (Eds.), *Forecast Verification: A Practitioner's Guide in Atmospheric Science* (pp. 77–94). (2nd ed.). Chichester, UK: John Wiley & Sons, Ltd.

Deser, C., Phillips, A. S., Alexander, M. A., & Smoliak, B. V. (2014). Projecting North American climate over the next 50 years: Uncertainty due to internal variability. *Journal of Climate*, *27*, 2271–2296.

Dijkstra, H. A., & Ghil, M. (2005). Low-frequency variability of the large-scale ocean circulation: a dynamical systems approach. *Reviews of Geophysics*, *43*, RG000122.

Doblas-Reyes, F., Andreu-Burillo, I., Chikamoto, Y., Garçia-Serrano, J., Guemas, V., Kimoto, M., et al. (2013). Initialized near-term regional climate change prediction. *Nature Communications*, *4*, 1715.

Doblas-Reyes, F., Garçia-Serrano, J., Lienert, F., Biescas, A. P., & Rodrigues, L. R. (2013). Seasonal climate predictability and forecasting: Status and prospects. *Wiley Interdisciplinary Reviews: Climate Change*, *4*, 245–268.

Doblas-Reyes, F., Hagedorn, R., & Palmer, T. (2005). The rationale behind the success of multi-model ensembles in seasonal forecasting. II. Calibration and combination. *Tellus A*, *57*, 234–252.

Eade, R., Smith, D., Scaife, A., Wallace, E., Dunstone, N., Hermanson, L., et al. (2014). Do seasonal-to-decadal climate predictions underestimate the predictability of the real world? *Geophysical Research Letters*, *41*, 5620–5628.

Ebert, E., Wilson, L., Weigel, A., Mittermaier, M., Nurmi, P., Gill, P., et al. (2013). Progress and challenges in forecast verification. *Meteorological Applications*, *20*, 130–139.

Feddersen, H., Navarra, A., & Ward, M. N. (1999). Reduction of model systematic error by statistical correction for dynamical seasonal predictions. *Journal of Climate*, *12*, 1974–1989.

Fyfe, J., Merryfield, W., Kharin, V., Boer, G., Lee, W. -S., & Von Salzen, K. (2011). Skillful predictions of decadal trends in global mean surface temperature. *Geophysical Research Letters*, *38*, GL049508.

Hagedorn, R., Doblas-Reyes, F., & Palmer, T. (2005). The rationale behind the success of multi-model ensembles in seasonal forecasting. I. Basic concept. *Tellus A*, *57*, 219–233.

Ho, C. K., Hawkins, E., Shaffrey, L., Bröcker, J., Hermanson, L., Murphy, J. M., et al. (2013). Examining reliability of seasonal to decadal sea surface temperature forecasts: The role of ensemble dispersion. *Geophysical Research Letters*, *40*, 5770–5775.

Hoskins, B. (2013). The potential for skill across the range of the seamless weather-climate prediction problem: a stimulus for our science. *Quarterly Journal of the Royal Meteorological Society, 139*, 573–584.

Johnson, C., & Bowler, N. (2009). On the reliability and calibration of ensemble forecasts. *Monthly Weather Review, 137*, 1717–1720.

Kharin, V., Boer, G., Merryfield, W., Scinocca, J., & Lee, W. -S. (2012). Statistical adjustment of decadal predictions in a changing climate. *Geophysical Research Letters, 39*, GL052647.

Kharin, V. V., & Zwiers, F. W. (2002). Climate predictions with multi-model ensembles. *Journal of Climate, 15*, 793–799.

Kharin, V. V., & Zwiers, F. W. (2003). Improved seasonal probability forecasts. *Journal of Climate, 16*, 1684–1701.

Krikken, F., Schmeits, M., Vlot, W., Guemas, V., & Hazeleger, W. (2016). Skill improvement of dynamical seasonal arctic sea ice forecasts. *Geophysical Research Letters, 43*, 5124–5132.

Krishnamurti, T., Kishtawal, C., LaRow, T. E., Bachiochi, D. R., Zhang, Z., Williford, C. E., et al. (1999). Improved weather and seasonal climate forecasts from multi-model superensemble. *Science, 285*, 1548–1550.

Kumar, A. (2009). Finite samples and uncertainty estimates for skill measures for seasonal prediction. *Monthly Weather Review, 137*, 2622–2631.

Kumar, A., Barnston, A. G., Peng, P., Hoerling, M. P., & Goddard, L. (2000). Changes in the spread of the variability of the seasonal mean atmospheric states associated with ENSO. *Journal of Climate, 13*, 3139–3151.

Kumar, A., Peng, P., & Chen, M. (2014). Is there a relationship between potential and actual skill? *Monthly Weather Review, 142*, 2220–2227.

Lovejoy, S. (2015). A voyage through scales, a missing quadrillion and why the climate is not what you expect. *Climate Dynamics, 44*, 3187–3210.

Maraun, D., Wetterhall, F., Ireson, A. M., Chandler, R. E., Kendon, E. J., Widmann, M., et al. (2010). Precipitation downscaling under climate change: Recent developments to bridge the gap between dynamical models and the end user. *Reviews of Geophysics, 48*.

Maraun, D., & Widmann, M. (2018). *Statistical Downscaling and Bias Correction for Climate Research.* Cambridge: Cambridge University Press.

Mason, S. J. (2012). Seasonal and longer-range forecasts. In I. T. Jolliffe & D. S. Stephenson (Eds.), *Forecast Verification: A Practitioner's Guide in Atmospheric Science* (pp. 203–220). (2nd ed.). Chichester, UK: John Wiley & Sons, Ltd.

Massonnet, F., Bellprat, O., Guemas, V., & Doblas-Reyes, F. (2016). Using climate models to estimate the quality of global observational data sets. *Science, 354*, 452–455.

Meehl, G. A., Goddard, L., Boer, G., Burgman, R., Branstator, G., Cassou, C., et al. (2014). Decadal climate prediction: An update from the trenches. *Bulletin of the American Meteorological Society, 95*, 243–267.

Messner, J. W., Mayr, G. J., Zeileis, A., & Wilks, D. S. (2014). Heteroscedastic extended logistic regression for postprocessing of ensemble guidance. *Monthly Weather Review, 142*, 448–456.

Molteni, F., Stockdale, T., Balmaseda, M., Balsamo, G., Buizza, R., Ferranti, L., et al. (2011). *The new ECMWF seasonal forecast system (System 4). European Centre for Medium-range Weather Forecasts.*

Murphy, J. (1990). Assessment of the practical utility of extended range ensemble forecasts. *Quarterly Journal of the Royal Meteorological Society, 116*, 89–125.

Nicolis, C. (2016). Error dynamics in extended-range forecasts. *Quarterly Journal of the Royal Meteorological Society, 142*, 1222–1231.

Ogallo, L., Bessemoulin, P., Ceron, J. -P., Mason, S., & Connor, S. J. (2008). Adapting to climate variability and change: The climate outlook forum process. *Bulletin of the World Meteorological Organization, 57*, 93–102.

Pena, M., & van den Dool, H. (2008). Consolidation of multi-model forecasts by ridge regression: Application to Pacific sea surface temperature. *Journal of Climate*, *21*, 6521–6538.

Prodhomme, C., Batté, L., Massonnet, F., Davini, P., Bellprat, O., Guemas, V., et al. (2016). Benefits of increasing the model resolution for the seasonal forecast quality in EC-earth. *Journal of Climate*, *29*, 9141–9162.

Prodhomme, C., Doblas-Reyes, F., Bellprat, O., & Dutra, E. (2016). Impact of land surface initialization on sub-seasonal to seasonal forecasts over Europe. *Climate Dynamics*, *47*, 919–935.

Raftery, A. E., Gneiting, T., Balabdaoui, F., & Polakowski, M. (2005). Using Bayesian model averaging to calibrate forecast ensembles. *Monthly Weather Review*, *133*, 1155–1174.

Rajagopalan, B., Lall, U., & Zebiak, S. E. (2002). Categorical climate forecasts through regularization and optimal combination of multiple GCM ensembles. *Monthly Weather Review*, *130*, 1792–1811.

Randall, D. A., Wood, R. A., Bony, S., Colman, R., Fichefet, T., Fyfe, J., et al. (2007). *Climate Models and Their Evaluation*. In: *Climate Change 2007: The Physical Science Basis* (pp. 589–662). *Contribution of Working Group I to the Fourth Assessment Report of the IPCC (FAR)*, Cambridge University Press.

Robertson, A. W., Lall, U., Zebiak, S. E., & Goddard, L. (2004). Improved combination of multiple atmospheric GCM ensembles for seasonal prediction. *Monthly Weather Review*, *132*, 2732–2744.

Royer, J.-F. (1993). Review of recent advances in dynamical extended range forecasting for the extratropics. In J. Shukla (Ed.), *Prediction of Interannual Climate Variations* (pp. 49–69). Berlin, Heidelberg: Springer.

Schefzik, R. (2017). Ensemble calibration with preserved correlations: Unifying and comparing ensemble copula coupling and member-by-member postprocessing. *Quarterly Journal of the Royal Meteorological Society*, *143*, 999–1008 17 pp.

Shukla, J. (1981). Dynamical predictability of monthly means. *Journal of the Atmospheric Sciences*, *38*, 2547–2572.

Siegert, S., Stephenson, D. B., Sansom, P. G., Scaife, A. A., Eade, R., & Arribas, A. (2016). A Bayesian framework for verification and recalibration of ensemble forecasts: How uncertain is NAO predictability? *Journal of Climate*, *29*, 995–1012.

Smith, L. A., Du, H., Suckling, E. B., & Niehorster, F. (2015). Probabilistic skill in ensemble seasonal forecasts. *Quarterly Journal of the Royal Meteorological Society*, *141*, 1085–1100.

Tippett, M. K., Kleeman, R., & Tang, Y. (2004). Measuring the potential utility of seasonal climate predictions. *Geophysical Research Letters*, *31*, L22201.

Van den Dool, H. (2007). *Empirical Methods in Short-Term Climate Prediction*. Oxford: Oxford University Press.

Van den Dool, H., & Rukhovets, L. (1994). On the weights for an ensemble-averaged 6–10-day forecast. *Weather and Forecasting*, *9*, 457–465.

Van Schaeybroeck, B., & Vannitsem, S. (2013). Reliable Probabilities Through Statistical Post-Processing of Ensemble Forecasts. In T. Gilbert, M. Kirkilionis, & G. Nicolis (Eds.), *Proceedings of the European conference on complex systems 2012, Springer proceedings on complexity, XVI* (pp. 347–352).

Van Schaeybroeck, B., & Vannitsem, S. (2015). Ensemble post-processing using member-by-member approaches: Theoretical aspects. *Quarterly Journal of the Royal Meteorological Society*, *141*, 807–818.

Vannitsem, S. (2015). The role of the ocean mixed layer on the development of the North Atlantic oscillation: A dynamical systems perspective. *Geophysical Research Letters*, *42*, 8615–8623.

Vannitsem, S., Demaeyer, J., De Cruz, L., & Ghil, M. (2015). Low-frequency variability and heat transport in a low-order nonlinear coupled ocean-atmosphere model. *Physica D: Nonlinear Phenomena*, *309*, 71–85.

Vannitsem, S., & Nicolis, C. (2008). Dynamical properties of Model Output Statistics forecasts. *Monthly Weather Review*, *136*, 405–419.

von Storch, H. (1999). On the use of inflation in statistical downscaling. *Journal of Climate*, *12*, 3505–3506.

Weigel, A. P., Liniger, M., & Appenzeller, C. (2008). Can multi-model combination really enhance the prediction skill of probabilistic ensemble forecasts? *Quarterly Journal of the Royal Meteorological Society*, *134*, 241–260.

Weigel, A. P., Liniger, M. A., & Appenzeller, C. (2009). Seasonal ensemble forecasts: Are recalibrated single models better than multimodels? *Monthly Weather Review*, *137*, 1460–1479.

Weisheimer, A., & Palmer, T. (2014). On the reliability of seasonal climate forecasts. *Journal of the Royal Society Interface*, *11*, 20131162.

Wilks, D. S. (2009). Extending logistic regression to provide full-probability-distribution MOS forecasts. *Meteorological Applications*, *16*, 361–368.

Wilks, D. S. (2018). Univariate ensemble postprocessing. In S. Vannitsem, D. Wilks, & J. Messner (Eds.), *Statistical Postprocessing of Ensemble Forecasts*. Elsevier.

ENSEMBLE POSTPROCESSING WITH R

11

Jakob W. Messner

Technical University of Denmark, Kongens Lyngby, Denmark

CHAPTER OUTLINE

Statistical Postprocessing of Ensemble Forecasts. https://doi.org/10.1016/B978-0-12-812372-0.00011-X

291

11.1 **INTRODUCTION**

As already pointed out in Hamill (2018, Chapter 7 of this book), computer software plays an integral role in the practical implementation of statistical ensemble postprocessing. Typical tasks for which software is required are data processing and analysis, model fitting, forecast generation and visualization, and forecast evaluation.

R is a free and open-source statistical software and programming language and its variety of convenient tools for data processing, graphics, and statistical modeling makes it very well suited for these tasks. It is mostly platform-independent and is free to download from the Comprehensive R Archive Network (CRAN[1]). Furthermore, the source code is fully open, which allows easy modifications and extensions. A big and active community steadily extends the functionality with numerous add-on packages that are also shared via CRAN.

The easy accessibility, the open-source philosophy, and the sharing facilities also make R an ideal environment for reproducible research. Packages that provide software for recent postprocessing methods also clearly facilitate the practical implementation of these methods. Furthermore, implementations of a wide range of other recent statistical methods can encourage researchers to investigate these methods for postprocessing.

This chapter provides a short introduction to some important and useful R functions for ensemble postprocessing. Code and results for four typical postprocessing examples are shown, including deterministic forecasting of temperature, univariate probabilistic forecasting of temperature, univariate ensemble postprocessing of precipitation, and bivariate probabilistic forecasting of temperature and precipitation. Each of these examples involves data processing, fitting of different postprocessing models, prediction, and forecast verification and comparison.

For the most part, the code is well explained and is kept as simple as possible. Nevertheless, a basic knowledge of R is expected. A comprehensive introduction to R can be found, for example, in R Development Core Team (2017). More detailed descriptions of specific functions can, for example, be accessed from within R using either ? or `help()` (i.e., `?lm` and `help(lm)` both open the help page for the function `lm()`). This chapter provides only brief descriptions of the different postprocessing and verification methods. More information on these methods can be found in Schefzik and Möller (2018, Chapter 4 of this book), Thorarinsdottir and Schuhen (2018, Chapter 6 of this book), and Wilks (2018, Chapter 3 of this book).

All examples in this chapter are fully reproducible and were computed on 64-bit Ubuntu Linux using R version 3.4.0 and the most current version of all used CRAN packages (2018-01-23). The following code can be used to install these packages:

```
R> install.packages(c("ensemblepp", "ensembleBMA", "crch", "gamlss",
+   "ensembleMOS",  "SpecsVerification", "scoringRules", "glmx", "ordinal",
+   "pROC", "mvtnorm"))
```

Throughout the chapter, code is written in `typewriter` font, function names are followed by parentheses (e.g., `function()`) and links to CRAN packages are in boldface (e.g., **ensemblepp**). Note that

---

[1]http://CRAN.R-project.org/.

package references can be found in Section 11.7.2. The used data sets and a copy of the entire code can be found in the **ensemblepp** package.

This chapter is organized as follows: A simple example on deterministic postprocessing of temperature in Section 11.2 introduces the typical workflow of statistical postprocessing in R. Sections 11.3 and 11.4 show examples of different univariate postprocessing (Wilks, 2018, Chapter 3 of this book) and verification (Thorarinsdottir & Schuhen, 2018, Chapter 6 of this book) methods on temperature and precipitation data, respectively. Finally, Section 11.5 extends these examples to bivariate postprocessing of temperature and precipitation (Schefzik & Möller, 2018, Chapter 4 of this book).

11.2 DETERMINISTIC POSTPROCESSING

Postprocessing of deterministic weather predictions with model output statistics (MOS) was proposed by Glahn and Lowry (1972) and has been common practice in weather forecasting for several decades. In principle, the different statistical ensemble postprocessing approaches that have been reviewed in the previous chapters can all be seen as an extension of this basic MOS approach for ensemble predictions. Although not an ensemble postprocessing technique in the strict sense, this section employs simple MOS to introduce the typical workflow of postprocessing in R, including data preparation and exploration, model fitting, prediction, and verification.

11.2.1 DATA

This and the following section employ the `temp` data set that is contained in the **ensemblepp** package. This data set contains 18–30 hour minimum temperature ensemble forecasts and corresponding 18–06 UTC observations at Innsbruck (Austria). The ensemble forecasts are taken from the second generation GEFS reforecast data set (Hamill et al., 2013) and span a time frame from 2000 to 2015.

The data set can be loaded into R with:

```
R> data("temp", package = "ensemblepp")
```

The commands

```
R> dim(temp)

[1] 2749 12

R> names(temp)

[1] "temp"     "tempfc.1" "tempfc.2" "tempfc.3" "tempfc.4" "tempfc.5"
[7] "tempfc.6" "tempfc.7" "tempfc.8" "tempfc.9" "tempfc.10" "tempfc.11"
```

reveal that `temp` consists of data for 2749 dates with temperature observations `temp` and 11 member temperature ensemble forecasts `tempfc.1` to `tempfc.11`.

The observation dates are stored in the row names and can be extracted with `rownames()`. With `as. Date()` this vector of date strings can be converted into a `"Date"` object, which allows more convenient date operations. The following code adds this date vector as an additional column to `temp` and retrieves the start and end dates:

```
R> temp$date <- as.Date(rownames(temp))
R> range(temp$date)
```

```
[1] "2000-01-02" "2016-01-01"
```

In the following, only winter data are considered. With `format()` the month can be extracted from `temp $date` to subset only data from December, January, and February:

```
R> temp <- temp[format(temp$date, "%m") %in% c("12", "01", "02"),]
```

This section will employ the ensemble mean as the deterministic forecast and the next section will additionally employ the ensemble standard deviation. Date-wise ensemble means and standard deviations can be computed with the `apply()` function:

```
R> temp$ensmean <- apply(temp[,2:12], 1, mean)
R> temp$enssd   <- apply(temp[,2:12], 1, sd)
```

Finally, the data are split into independent training (before 2010-03-01) and test data (after 2010-03-01):

```
R> temptrain <- temp[temp$date <  "2010-03-01",]
R> temptest  <- temp[temp$date >= "2010-03-01",]
```

These data frames consist of three additional columns (dates, ensemble mean, and standard deviation) and have 417 and 253 rows:

```
R> dim(temptrain)
```

```
[1] 417 15
```

```
R> dim(temptest)
```

```
[1] 253 15
```

```
R> names(temp)
```

```
 [1] "temp"     "tempfc.1"  "tempfc.2"  "tempfc.3"  "tempfc.4"  "tempfc.5"
 [7] "tempfc.6" "tempfc.7"  "tempfc.8"  "tempfc.9"  "tempfc.10" "tempfc.11"
[13] "date"     "ensmean"   "enssd"
```

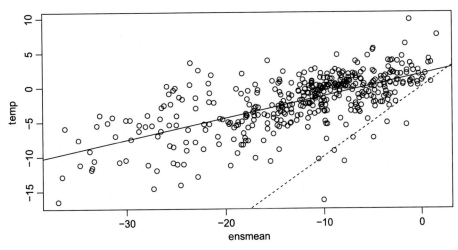

FIG. 11.1

Scatterplot of temperature observations by corresponding ensemble mean forecasts. Additionally, the *dashed line* shows the 1:1 relationship and the *solid line* shows the MOS regression fit.

A scatterplot of observations versus the ensemble mean forecast, which is obtained with

```
R> plot(temp ~ ensmean, data = temptrain)
R> abline(0, 1, lty = 2)
```

and is shown in Fig. 11.1, reveals a clear cold bias with the majority of points lying above the dashed diagonal line. Furthermore this cold bias is clearly stronger for lower temperatures. The syntax temp~ensmean is called formula and is often used in R to specify dependent (predictand) and independent (predictor) variables. More details on formulae can be found in R Development Core Team (2017).

11.2.2 MODEL FITTING

MOS (Glahn & Lowry, 1972) is a common approach to correcting systematic forecast biases in deterministic predictions. Least-squares regression is applied on training data to estimate a linear relationship between predictions and corresponding observations. This linear relationship can then be used to correct future forecasts.

In R linear regression can easily be fitted with lm() using the same arguments as for plot():

```
R> MOS <- lm(temp ~ ensmean, data = temptrain)
R> abline(MOS)
```

where abline() adds the regression line to the scatterplot in Fig. 11.1.

MOS is a fitted model object of class "lm" and contains coefficient estimates and further information on the regression fit. summary() can be used to extract parts of this information such as coefficient estimates with standard errors and P values, residual standard errors, R^2 or F-statistics (see e.g., Wilks, 2011):

```
R> summary(MOS)
Call:
lm(formula = temp ~ ensmean, data = temptrain)

Residuals:
     Min      1Q  Median      3Q     Max
 -15.0234 -1.5003  0.1596  1.7068  9.1260

Coefficients:
            Estimate Std. Error t value Pr(>|t|)
(Intercept)  1.89614    0.26189    7.24 2.19e-12 ***
ensmean      0.31579    0.01768   17.86  < 2e-16 ***
---
Signif. codes: 0 '***' 0.001 '**' 0.01 '*' 0.05 '.' 0.1 ' ' 1

Residual standard error: 2.953 on 415 degrees of freedom
Multiple R-squared:  0.4346,        Adjusted R-squared: 0.4332
F-statistic:    319 on 1 and 415 DF,  p-value: < 2.2e-16
```

For an unbiased ensemble mean forecast, the intercept and ensemble mean coefficient should be close to 0 and 1, respectively. The smaller ensemble mean coefficient means that for a 1 degree increase in the ensemble mean forecast only a 0.316 degree increase in the observed temperature is expected. This relationship can also be seen in Fig. 11.1.

summary() is a common *generic function*, which can be applied to different objects and where the output depends on the object class. For example, summary() generates different outputs for, for example, "lm" or "data.frame" objects. Other typical generic functions for model fits are coef(), predict(), or plot(). See R Development Core Team (2017) for more details on object orientated programming in R.

11.2.3 PREDICTION

With the linear regression fit from the training data, MOS predictions can be generated by applying the regression equation to independent (future) ensemble mean forecasts. This can be done conveniently predict() with the tesdata set as newdata:

```
R> fcMOS <- predict(MOS, newdata = temptest)
```

A plot of the first 20 dates with corresponding ensemble mean forecasts and observations can be generated with

```
R> plot(fcMOS[1:20], type = "l", lty = 2, ylab = "2m temperature",
+    xlab = "date", xaxt = "n", ylim = c(-35, 10))
```

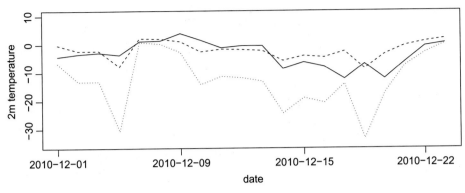

FIG. 11.2

Forecast example for dates 2010-12-01 to 2010-12-23. The raw ensemble mean forecast is shown as *dotted*, the MOS forecast as *dashed*, and the corresponding observations as a *solid line*.

```
R> axis(1, at = seq(1,20,6), temptest$date[seq(1, 20, 6)])
R> lines(temptest$temp[1:20])
R> lines(temptest$ensmean[1:20], lty = 3)
```

where the plot argument `xaxt = "n"` and the `axis()` function create an axis of dates instead of numerical indices. This plot (Fig. 11.2) shows for these example dates, that the MOS model corrects the strong cold bias well in the ensemble mean.

11.2.4 VERIFICATION

Fig. 11.2 already indicates a clear advantage of MOS compared with the raw ensemble mean. Verification scores such as bias (BIAS), mean absolute error (MAE), or root mean squared error (RMSE) allow a quantitative and more comprehensive assessment of forecast quality. The following code creates a table of these scores for the raw ensemble mean prediction and the MOS forecast.

```
R> rbind(raw = c(BIAS = mean(temptest$ensmean-temptest$temp),
+            MAE = mean(abs(temptest$ensmean - temptest$temp)),
+            RMSE = sqrt(mean((temptest$ensmean - temptest$temp)^2))),
+       MOS = c(BIAS = mean(fcMOS-temptest$temp),
+            MAE = mean(abs(fcMOS - temptest$temp)),
+            RMSE = sqrt(mean((fcMOS - temptest$temp)^2))))
         BIAS       MAE       RMSE
raw -9.4870258  9.531735  11.022110
MOS  0.3179221  2.346333   3.202392
```

A clear advantage of MOS can be seen here in all three scores (smaller absolute values).

11.3 **UNIVARIATE POSTPROCESSING OF TEMPERATURE**

The previous section has shown a very simple approach to calibration of deterministic (ensemble mean) forecasts based on past data. However, ensemble spread information was neglected so that the resulting forecasts were deterministic and so did not provide any case-dependent information on forecast uncertainty. This section extends the example from the previous section and applies some of the univariate ensemble postprocessing and verification methods that have been reviewed in Wilks (2018, Chapter 3 of this book) and Thorarinsdottir and Schuhen (2018, Chapter 6 of this book). More specifically, non-homogeneous Gaussian regression (NGR; Gneiting, Raftery, Westveld, & Goldman, 2005), Bayesian model averaging (BMA; Raftery, Gneiting, Balabdaoui, & Polakowski, 2005), and two ensemble dressing approaches (Bröcker & Smith, 2008; Wang & Bishop, 2005) are applied and compared with each other. Logistic regression and its variants are more common for precipitation forecasts and are therefore presented in the next section.

11.3.1 **DATA**

This section uses the same data as the previous section. However, the functions `ensembleBMA()` and `ensembleMOS()`, which are used to fit BMA and NGR, require the data to have a specific `"ensemble-Data"` structure. Such `"ensembleData"` objects can be created with the `ensembleData()` function from the **ensembleBMA** package.

```
R> library("ensembleBMA")
R> temptrain_eD <- ensembleData(forecasts = temptrain[,2:12],
+      dates = temptrain$date, observations = temptrain$temp,
+      forecastHour = 24, initializationTime = "00", exchangeable = rep(1, 11))
R> temptest_eD <- ensembleData(forecasts = temptest[,2:12],
+      dates = temptest$date, observations = temptest$temp,
+      forecastHour = 24, initializationTime = "00", exchangeable = rep(1, 11))
```

These objects `temptrain_eD` and `temptest_eD` contain the same data as `temptrain` and `temptest` together with additional information such as forecast lead time or initialization time. In the following the GEFS reforecast are assumed to be exchangeable, which is specified with `exchangeable = rep(1, 11)` (i.e., all 11 members are in group 1).

The previous section already revealed the cold bias in the ensemble mean. The calibration of the full ensemble is commonly assessed with a verification rank histogram (e.g., Wilks, 2011), which can be generated with `rank()` and `hist()`:

```
R>    rank <- apply(temptrain[, 1:12], 1, rank)[1,]
R>    hist(rank, breaks = 0:12 + 0.5, main = "Verification Rank Histogram")
```

In this verification rank histogram, which is shown in the left panel of Fig. 11.3, the cold bias can be seen by the large fraction of observations that are greater than the highest ensemble member (peak at rank 12). Furthermore, almost all remaining observations have rank 1, which indicates the common problem of underdispersion (Buizza, 2018, Chapter 2 of this book).

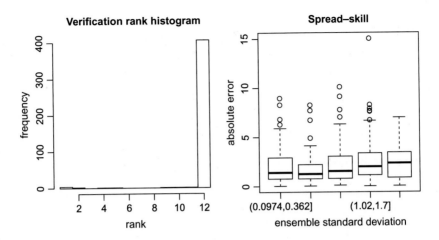

FIG. 11.3

Verification rank histogram (*left*) and spread skill relationship (*right*) of the temperature ensemble forecast.

Although the ensemble is clearly uncalibrated, it still might contain useful information on the forecast uncertainty if there is a clear relationship between forecast error and ensemble spread. This spread-skill relationship can, for example, be assessed in a boxplot of the absolute forecast error for different intervals of ensemble standard deviation:

```
R> sdcat <- cut(temptrain$enssd,
+     breaks = quantile(temptrain$enssd, seq(0, 1, 0.2)))
R> boxplot(abs(residuals(MOS)) ~ sdcat, ylab = "absolute error",
+     xlab = "ensemble standard deviation", main = "Spread-Skill")
```

This plot, which is shown in the right panel of Fig. 11.3, shows a slight positive relationship between the ensemble spread and the absolute error of the deterministic MOS forecasts.

11.3.2 MODEL FITTING

Fig. 11.3 shows that, although the ensemble is uncalibrated, it contains forecast uncertainty information in the ensemble spread. In the following, different implementations of ensemble postprocessing methods are presented to exploit this spread-skill relationship and generate calibrated probabilistic temperature forecasts.

Nonhomogeneous Gaussian regression

Nonhomogeneous Gaussian regression (NGR; Gneiting et al., 2005) is one of the most popular postprocessing approaches for temperature ensemble forecasts. It is closely related to linear regression MOS, but assumes the observations to be normally distributed around a mean regression fit with a standard deviation that is a function of the ensemble spread. NGR can be formulated as

$$T \sim \mathcal{N}(\mu, \sigma) \tag{11.1}$$

$$\mu = \beta_0 + \beta_1 m \tag{11.2}$$

$$\log(\sigma) = \gamma_0 + \gamma_1 s \tag{11.3}$$

where T is the observed temperature, m and s are the ensemble mean and standard deviation, respectively, and $\beta_0, \beta_1, \gamma_0, \gamma_1$ are regression coefficients. The logarithm in Eq. (11.3) is a device to ensure positive values for σ (e.g., Messner, Mayr, Wilks, & Zeileis, 2014). Alternatively also σ or σ^2 can be modeled, for example,

$$\sigma^2 = \gamma_0 + \gamma_1 s^2 \qquad (11.4)$$

with γ_0 and γ_1 restricted to be positive (e.g., Gneiting et al., 2005). Note that for exchangeable ensemble members Eqs. (11.1), (11.2), (11.4) are equivalent to Eqs. (3.1)–(3.5) in Wilks (2018, Chapter 3 of this book).

Estimation of the model coefficients is usually carried out either with maximum likelihood or minimum CRPS estimation (Gneiting et al., 2005). In R, maximum likelihood estimation of model (11.1)–(11.3) can be performed with crch() from **crch** or gamlss() from **gamlss**. The interface of both functions is very similar to lm(). The only difference is that in addition to the predictor variables for the mean μ, predictor variables for σ have to be specified. In crch() the predictor variables for μ and σ are separated by "|" in a two-part formula (Zeileis & Croissant, 2010). In gamlss() the predictor variables for σ are specified in a second formula sigma.formula:

```
R> library("crch")
R> NGR_crch <- crch(temp ~ ensmean | enssd, data = temptrain)
R> library("gamlss")
R> NGR_gamlss <- gamlss(temp ~ ensmean, sigma.formula = ~ enssd,
+     data = temptrain)
```

With slight modifications and additional arguments crch() can also be employed to fit models with minimum CRPS estimation and the σ parameterization (Eq. 11.4):

```
R> NGR_crch2 <- crch(temp ~ ensmean | I(enssd^2), data = temptrain,
+     link.scale = "quad", type = "crps")
```

Another R implementation of this model is provided by fitMOS() from **ensembleMOS**. It has a slightly different interface and does not use the formula notation. Furthermore, it requires the input data to be a "ensembleData" object such as temptrain_eD or temptrain_eD (see Section 11.3.1).

```
R> library("ensembleMOS")
R> NGR_ensembleMOS <- fitMOS(temptrain_eD, model = "normal")
```

crch and **gamlss** also provide methods for typical extractor functions such as summary() or coef(). In NGR_ensembleMOS the coefficients can be accessed with:

```
R> coef_ensembleMOS <- c(NGR_ensembleMOS$a, 11 * NGR_ensembleMOS$B[1],
+     NGR_ensembleMOS$c, NGR_ensembleMOS$d)
```

where NGR_ensembleMOS$B is actually a vector of coefficients for each ensemble member. But because the ensemble members are assumed to be exchangeable all coefficients are equal to 1/11 of the coefficient for the ensemble mean. A comparison of the coefficients from the different model fits shows that `NGR_crch` and `NGR_gamlss` and `NGR_crch2` and `NGR_ensembleMOS` are equivalent, respectively:

```
R> rbind(crch = coef(NGR_crch),
+      gamlss = c(coef(NGR_gamlss), coef(NGR_gamlss, what = "sigma")),
+      ensembleMOS = coef_ensembleMOS, crch2 = coef(NGR_crch2))

              (Intercept)   ensmean   (scale)_(Intercept) (scale)_enssd
crch           1.889426  0.3186283          1.032608       0.03955893
gamlss         1.889460  0.3186099          1.032671       0.03950481
ensembleMOS    2.032526  0.3197697          6.185533       0.29298578
crch2          2.032560  0.3197728          6.185556       0.29299460
```

Therefore, only `NGR_crch` and `NGR_crch2` will be used in the following. The clear difference between these models in the standard deviation coefficients (`(scale)_(Intercept)` and `(scale)_enssd`) originates from the different parameterization in Eqs. (11.3), (11.4).

BMA and other ensemble dressing approaches

BMA (e.g., Raftery et al., 2005) and other ensemble dressing methods (e.g. Bröcker & Smith, 2008; Wang & Bishop, 2005) are also common postprocessing techniques for continuous predictands such as temperature. Unlike NGR, the continuous predictive distributions are not of a specific parametric form, but are (weighted) mixtures of component distributions centered at the corrected ensemble forecasts. See Wilks (2018, Chapter 3 of this book) for details on these models.

In R, Gaussian BMA (Raftery et al., 2005) can be estimated with `fitBMA` from **ensembleBMA** with a similar interface to `ensembleMOS()` and input data of class `"ensembleData"`.

```
R> BMA <- fitBMA(temptrain_eD, model = "normal")
```

Simpler ensemble dressing approaches such as that described by Wang and Bishop (2005) do not have direct R implementations. However, the model that is specified by Eq. (3.48) in Wilks (2018, Chapter 3 of this book) can easily be applied manually by employing the deterministic MOS model from Section 11.2:

```
R> smuy2 <- var(MOS$residuals)
R> st2 <- mean(temptrain$enssd^2)
R> dress <- sqrt(smuy2 - (1 + 1/8) * st2)
```

In the preceding BMA model the exchangeable members have equal weights and a common component distribution variance. Therefore, both BMA and ensemble dressing have only three model parameters: two to correct the ensemble-member forecasts and one component distribution standard deviation. Therefore, the parameters of these models are comparable:

```
R> rbind(BMA = c(BMA$biasCoefs[,1], BMA$sd),
+      ensdress = c(coef(MOS), sd = dress))

            (Intercept)   ensmean        sd
BMA          1.76468   0.3051412   2.909911
ensdress     1.89614   0.3157856   2.410527
```

Affine kernel dressing (AKD; Bröcker & Smith, 2008) is another ensemble dressing approach (see also Wilks, 2018, Chapter 3 of this book), which can be fitted by `FitAkdParameters()` from **SpecsVerification**. It expects input of a matrix of ensemble forecasts and a vector of observations and returns the model coefficients (see Eqs. 3.50, 3.51 in Wilks, 2018, Chapter 3 of this book).

```
R> library("SpecsVerification")
R> (AKD <- FitAkdParameters(ens = as.matrix(temptrain[,2:12]),
+      obs = temptrain$temp))

           a          r1          r2          s1          s2
  0.311741683  2.031694139  0.007992924  14.405440887  4.787428548
```

11.3.3 PREDICTION

Probabilistic forecasts can be expressed in multiple ways. Full continuous predictive distributions can be shown as, for example, probability density plots, but often these are difficult to display, especially when regarding multiple forecasts. Therefore, probabilities to fall below or over certain thresholds or predictive quantiles (prediction intervals) are often extracted. This section presents the generation of these different kinds of forecasts from the models that were fitted previously.

For NGR these forecasts can be based on the distribution functions provided by R (i.e., `dnorm()`, `pnorm()`, `qnorm()` for densities, probabilities, and quantiles of the normal distribution, respectively). Therefore, first the predictive means and standard deviations have to be derived with `predict()` analogously to the linear regression in Section 11.2. Forecast mean μ and standard deviation σ are derived by `type = "location"` and `type = "scale"`, respectively:

```
R> mean_NGR   <- predict(NGR_crch,  newdata = temptest, type = "location")
R> sd_NGR     <- predict(NGR_crch,  newdata = temptest, type = "scale")
R> mean_NGR2  <- predict(NGR_crch2, newdata = temptest, type = "location")
R> sd_NGR2    <- predict(NGR_crch2, newdata = temptest, type = "scale")
```

From these values, densities and probabilities can easily be produced with

```
R> x <- seq(-10, 10, 0.1)
R> pdf_NGR  <- dnorm(x, mean_NGR[1], sd_NGR[1])
R> cdf_NGR  <- pnorm(0, mean_NGR, sd_NGR)
```

This code generates the predictive density for the first date in the test data, evaluated on values from −10 to 10 (`pdf_NGR`), and a series of probability forecasts for freezing temperatures (`cdf_NGR`). Predictive quantiles could be derived similarly with `qnorm()`, but there is also a `type = "quantile"` option for `predict()`:

```
R> quant_NGR <- predict(NGR_crch, newdata = temptest, type = "quantile",
+      at = c(0.25, 0.5, 0.75))
```

which generates the predictive quartiles and median for all dates in the test data.

The predictive distributions of BMA and ensemble dressing are mixtures of normal distributions that are centered at the corrected ensemble forecasts. Therefore, the corrected ensemble forecasts and standard deviation of the component distributions have to be derived first. **ensembleBMA** does not have a direct function to derive the corrected ensemble, but the correction can easily be applied with matrix multiplication `%*%` that is applied to each of the ensemble members with `apply()`:

```
R> corrected_BMA  <- apply(temptest[,2:12], 2,
+     function(x) BMA$biasCoefs[,1] %*% rbind(1, x))
```

Similarly, the correction is also applied for ensemble dressing:

```
R> corrected_dress <- apply(temptest[,2:12], 2,
+     function(x) coef(MOS) %*% rbind(1, x))
```

For AKD, **SpecsVerification** provides the `DressEnsemble()` function to derive the corrected ensemble and the component distribution standard deviation:

```
R> AKDobj <- DressEnsemble(ens = as.matrix(temptest[2:12]),
+     dressing.method = "akd", parameters = as.list(AKD))
```

This line of code returns a list with matrices of corrected ensemble forecasts and component distribution standard deviations.

To use some of the functions from **SpecsVerification**, a similar object for the basic ensemble dressing model is created by:

```
R> dressobj <- list(ens = corrected_dress,
+     ker.wd = matrix(dress, nrow = nrow(corrected_dress), ncol = 11))
```

ensembleBMA provides direct functions to predict probabilities and quantiles of `"fitBMA"` objects. For ensemble dressing and AKD, `pnorm()` can be employed to derive probability forecasts:

```
R> cdf_BMA   <- cdf(BMA, temptest_eD, values = 0)
R> cdf_AKD   <- rowMeans(pnorm(0,   AKDobj$ens,   AKDobj$ker.wd))
R> cdf_dress <- rowMeans(pnorm(0, dressobj$ens, dressobj$ker.wd))
```

Predictive probability density functions (PDFs) could be derived similarly for probabilities with `dnorm()` instead of `pnorm()`. Alternatively, the `density()` function can be employed. Furthermore, there is a method for the generic `plot()` function to generate density plots for `"fitBMA"` models.

The following code creates predictive PDF plots for the first date in the test data set for NGR, BMA, ensemble dressing, and AKD:

```
R> par(mfrow = c(1,4))
R> plot(x, pdf_NGR, type = "l", xlab = "Temperature", ylab = "Density",
+     lwd = 3, main = "NGR")
R> abline(v = temptest$temp[1], col = "orange", lwd = 3)
```

```
R> plot(BMA, temptest_eD[1,])
R> title(main = "BMA")
R> plot(density(dressobj$ens[1,], bw = dressobj$ker.wd[1, 1]),
+     xlab= "Temperature", main = "ensemble dressing", lwd = 3)
R> abline(v = temptest$temp[1], col = "orange", lwd = 3)
R> plot(density(AKDobj$ens[1,], bw = AKDobj$ker.wd[1, 1]),
+     xlab = "Temperature", main = "AKD", lwd = 3)
R> abline(v = temptest$temp[1], col = "orange", lwd = 3)
```

These density plots are shown in Fig. 11.4 and look very similar for all forecast models.

Full predictive PDF plots are difficult to display for a series of days. Therefore, quantile or interval predictions are often shown. An example plot of median and quartile predictions for NGR, shown in the left panel of Fig. 11.5, can be created with

```
R> plot(quant_NGR[1:20, 2], type = "l", lty = 2, ylab = "2m temperature",
+     xlab = "date", xaxt = "n", ylim = c(-15, 10))
R> axis(1, at = seq(1, 20, 6), temptest$date[seq(1, 20, 6)])
R> polygon(c(1:20, 20:1), c(quant_NGR[1:20, 1], quant_NGR[20:1, 3]),
+     col = gray(0.1, alpha = 0.1), border = FALSE)
R> lines(temptest$temp[1:20])
```

where `polygon()` adds the shaded area.

Alternatively to quantiles, threshold probabilities can be plotted for a series of dates:

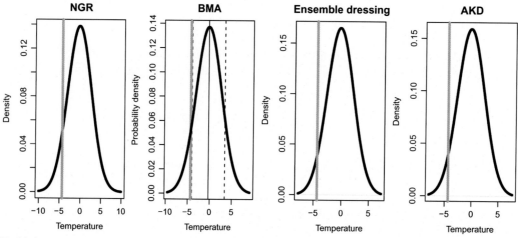

FIG. 11.4

Predictive densities of NGR, BMA, ensemble dressing, and AKD for 2010-12-01. *Black solid lines* show the predictive densities and *vertical orange lines* show the verifying observations. For BMA, the predictive median (*thin vertical line*) and 0.1 and 0.9 quantiles (*dashed lines*) are also shown.

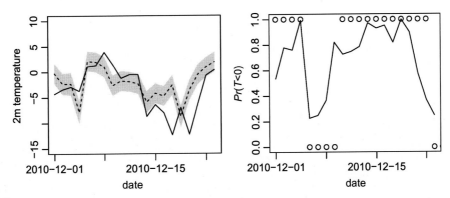

FIG. 11.5

Left: NGR median prediction (*dashed line*) and predictive interquartile range (*gray shading*). Observations are shown as the *solid line*. *Right*: NGR probability of freezing temperatures (*solid line*) and corresponding observations (*circles* at 1 for occurrence and at 0 for nonoccurrence).

```
R> plot(cdf_NGR[1:20], type = "l", ylab = "Pr(T<0)",
+    xlab = "date", xaxt = "n", ylim = c(0, 1))
R> axis(1, at = seq(1, 20, 6), temptest$date[seq(1, 20, 6)])
R> points(temptest$temp<0)
```

which is shown in the right panel of Fig. 11.5.

11.3.4 **VERIFICATION**

A variety of different approaches is available to measure the performance of probabilistic ensemble forecasts, which have been reviewed in Thorarinsdottir and Schuhen (2018, Chapter 6 of this book). This section shows some of the most common verification measures for continuous probabilistic forecasts, namely the continuous ranked probability score (CRPS), the ignorance score, and probability integral transform (PIT) histograms. It is also common to evaluate threshold probabilities of these forecasts with binary or categorical verification measures but we omit this here and refer to Section 11.4.4 for details on these measures.

The CRPS (see e.g., Thorarinsdottir & Schuhen, 2018, Chapter 6 of this book) is one of the most common single-number scores for continuous probabilistic forecasts. For certain parametric predictive distributions such as the Gaussian distribution, `crps()` from **scoringRules** has implementations of analytical CRPS expressions. Analytic CRPS derivations for Gaussian mixture distributions are implemented in `crps()` from **ensembleBMA** and `DressCrps()` from **specsVerification**.

The following code creates a matrix of CRPS values with the different methods in the columns and the different dates in the rows:

```
R> library("scoringRules")
R> crps_all <- cbind(
+    NGR1 = scoringRules::crps(temptest$temp, family = "normal",
```

```
+         mean = mean_NGR, sd = sd_NGR),
+     NGR2 = scoringRules::crps(temptest$temp, family = "normal",
+         mean = mean_NGR2, sd = sd_NGR2),
+     BMA   = ensembleBMA::crps(BMA, temptest_eD)[, 2],
+     dress = DressCrps(dressobj, temptest$temp),
+     AKD   = DressCrps(AKDobj, temptest$temp))
```

Note that the syntax `packagename::crps()` is used here because **ensembleBMA** and **scoringRules** both have a function `crps()` so that the package has to be specified from which the function should be taken. `ensembleBMA::crps()` returns a two-column matrix of raw ensemble and BMA CRPS where here only the BMA column is taken with `[, 2]`.

Commonly the CRPS is averaged over the test data set, which can be derived, for example, with `colMeans()`. To estimate the sampling distribution of this mean CRPS, a bootstrapping approach (Efron, 1979) is used where average CRPS values are computed on 250 bootstrap samples. The following simple function draws 250 samples with replacement, computes the average CRPS on each of them, and returns a matrix with the different postprocessing methods in the columns and the 250 average CRPS values in the rows.

```
R> bootmean <- function(scores, nsamples = 250) {
+     boot <- NULL
+     for(i in 1:nsamples) {
+         bindex <- sample(nrow(scores), replace = TRUE)
+         boot <- rbind(boot, colMeans(scores[bindex,]))
+     }
+     boot
+ }
```

These sampling distributions can then be plotted in a boxplot:

```
R> boxplot(bootmean(crps_all), ylab = "CRPS")
```

which is shown in the left panel of Fig. 11.6.

A second common single number score is the ignorance or logarithmic score, which is the predictive log-density evaluated at the observation (Thorarinsdottir & Schuhen, 2018, Chapter 6 of this book). Therefore, its computation is similar to that of density forecasts and can be performed with `dnorm()`:

```
R> ign_all <- cbind(
+     NGR1  = -dnorm(temptest$temp, mean_NGR, sd_NGR, log = TRUE),
+     NGR2  = -dnorm(temptest$temp, mean_NGR2, sd_NGR2, log = TRUE),
+     BMA   = -log(rowSums(BMA$weights * dnorm(temptest$temp, corrected_BMA, BMA$sd))),
+     dress = -log(rowMeans(dnorm(temptest$temp, dressobj$ens, dressobj$ker.wd))),
+     AKD   = -log(rowMeans(dnorm(temptest$temp, AKDobj$ens, AKDobj$ker.wd))))
```

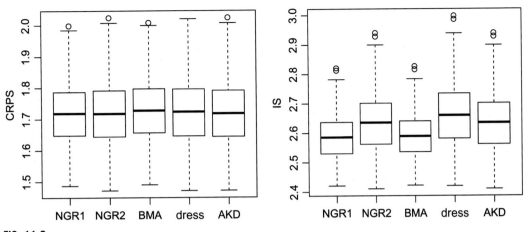

FIG. 11.6

Sampling distributions for continuous ranked probability score (CRPS; *left*) and ignorance score (IS; *right*) for NGR, BMA, ensemble dressing, and AKD. The *horizontal lines* mark the medians and the *boxes* the interquartile range of 250 values from bootstrapping. The *whiskers* show the most extreme values that are less than 1.5 times the length of the box away from the box, and *empty circles* are plotted for values that are outside the *whiskers*. NGR1 and NGR2 have been derived using Eqs. (11.3), (11.4), respectively.

Similar to the CRPS, a box plot of the sampling distributions is created with

```
R> boxplot(bootmean(ign_all), ylab = "IS")
```

and is shown in the right panel of Fig. 11.6.

While the CRPS in Fig. 11.6 shows only small differences between the different forecast methods, the ignorance score indicates slightly better performance for the NGR and BMA models.

PIT histograms are the continuous counterpart of verification rank histograms (e.g., left side of Fig. 11.3) and can be used to assess the reliability (i.e., calibration) of continuous probabilistic forecasts (Thorarinsdottir & Schuhen, 2018, Chapter 6 of this book). PITs are the predictive cumulative distribution functions evaluated at the observations so that their computation is similar to that for the threshold probabilities. For BMA the function `pit()` from **ensembleBMA** can be used. For the other methods the PITs are derived with `pnorm()`:

```
R> pit <- cbind(
+    NGR1  = pnorm(temptest$temp, mean_NGR, sd_NGR),
+    NGR2  = pnorm(temptest$temp, mean_NGR2, sd_NGR2),
+    BMA   = pit(BMA, temptest_eD),
+    dress = rowMeans(pnorm(temptest$temp, dressobj$ens, dressobj$ker.wd)),
+    AKD   = rowMeans(pnorm(temptest$temp, AKDobj$ens, AKDobj$ker.wd)))
```

PIT histograms can then be plotted with `hist()`

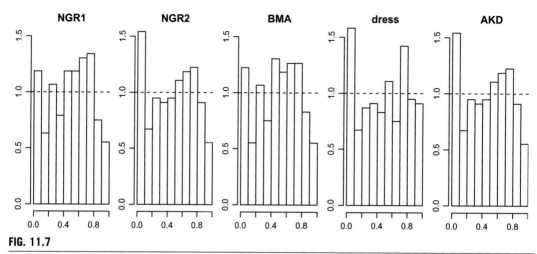

FIG. 11.7

PIT histograms for NGR, BMA, ensemble dressing, and AKD. Perfect calibration is shown as *dashed horizontal line*.

```
R> par(mfrow = c(1, ncol(pit)))
R> for(model in colnames(pit)){
+     hist(pit[, model], main = model, freq = FALSE,
+        xlab = "", ylab = "", ylim = c(0, 1.6))
+     abline(h = 1, lty = 2)
+  }
```

These PIT histograms in Fig. 11.7 show that compared with the raw ensemble (Fig. 11.3) the postprocessed forecasts are better calibrated. However, the wave-like form of the histograms indicates that the assumed symmetry of the Gaussian distribution might not be completely appropriate for this data set.

11.4 POSTPROCESSING OF PRECIPITATION

The previous two sections have shown postprocessing of ensemble temperature forecasts, where all of the methods used assume Gaussian distributions either directly or for the components of mixture distributions. Meteorological variables such as precipitation or wind speed are nonnegative and therefore clearly non-Gaussian so that these postprocessing techniques cannot be applied directly. Ensemble postprocessing for precipitation has been an active field of research and various methods have been proposed (e.g., Hamill, Whitaker, & Wei, 2004; Scheuerer, 2014; Scheuerer & Hamill, 2015 a; Sloughter, Raftery, Gneiting, & Fraley, 2007; Stauffer, Umlauf, Messner, Mayr, & Zeileis, 2017; Wilks, 2009). A review of these methods can be found in Wilks (2018, Chapter 3 of this book). This section shows R implementations of some of these methods, namely censored nonhomogeneous logistic regression, BMA with discrete-gamma mixture component distributions, and several variants of logistic regression.

11.4.1 DATA

This section uses the `rain` data set that is also included in the **ensemblepp** package and can be loaded with:

```
R> data("rain", package = "ensemblepp")
```

It contains 12-hour accumulated precipitation observations and corresponding 18–30 hours ensemble forecasts at Innsbruck (Austria). Similar to `temp`, the ensemble forecasts come from the second generation GEFS reforecast data set (Hamill et al., 2013) and span a time frame from 2000 to 2015.

```
R> dim(rain)

[1] 2749  12

R> names(rain)

 [1] "rain"     "rainfc.1"  "rainfc.2"  "rainfc.3"  "rainfc.4"  "rainfc.5"
 [7] "rainfc.6"  "rainfc.7"  "rainfc.8"  "rainfc.9"  "rainfc.10" "rainfc.11"
```

It is common to transform precipitation data prior to applying postprocessing methods. Common choices are the square root (e.g., Stauffer et al., 2017; Wilks, 2009) or the cube root (e.g., Schmeits & Kok, 2010; Sloughter et al., 2007). For an easy comparison among the methods, only a square root transformation is used here.

```
R> rain <- sqrt(rain)
```

Again, only data from December, January, and February are considered:

```
R> rain$date <- as.Date(rownames(rain))
R> rain <- rain[format(rain$date, "%m") %in% c("12", "01", "02"),]
```

Furthermore, as was done for the temperature data in Section 11.2.1, ensemble means and standard deviations are computed and the data are split into training and test data.

```
R> rain$ensmean <- apply(rain[,2:12], 1, mean)
R> rain$enssd   <- apply(rain[,2:12], 1, sd)
R> raintrain <- rain[rain$date < "2010-03-01",]
R> raintest  <- rain[rain$date > "2010-03-01",]
```

A scatterplot of precipitation by ensemble mean forecasts, a verification rank histogram, and a spread-skill plot can be created similarly to those in Sections 11.2 and 11.3. The following code also creates a histogram of precipitation observations:

```
R> par(mfrow = c(2,2))
R> plot(rain~ensmean, raintrain, col = gray(0.2, alpha = 0.4),
+     main = "Scatterplot")
R> abline(0, 1, lty = 2)
```

```
R> rank <- apply(raintrain[,1:12], 1, rank)[1,]
R> hist(rank, breaks = 0:12 + 0.5, main = "Verification Rank Histogram")
R> sdcat <- cut(raintrain$enssd, quantile(raintrain$enssd, seq(0, 1, 0.2)))
R> boxplot(abs(rain-ensmean)~sdcat, raintrain, ylab = "absolute error",
+    xlab = "ensemble standard deviation", main = "Spread-Skill")
R> hist(rain$rain, xlab = "square root of precipitation", main = "Histogram")
```

These are plotted in Fig. 11.8 and show that, compared with temperature, the precipitation ensemble means have a weaker correlation with the observations, but the ensembles are better calibrated and exhibit a stronger spread-skill relationship. The histogram shows the non-Gaussian distribution of precipitation and the large number of zero precipitation events.

11.4.2 MODEL FITTING

Precipitation data are nonnegative, non-Gaussian, and usually contain a large number of zeros. Therefore, the methods that were used in the previous section should not be applied directly. This section introduces implementations of some postprocessing methods that are better suited for precipitation. Continuous probabilistic precipitation forecasts can be derived by non-Gaussian nonhomogeneous regression methods, BMA with non-Gaussian component distributions, or extended logistic regression.

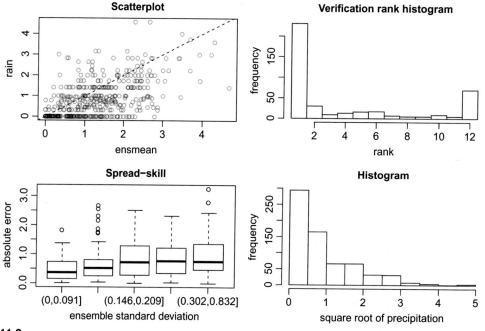

FIG. 11.8

Diagnostic plots for 18–30 hours accumulated precipitation ensemble forecasts. *Upper left*: Scatterplot of 12-hour accumulated precipitation by ensemble mean forecast. 1:1 relationship is shown as *dashed line*. *Upper right*: Verification rank histogram. *Bottom left*: Spread skill relationship. *Bottom right*: Histogram of transformed precipitation observations.

Often, distribution assumptions are avoided by using other logistic regression approaches to predict only probabilities to fall below or above one or several thresholds.

Nonhomogeneous regression

Several variants of nonhomogeneous regression for precipitation have been proposed in the literature, which mainly differ in the assumed predictand distribution. For example, Scheuerer (2014) proposed generalized extreme value (GEV) distributions, Messner, Mayr, et al. (2014) proposed logistic distributions, and Scheuerer and Hamill (2015a) proposed shifted gamma distributions. All of these approaches account for the large number of zero precipitation events by censoring the data. Therefore, precipitation r is modeled as a latent variable r^* that can also take on negative values, but where these negative values are only observed as 0.

$$r = \begin{cases} 0 & r^* \leq 0 \\ r^* & r^* > 0 \end{cases} \tag{11.5}$$

Unfortunately, no direct R implementations of the censored GEV and censored shifted gamma models are available. Therefore, only the censored nonhomogeneous logistic regression (cNLR; Messner, Mayr, et al., 2014) is considered in the following. This model can be formulated similarly to NGR in Eqs. (11.1)–(11.3), but with the logistic instead of the Gaussian distribution in Eq. (11.1):

$$r^* \sim \mathcal{L}(\mu, \sigma) \tag{11.6}$$

$$\mu = \beta_0 + \beta_1 m \tag{11.7}$$

$$\log(\sigma) = \gamma_0 + \gamma_1 s \tag{11.8}$$

where $\mathcal{L}(\mu, \sigma)$ is the logistic distribution with mean μ and scale σ.

Maximum likelihood (and minimum CRPS) estimation of this model can be carried out with `crch()` (see also Section 11.3) by setting the left censoring point `left = 0` and `dist = "logistic"`:

```
R> cNLR <- crch(rain ~ ensmean | enssd, data = raintrain, left = 0,
+      dist = "logistic")
```

As an alternative to `crch()`, maximum likelihood estimation of this model is also supported by `gamlss()`.

Bayesian model averaging

Sloughter et al. (2007) proposed a BMA variant that is tailored to quantitative precipitation forecasts. Instead of Gaussian distributions, the component distributions are mixtures of gamma distributions and point masses at zero (Wilks, 2018, Chapter 3 of this book). An R implementation of this model is provided by `fitBMA()` from **ensembleBMA**. Therefore, the data have to be converted first into `"ensembleData"` objects:

```
R> raintrain_eD <- ensembleData(forecasts = raintrain[,2:12],
+      dates = raintrain$date, observations = raintrain$rain,
+      forecastHour = 24, initializationTime = "00", exchangeable = rep(1, 11))
R> raintest_eD <- ensembleData(forecasts = raintest[,2:12],
+      dates = raintest$date, observations = raintest$rain,
+      forecastHour = 24, initializationTime = "00", exchangeable = rep(1, 11))
```

As was the case for Gaussian BMA in Section 11.3, model fitting is carried out with `fitBMA()` but with `model = "gamma0"`. By default, the data are internally cube root transformed prior to BMA fitting. However, because the data here have already been square root transformed, the power transformation is switched off by setting `control = controlBMAgamma0(power = 1)`:

```
R> gBMA <- fitBMA(raintrain_eD, model = "gamma0",
+     control = controlBMAgamma0(power = 1))
```

Logistic regression

Often, it is sufficient to only predict probabilities of precipitation to fall below or above certain thresholds. A typical example is precipitation occurrence forecasts with a threshold of zero. By modeling only these threshold probabilities instead of full continuous distributions, specific distribution assumptions can be avoided. Hamill et al. (2004) proposed postprocessing ensemble forecasts with logistic regression, which is a member of the generalized linear model (GLM; Nelder & Wedderburn, 1972) family and is a common regression method for binary outcomes.

In R GLMs are usually fitted with `glm()`, which has a similar interface to `lm()`. For logistic regression, the predictand has to be a binary variable (`TRUE` and `FALSE` or 0 and 1), which can, for example, be created by relational operators such as `">"`. Furthermore, `family` has to be set to `binomial()`:

```
R> logreg <- glm(rain > 0 ~ ensmean, data = raintrain, family = binomial())
```

The preceding example code uses the ensemble mean as the sole predictor, because various studies have shown that including the ensemble standard deviation as an additional predictor does not improve the forecasts (e.g., Hamill et al., 2004; Wilks & Hamill, 2007). Messner, Zeileis, Mayr, and Wilks (2014) pointed out that the ensemble spread cannot directly affect the forecast uncertainty when used as a standard predictor variable and that the ensemble spread can be exploited more efficiently when used as a predictor variable for the scale of the logistic function (similar to Eqs. 11.6–11.8 or Eq. 3.13 in Wilks, 2018, Chapter 3 of this book). This approach can be carried out in `hetglm()` from **glmx** where, similar to `crch()`, the predictor variables are specified with a formula that separates location and scale predictors by `"|"`.

```
R> library("glmx")
R> hlogreg <- hetglm(rain > 0 ~ ensmean | enssd, data = raintrain,
+     family = binomial())
```

Often, more than one threshold is of interest, for example, the climatological deciles (quantiles with probabilities from 0.1 to 0.9):

```
R> q <- unique(quantile(raintrain$rain, seq(0.1, 0.9, 0.1)))
```

With `unique()`, quantiles of equal values are merged (here, the 0.1, 0.2, and 0.3 quantiles are zero).

As a simple approach, logistic regression can be fitted for each threshold separately. The following code fits these separate (heteroscedastic) logistic regressions and writes the fitted model objects in lists:

```
R> logreg2 <- hetlogreg2 <- list()
R> for(i in 1:length(q)){
```

```
+    logreg2[[i]] <- glm(rain <= q[i] ~ ensmean, data = raintrain,
+       family = binomial())
+    hetlogreg2[[i]] <- hetglm(rain <= q[i] ~ ensmean | enssd,
+       data = raintrain, family = binomial())
+  }
```

A problem with this approach is that the regression lines for the different thresholds can cross, which will lead to nonsense negative probabilities for intervals between these thresholds (see e.g., Fig. 3.4 in Wilks, 2018, Chapter 3 of this book). Wilks (2009) proposed avoiding this problem with an extended logistic regression that fits a single equation for all thresholds, with the threshold as additional predictor variable (see also Wilks, 2018, Chapter 3 of this book). In addition to avoiding negative probabilities, this extended logistic regression can also provide full continuous predictive distributions. Messner, Zeileis, et al. (2014) further extended this approach to efficiently exploit the ensemble spread, similar to the heteroscedastic logistic regression (hetglm()) or nonhomogeneous regression (Eqs. 11.6–11.8). hxlr() from **crch** provides an implementation of this heteroscedastic extended logistic regression (HXLR) with a similar interface to crch() or hetglm():

```
R> HXLR <- hxlr(rain ~ ensmean | enssd, data = raintrain, thresholds = q)
```

Proportional-odds or ordered logistic regression (OLR; e.g., Hemri, Haiden, & Pappenberger, 2016; Messner, Mayr, et al., 2014) is another closely related logistic regression approach, which is applicable to a finite collection of thresholds, and avoids crossing of regression lines by restricting them to be parallel. Implementations of OLR can be found in polr() from **MASS** or clm() from **ordinal**. For these functions, the predictand has to be an ordered factor which can, for example, be created by cut():

```
R> raintrain$raincat <- cut(raintrain$rain, c(-Inf, q, Inf))
```

clm() also has an option to specify predictor variables for the scale with a second formula scale:

```
R> library("ordinal")
R> OLR <- clm(raincat ~ ensmean, scale = ~ enssd, data = raintrain)
```

11.4.3 PREDICTION

Predictions for the models that have been fitted in the previous subsection are created similarly to the temperature predictions in Section 11.3.3. However, the binary and ordered logistic regression models do not support density or quantile predictions, and only probability forecasts can be derived.

For cNLR, location and scale parameters of the predictive logistic distribution can be derived with predict():

```
R> location_cNLR  <- predict(cNLR, newdata = raintest, type = "location")
R> scale_cNLR     <- predict(cNLR, newdata = raintest, type = "scale")
```

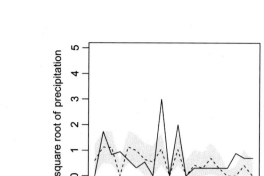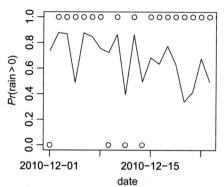

FIG. 11.9

Left: cNLR median prediction (*dashed line*) and predictive interquartile range (*gray shading*). Observations are shown as *solid line*. *Right*: cNLR probability of precipitation (*solid line*) and corresponding observations (*circles* at 1 for occurrence and at 0 for nonoccurrence).

Density, quantiles, and probabilities are then derived similarly to the previous temperature forecasts, but with distribution functions for the censored logistic distribution (dclogis(), pclogis(), qclogis() from **crch**). For example, the prediction intervals or probability-of-precipitation plots for cNLR in Fig. 11.9 are created similarly to Fig. 11.5:

```
R> par(mfrow = c(1,2))
R> quant_cNLR <- predict(cNLR, newdata = raintest, type = "quantile",
+     at = c(0.25, 0.5, 0.75))
R> plot(quant_cNLR[1:20, 2], type = "l", lty = 2, ylim = c(0, 5),
+     ylab = "square root of precipitation", xlab = "date", xaxt = "n")
R> axis(1, at = seq(1, 20, 6), raintest$date[seq(1, 20, 6)])
R> polygon(c(1:20, 20:1), c(quant_cNLR[1:20, 1], quant_cNLR[20:1, 3]),
+     col = gray(0.1, alpha = 0.1), border = FALSE)
R> lines(raintest$rain[1:20])

R> cdf_cNLR <- sapply(q, function(q) plogis(q, location_cNLR, scale_cNLR))
R> plot(1 - cdf_cNLR[1:20], type = "l", ylab = "Pr(rain>0)",
+     xlab = "date", xaxt = "n", ylim = c(0,1))
R> axis(1, at = seq(1, 20, 6), temptest$date[seq(1, 20, 6)])
R> points(raintest$rain>0)
```

Predictive densities of cNLR, BMA, and HXLR, similar to Fig. 11.4 can be plotted with:

```
R> par(mfrow = c(1,3))
R> x <- c(0, seq(1e-8, 6.5, 0.1))
R> ## cNLR
R> plot(x, dclogis(x, location_cNLR[2], scale_cNLR[2], left = 0), type = "l",
+     lwd = 3, xlab = "Square root of precipitation", ylab = "PDF", main = "cNLR")
```

```
R> abline(v = raintest$rain[2], lwd = 3, col = "orange")
R> ## BMA
R> plot(gBMA, raintest_eD[2,])
R> title(main = "BMA")
R> ## HXLR
R> location_HXLR <- predict(HXLR, newdata = raintest, type = "location")
R> scale_HXLR   <- predict(HXLR, newdata = raintest, type = "scale")
R> plot(x, dclogis(x, location_HXLR[2], scale_HXLR[2], left = 0), type = "l",
+     lwd = 3, xlab = "Square root of precipitation", ylab = "PDF", main = "HXLR")
R> abline(v = raintest$rain[2], lwd = 3, col = "orange")
```

These predictive densities, which are shown in Fig. 11.10, all have a discrete probability at 0. However, while the BMA gamma distribution has close to zero probability for very small precipitation amounts, a smoother transition from zero to positive amounts can be seen for the censored logistic distributions of cNLR and HXLR.

The verification in the next section mainly concentrates on threshold probability forecasts. For BMA, these forecasts can be generated with cdf():

```
R> cdf_gBMA <- cdf(gBMA, raintest_eD, values = q)
```

For binary (heteroscedastic) logistic regression, these probabilities are computed with predict() where type="response" has to be set. The following code uses sapply() to predict the logistic regression models for all thresholds and create a matrix of cumulative probabilities for all thresholds.

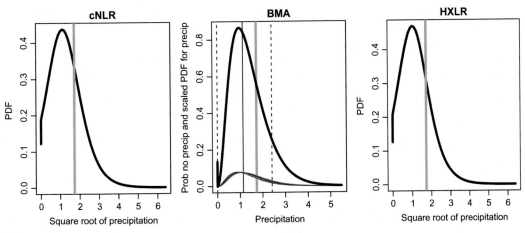

FIG. 11.10

Predictive densities of cNLR, BMA, and HXLR for 2010-12-01. *Black solid lines* show the predictive densities and *vertical orange lines* show the verifying observations. For BMA also the predictive median (*thin vertical line*), 0.1 and 0.9 quantiles (*dashed lines*), and the member distributions (*colored curves*) are shown. Note that the BMA plot method ignores the xlab argument so that the x-axis label for the middle (BMA) plot incorrectly states "Precipitation" when it should be the "Square root of precipitation."

```
R> cdf_logreg <- sapply(logreg2, function(mod)
+     predict(mod, newdata = raintest, type = "response"))
R> cdf_hetlogreg <- sapply(hetlogreg2, function(mod)
+     predict(mod, newdata = raintest, type = "response"))
```

predict() is also used for threshold probability forecasts of HXLR and OLR with type = "cumprob" and type = "cum.prob", respectively.

```
R> cdf_HXLR <- predict(HXLR, newdata = raintest, type = "cumprob")
R> cdf_OLR <- predict(OLR, newdata = raintest, type = "cum.prob")$cprob1[,-7]
```

Finally, the forecast probabilities from the different methods are combined to a list for a more convenient verification in the following:

```
R> CDF <- list(cNLR = cdf_cNLR, BMA = cdf_gBMA, logreg = cdf_logreg,
+     hlogreg = cdf_hetlogreg, HXLR = cdf_HXLR, OLR = cdf_OLR)
```

11.4.4 VERIFICATION

cNLR, BMA, and HXLR provide full continuous predictive distributions. The other logistic regression methods can only predict threshold probabilities. Therefore, this section compares the different methods, mainly on scores for binary or categorical predictions. CRPS, ignorance score, or PIT histograms for cNLR, BMA, and HXLR can be derived similarly to Section 11.3.4.

In the following, binary verification methods are shown only on precipitation occurrence (threshold zero; first columns of CDF matrices). Other thresholds can be evaluated similarly.

The Brier score is one of the most common verification measures for binary probabilistic forecasts. Essentially, it is the squared error of probability forecasts where the observations are considered as 0 (not occurred) or 1 (occurred). The following code derives the Brier scores of all methods and combines them in a matrix with the methods in the columns and the different dates in the rows:

```
R> brier_all <- NULL
R> for(n in names(CDF)) {
+     brier_all <- cbind(brier_all, ((raintest$rain <= 0) - CDF[[n]][,1])^2)
+ }
R> colnames(brier_all) <- names(CDF)
```

Similar to the CRPS and ignorance score in Section 11.3.4, the sampling distributions of the mean Brier scores can be illustrated in box plots of bootstrapped means:

```
R> boxplot(bootmean(brier_all), las = 2, ylab = "Brier score")
```

This plot, which is shown in the left panel of Fig. 11.11, indicates only small differences between the forecast methods with BMA having a slightly worse performance than the other approaches.

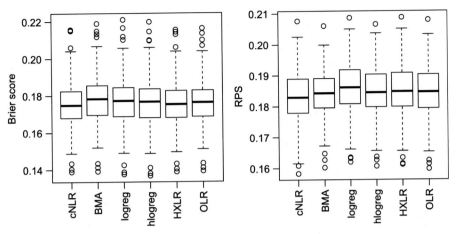

FIG. 11.11

Bootstrapped sampling distributions for Brier scores for probability of no precipitation (*left*) and ranked probability score (*right*) for the different precipitation postprocessing methods. *Boxes, whiskers,* and *circles* have the same meaning as in Fig. 11.6.

Often, the Brier score is decomposed into reliability, resolution, and uncertainty (see e.g., Thorarinsdottir & Schuhen, 2018). This decomposition is implemented in `brier()` from **verification** or `BrierDecomp()` from **SpecsVerification**:

```
R> sapply(CDF, function(x) BrierDecomp(x[,1], y = (raintest$rain <= 0))[1,])
            cNLR         BMA      logreg     hlogreg        HXLR         OLR
REL 0.004704628 0.009858622 0.003886837 0.004187534 0.006635748 0.00464618
RES 0.043203596 0.043641378 0.040695738 0.041052569 0.043469168 0.03907385
UNC 0.214751051 0.214751051 0.214751051 0.214751051 0.214751051 0.21475105
```

`BrierDecomp()` returns a matrix with the Brier composition in the first and its estimated standard deviations in the second line. Therefore, the first line is subset with `[1,]`. This decomposition reveals that the high Brier score of BMA can mainly be attributed to its large (i.e., relatively poor) reliability.

A more comprehensive evaluation of binary forecasts is provided by reliability diagrams (Wilks, 2011), for which implementations are provided by `reliability.plot()` from **verification** or `ReliabilityDiagram()` from **SpecsVerification**. Reliability diagrams for all method can be created by following code and are shown in Fig. 11.12:

```
R> par(mfrow=c(2,3))
R> for(n in 1:length(names(CDF))){
+    ReliabilityDiagram(1 - CDF[[n]][,1], (raintest$rain > 0), plot=TRUE)
+    par(mfg = c((n-1) %/% 3 + 1, (n-1) %% 3 + 1))
+    title(main = names(CDF)[n])
+ }
```

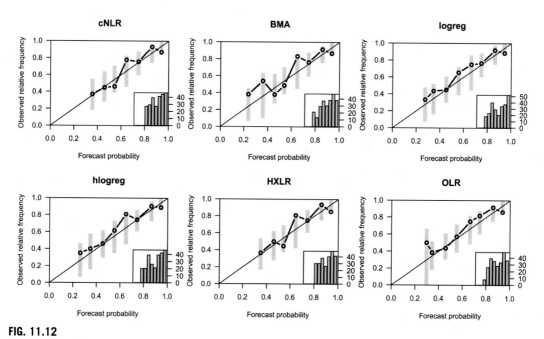

FIG. 11.12

Reliability diagrams for probability of precipitation (i.e., $Pr(r > 0)$) from the different postprocessing methods. The calibration functions are shown as *circles connected by lines*. *Gray bars* show consistency bars from bootstrap resampling (Bröcker & Smith, 2007).

The command $par(mfg = c((n-1) \%/\% 3 + 1, (n-1) \%\% 3 + 1)$ is required to arrange the diagrams in the 6 panels because ReliabilityDiagram() resets the graphical parameters.

The reliability diagrams show good calibration for all of the forecast methods (calibration function close to diagonal line). In agreement with the reliability from the Brier score decomposition, BMA deviates most from the diagonal, but still lies within the consistency range.

Receiver operating characteristic (ROC) diagrams are another common graphical verification tool for binary predictions (Wilks, 2011). R provides various functions to create ROC diagrams such as roc.plot() from **verification**, plot() from **ROCR**, or roc() from **pROC**. The following code creates ROC diagrams for all methods and adds the area under the curve (AUC) to these plots:

```
R> library("pROC")
R> par(mfrow = c(2, 3))
R> for(n in names(CDF)){
+     rocplot <- roc((raintest$rain > 0) ~ I(1 - CDF[[n]][, 1]), plot=TRUE,
+       main = n)
+     text(0.2, 0.2, paste("AUC =", round(rocplot$auc, digits = 4)))
+ }
```

These ROC diagrams, which are shown in Fig. 11.13, look very similar for all methods, but BMA has the best (highest) AUC.

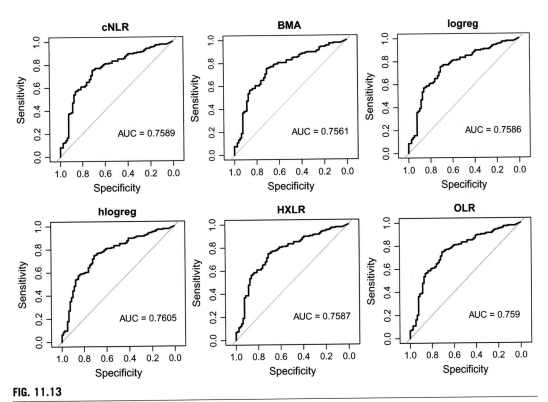

FIG. 11.13

Receiver operator characteristic (ROC) diagrams for probability of precipitation (i.e., $Pr(r > 0)$) from the different postprocessing methods. The areas under the curves (AUC) are displayed in lower right corners of the respective plots.

The Brier score is a well-suited verification measure for single-threshold probability forecasts. For two or more thresholds, the Brier score can be generalized to the ranked probability score (RPS). Essentially the RPS is the mean Brier score over all thresholds (here we use the same thresholds that have been used for fitting in Section 11.4.2, that is, climatological deciles with probabilities $0.1, 0.2, \ldots, 0.9$) and can, for example, be computed by:

```
R> cdf_obs <- sapply(q, function(q) raintest$rain <= q)
R> rps_all <- NULL
R> for(n in names(CDF)) {
+    rps_all <- cbind(rps_all,
+    rowSums((CDF[[n]] - cdf_obs)^2)/(ncol(cdf_HXLR) - 1))
+ }
R> colnames(rps_all) <- names(CDF)
```

A boxplot of the sampling distribution from bootstrapped RPS means can be created by the following code and is shown in the right panel of Fig. 11.11.

```
R> boxplot(bootmean(rps_all), las = 2, ylab = "RPS")
```

In agreement with the Brier score, cNLR performs well. The relatively poor performance of the standard logistic regression (logreg), especially in comparison with heteroscedastic logistic regression (hlogreg) indicates an advantage of exploiting the ensemble spread.

11.5 MULTIVARIATE POSTPROCESSING OF TEMPERATURE AND PRECIPITATION

Sections 11.3 and 11.4 performed univariate ensemble postprocessing on temperature and precipitation data examples, respectively. This section extends these examples to bivariate ensemble postprocessing of these two variables.

Schefzik and Möller (2018) have reviewed various multivariate postprocessing approaches. This chapter performs a parametric multivariate Gaussian copula approach (Möller, Lenkoski, & Thorarinsdottir, 2013) and the nonparametric ensemble copula coupling (ECC; Schefzik, Thorarinsdottir, & Gneiting, 2013) and Schaake shuffle (Clark, Gangopadhyay, Hay, Rajagopalan, & Wilby, 2004) approaches.

11.5.1 DATA

This section employs the same data as the previous sections. Fig. 11.14 shows a plot of precipitation by temperature, which is created by:

```
R> plot(raintrain$rain~temptrain$temp, xlab = "temperature",
+    ylab = "precipitation")
R> abline(lm(raintrain$rain~temptrain$temp))
```

This plot shows a weak but positive relationship between these two variables, which can be explained by cold temperatures that mainly occur on clear nights with little or no precipitation. In the following, multivariate methods are employed to capture this correlation.

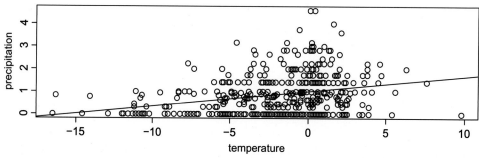

FIG. 11.14

Innsbruck 3-day accumulated precipitation by temperature. The *solid line* shows the linear regression fit.

11.5.2 **MODEL FITTING**

The copula-based multivariate postprocessing approaches that are used in the following are all based on univariate postprocessing of the marginal predictive distributions. Here we use NGR (Eqs. 11.1–11.3) for temperature and cNLR (Eqs. 11.6–11.8) for precipitation, which have been fitted in Sections 11.3 and 11.4.

In the Gaussian copula approach, the Gaussian quantile function (qnorm()) and the marginal predictive CDFs are used to transform the data into multivariate Gaussian samples. For censored distributions, the CDF cannot be derived exactly at values on the censoring point (zero precipitation). Therefore, the clogispit() function, which can be found in Section 11.7.1, derives the marginal CDF for these data by random sampling. Thus, the Gaussian copula can be applied to the training data with:

```
R> trtemp <- qnorm(pnorm(temptrain$temp,
+     fitted(NGR_crch, type = "location"),
+     fitted(NGR_crch, type = "scale")))
R> trrain <- qnorm(clogispit(raintrain$rain,
+     fitted(cNLR, type = "location"),
+     fitted(cNLR, type = "scale")))
```

In the Gaussian copula approach, these transformed data are assumed to follow a multivariate Gaussian distribution with a covariance matrix that can be derived with cov():

```
R> covmatrix <- cov(cbind(trtemp, trrain))
```

The nonparametric ECC and the Schaake shuffle do not require any prior model fitting and are directly applied in the prediction.

11.5.3 **PREDICTION**

Multivariate predictions are commonly provided as scenarios or ensembles. The number of scenarios in the ECC approach has to match the size of the postprocessed ensemble. Thus, to compare the different approaches, nscen = 11 scenarios are generated for all approaches.

```
R> nscen <- 11
```

For the Gaussian copula approach, first nscen bivariate Gaussian random samples have to be drawn, which, for example, can be done with rmvnorm() from **mvtnorm**. These random samples are then transformed back to the marginal distributions by the marginal quantile functions (qnorm() and qclogis()) and the Gaussian CDF (pnorm()). The following code generates temperature and precipitation scenario predictions for the full test data set:

```
R> library("mvtnorm")
R> GCtemp <- GCrain <- NULL
R> for(i in 1:nrow(temptest)) {
+     sim <- rmvnorm(nscen, c(0, 0), sigma = covmatrix)
```

```
+    GCtemp <- rbind(GCtemp, qnorm(pnorm(sim[,1]), mean_NGR[i], sd_NGR[i]))
+    GCrain <- rbind(GCrain,
+      qclogis(pnorm(sim[,2]), location_cNLR[i], scale_cNLR[i], left = 0))
+  }
```

ECC and Schaake shuffle are both based on univariate scenarios drawn from the marginal predictive distributions, which can easily be generated by the random number functions for the respective distributions (rnorm() and rclogis()):

```
R> UNIVtemp <- UNIVrain <- NULL
R> for(i in 1:nrow(temptest)) {
+    UNIVtemp <- rbind(UNIVtemp, rnorm(nscen, mean_NGR[i], sd_NGR[i]))
+    UNIVrain <- rbind(UNIVrain,
+      rclogis(nscen, location_cNLR[i], scale_cNLR[i], left = 0))
+  }
```

Subsequently, the rank order of these scenarios are adjusted to the rank order of the raw ensemble (ECC) or the rank order of sampled past data (Schaake shuffle). This ordering can be carried out with sort() and rank():

```
R> ECCtemp <- ECCrain <- NULL
R> for(i in 1:nrow(temptest)) {
+    ECCtemp <- rbind(ECCtemp, sort(UNIVtemp[i,])[rank(temptest[i, 2:12])])
+    ECCrain <- rbind(ECCrain, sort(UNIVrain[i,])[rank(raintest[i, 2:12])])
+  }
```

```
R> SStemp <- SSrain <- NULL
R> for(i in 1:nrow(temptest)) {
+    ind <- sample(nrow(temptrain), nscen)
+    SStemp <- rbind(SStemp, sort(UNIVtemp[i,])[rank(temptrain$temp[ind])])
+    SSrain <- rbind(SSrain, sort(UNIVrain[i,])[rank(raintrain$rain[ind])])
+  }
```

11.5.4 VERIFICATION

Thorarinsdottir and Schuhen (2018, Chapter 6 of this book) have reviewed different verification measures for multivariate probabilistic forecasts. This section compares the bivariate ensemble temperature and precipitation forecasts with the common energy score (Gneiting, Stanberry, Grimit, Held, & Johnson, 2008), the variogram score (Scheuerer & Hamill, 2015b), and multivariate rank histograms (Gneiting et al., 2008).

The energy and variogram scores can be computed by es_sample() and vs_sample() from **scoringRules**, respectively. The following code derives both scores for all dates in the test data set:

```
R> es <- vs <- NULL
R> for(i in 1:nrow(temptest)) {
```

```
+    obs <- c(temptest$temp[i], raintest$rain[i])
+    es <- rbind(es, c(
+      UNIV = es_sample(obs, rbind(UNIVtemp[i,], UNIVrain[i,])),
+      ECC  = es_sample(obs, rbind(ECCtemp[i,], ECCrain[i,])),
+      GC   = es_sample(obs, rbind(GCtemp[i,], GCrain[i,])),
+      SS   = es_sample(obs, rbind(SStemp[i,], SSrain[i,]))))
+    vs <- rbind(vs, c(
+      UNIV = vs_sample(obs, rbind(UNIVtemp[i,], UNIVrain[i,])),
+      ECC  = vs_sample(obs, rbind(ECCtemp[i,], ECCrain[i,])),
+      GC   = vs_sample(obs, rbind(GCtemp[i,], GCrain[i,])),
+      SS   = vs_sample(obs, rbind(SStemp[i,], SSrain[i,]))))
+
+  }
```

Similar to, for example, Fig. 11.6, the sampling distributions of the mean scores can be depicted by boxplots of bootstrapped means:

```
R> par(mfrow = c(1, 2))
R> boxplot(bootmean(es), ylab = "ES")
R> boxplot(bootmean(vs), ylab = "VS")
```

Fig. 11.15 shows only small differences among the different approaches. The Gaussian copula approach has a slightly better performance than the other methods in both scores. The Schaake shuffle is slightly better than the univariate scenarios and ECC in the Variogram score. The very little difference between the univariate scenarios and ECC indicates that the raw ensembles do not reflect the correlation between temperature and precipitation well.

Multivariate verification rank histograms are another common verification tool for multivariate forecasts. These diagnostic plots are similar to univariate verification rank histograms but use multivariate rank orders.

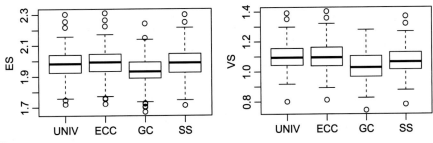

FIG. 11.15

Bootstrap sampling distributions for the energy (*left*) and variogram score (*right*) of the different multivariate postprocessing methods.

A function to compute multivariate ranks is provided in the supplemental material of Thorarinsdottir, Scheuerer, and Heinz (2016) and can also be found in Section 11.7.1. With the function `mv.rank()` multivariate ranks for the test data set can be derived by:

```
R> r <- NULL
R> for(i in 1:nrow(temptest)) {
+     obs <- c(temptest$temp[i], raintest$rain[i])
+     r <- rbind(r, c(
+     UNIV = mv.rank(cbind(obs, rbind(UNIVtemp[i,], UNIVrain[i,])))[1],
+     ECC  = mv.rank(cbind(obs, rbind(ECCtemp[i,], ECCrain[i,])))[1],
+     GC   = mv.rank(cbind(obs, rbind(GCtemp[i,], GCrain[i,])))[1],
+     SS   = mv.rank(cbind(obs, rbind(SStemp[i,], SSrain[i,])))[1]))
+ }
```

A multivariate verification rank histogram is then plotted in Fig. 11.16:

```
R> par(mfrow = c(1, 4))
R> for(n in colnames(r)){
+     hist(r[,n], freq = FALSE, breaks = 0:12 + 0.5, main = n, ylim = c(0, 0.12))
+     abline(h = 1/11, lty = 2)
+ }
```

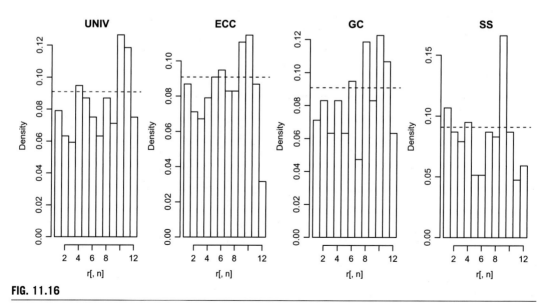

FIG. 11.16

Multivariate verification rank histograms for the different multivariate postprocessing approaches.

In agreement with the Energy score, the multivariate rank histograms show only small differences between the different methods.

11.6 SUMMARY AND DISCUSSION

This chapter introduced the statistical software and programming language R in four typical postprocessing examples, namely deterministic postprocessing of temperature, univariate ensemble postprocessing of temperature, univariate ensemble postprocessing of precipitation, and bivariate postprocessing of temperature and precipitation. These examples involved different postprocessing and verification methods, but clearly there are many other tasks and useful R functions for ensemble postprocessing that could not all be covered by this chapter.

Hamill (2018, Chapter 7 of this book) encouraged the use of open source software and common repositories for code and data. The CRAN repository provides a well-established platform to share and access open-source R code and ensemble postprocessing packages such as **ensembleBMA**, **crch**, **scoringRules**, or **SpecsVerification** already use these facilities. This chapter showed that these and other R functions and packages allow for easy testing and comparison of both well known and newly developed postprocessing methods.

11.7 APPENDICES

11.7.1 APPENDIX A: CODE FOR SOME FUNCTIONS USED IN THIS CHAPTER

This appendix provides code for some of the functions that are used in this chapter.

PITs of censored distributions, which are, for example, used in Section 11.5.2, cannot be derived exactly at the censoring point (zero precipitation). To get uniformly distributed PITs, random values are drawn for censored data. These random values correspond to negative latent (nonobserved) values. The following function computes these PITs for the censored logistic distribution:

```
R> clogispit <- function(obs, location, scale, left = 0) {
+     pit <- pclogis(obs, location, scale, left = left)
+     pit[obs <= 0] <- runif(sum(obs <= 0), 0, pit[obs <= 0])
+     return(pit)
+   }
```

The following function is taken from the supplemental material of Thorarinsdottir et al. (2016) and can be used to rank multivariate data:

```
R> mv.rank <- function(x) {
+     d <- dim(x)
+     x.prerank <- numeric(d[2])
+     for(i in 1:d[2]) {
+       x.prerank[i] <- sum(apply(x<=x[,i],2,all))
```

```
+   }
+   x.rank <- rank(x.prerank,ties="random")
+   return(x.rank)
+ }
```

11.7.2 APPENDIX B: AVAILABLE R PACKAGES FOR ENSEMBLE POSTPROCESSING

This appendix lists available packages that are particularly useful for ensemble postprocessing. More information on these packages can be found in their respective package documentations.

Available data sets and data processing

Example data of ensemble forecasts and corresponding observations can be found in **ensemblepp** (Messner, 2017), **crch** (Messner, 2016), **ensembleMOS** (Yuen, Gneiting, Thorarinsdottir, & Fraley, 2017), **ensembleBMA** (Fraley, Raftery, Sloughter, Gneiting, & University of Washington., 2018), and **SpecsVerification** (Siegert, 2017), and can be used conveniently to test existing or new ensemble postprocessing methods. Data that are stored in text format can be loaded with the base R functions `read.table()` and `read.csv()`, **RNetCDF** (Michna & Woods, 2016) and **rNOMADS** (Bowman & Lees, 2015) provide functions to load netCDF and grib files, respectively.

Infrastructure to work conveniently with time series data is provided by **zoo** (Zeileis & Grothendieck, 2005), **chron** (James & Hornik, 2018), **tseries** (Trapletti & Hornik, 2017), or **xts** (Ryan & Ulrich, 2017). Spatial data can be handled, for example, with **sp** (Bivand, Pebesma, & Gomez-Rubio, 2013), **raster** (Hijmans, 2017), or **rgdal** (Bivand, Keitt, & Rowlingson, 2017). See also the respective CRAN task views Hyndman (2017) and Bivand (2017) for an overview of available packages for time series and spatial data.

Ensemble postprocessing models

The most important R packages and functions for ensemble postprocessing have already been covered by this chapter (**ensembleBMA** (Fraley et al., 2018), **crch** (Messner, 2016), gamlss (Rigby & Stasinopoulos, 2005), **ensembleMOS** (Yuen et al., 2017), **SpecsVerification** (Siegert, 2017), **glmx** (Zeileis, Koenker, & Doebler, 2015), **ordinal** (Christensen, 2015)), but there are clearly many more packages that provide other useful postprocessing functions, for example, **quantreg** (Koenker, 2017) for quantile regression, **ensAR** (Groß & Möller, 2016) for autoregressive postprocessing (Möller & Groß, 2016), **INLA** or **bamlss** (Umlauf, Klein, & Zeileis, 2017) for Bayesian model fitting, **RandomFields** (Schlather, Malinowski, Menck, Oesting, & Strokorb, 2015) and **SpatialExtremes** (Ribatet, 2018) to simulate max-stable processes (Friederichs, Buschow, & Wahl, 2018, Chapter 5 of this book), or **CDVine** (Brechmann & Schepsmeier, 2013) for C-Vine copulae (Hemri, 2018, Chapter 8 of this book).

Verification

The forecast verification in this chapter is mainly based on basic R functions and functions from **scoringRules** (Jordan, Krueger, & Lerch, 2017) and **SpecsVerification** (Siegert, 2017). Other very useful verification functions can be found in **verification** (NCAR—Research Applications Laboratory, 2015), **easyVerification** (MeteoSwiss, 2017), or **SpatialVx** (Gilleland, 2017).

REFERENCES

Bivand, R. (2017). *CRAN task view: Analysis of spatial data.* https://CRAN.R-project.org/view=Spatial. Accessed 23 January 2018.

Bivand, R., Keitt, T., & Rowlingson, B. (2017). *rgdal: Bindings for the geospatial data abstraction library. R package version 1.2-16.* https://CRAN.R-project.org/package=rgdal. Accessed 23 January 2018.

Bivand, R. S., Pebesma, E., & Gomez-Rubio, V. (2013). *Applied Spatial Data Analysis With R.* New York: Springer. http://www.asdar-book.org/.

Bowman, D. C., & Lees, J. M. (2015). Near real time weather and ocean model data access with rNOMADS. *Computers & Geosciences, 78,* 88–95.

Brechmann, E. C., & Schepsmeier, U. (2013). Modeling dependence with C- and D-vine copulas: the R package CDVine. *Journal of Statistical Software, 52,* 1–27.

Bröcker, J., & Smith, L. A. (2007). Increasing the reliability of reliability diagrams. *Weather and Forecasting, 22,* 651–661.

Bröcker, J., & Smith, L. A. (2008). From ensemble forecasts to predictive distribution functions. *Tellus A, 60,* 663–678.

Buizza, R. (2018). Ensemble forecasting and the need for calibration. In S. Vannitsem, D. S. Wilks, & J. W. Messner (Eds.), *Statistical Postprocessing of Ensemble Forecasts.* Elsevier.

Christensen, R. H. B. (2015). *Ordinal—Regression models for ordinal data. R package version 2015. 6-28.* https://cran.r-project.org/package=ordinal. Accessed 23 January 2018.

Clark, M., Gangopadhyay, S., Hay, L., Rajagopalan, B., & Wilby, R. (2004). The Schaake shuffle: a method for reconstructing spacetime variability in forecasted precipitation and temperature fields. *Journal of Hydrometeorology, 5,* 243–262.

Efron, B. (1979). Bootstrap methods: another look at the jackknife. *The Annals of Statistics, 7,* 1–26.

Fraley, C., Raftery, A. E., Sloughter, J. M., & Gneiting, T., & *University of Washington.* (2017). *ensembleBMA: Probabilistic forecasting using ensembles and Bayesian model averaging. R package version 5.1.5.* https://CRAN.R-project.org/package=ensembleBMA. Accessed 23 January 2018.

Friederichs, P., Buschow, S., & Wahl, S. (2018). Post-processing for extreme events. In S. Vannitsem, D. S. Wilks, & J. W. Messner (Eds.), *Statistical Postprocessing of Ensemble Forecasts.* Elsevier.

Gilleland, E. (2017). *SpatialVx: Spatial forecast verification. R package version 0.6-1.* https://CRAN.R-project.org/package=SpatialVx. Accessed 23 January 2018.

Glahn, H., & Lowry, D. (1972). The use of model output statistics (MOS) in objective weather forecasting. *Journal of Applied Meteorology, 11,* 1203–1211.

Gneiting, T., Raftery, A. E., Westveld, A. H., & Goldman, T. (2005). Calibrated probabilistic forecasting using ensemble model output statistics and minimum CRPS estimation. *Monthly Weather Review, 133,* 1098–1118.

Gneiting, T., Stanberry, L. I., Grimit, E. P., Held, L., & Johnson, N. A. (2008). Assessing probabilistic forecasts of multivariate quantities, with an application to ensemble predictions of surface winds. *TEST, 17,* 211.

Groß, J., & Möller, A. (2016). *ensar: Autoregressive postprocessing methods for ensemble forecasts. R package version 0.0.0.9000.* https://github.com/JuGross/ensAR. Accessed 23 January 2018.

Hamill, T. M. (2018). Practical aspects of statistical postprocessing. In S. Vannitsem, D. S. Wilks, & J. W. Messner (Eds.), *Statistical Postprocessing of Ensemble Forecasts.* Elsevier.

Hamill, T. M., Bates, G. T., Whitaker, J. S., Murray, D. R., Fiorino, M., & Galarneau, T. J., Jr. (2013). NOAA's second-generation global medium-range ensemble reforecast dataset. *Bulletin of the American Meteorological Society, 94,* 1553–1565.

Hamill, T. M., Whitaker, J. S., & Wei, X. (2004). Ensemble reforecasting: improving medium-range forecast skill using retrospective forecasts. *Monthly Weather Review, 132,* 1434–1447.

Hemri, S. (2018). Application of postprocessing for hydrological forecasts. In S. Vannitsem, D. S. Wilks, & J. W. Messner (Eds.), *Statistical Postprocessing of Ensemble Forecasts*. Elsevier.

Hemri, S., Haiden, T., & Pappenberger, F. (2016). Discrete postprocessing of total cloud cover ensemble forecasts. *Monthly Weather Review, 144*, 2565–2577.

Hijmans, R. J. (2017). *raster: Geographic data analysis and modeling. R package version 2.6-7*. https://CRAN.R-project.org/package=raster. Accessed 23 January 2018.

Hyndman, R. J. (2017). *CRAN task view: Time series analysis*. https://CRAN.R-project.org/view=TimeSeries. Accessed 23 January 2018.

James, D., & Hornik, K. (2018). *chron: Chronological objects which can handle dates and times. R package version 2.3-52*. https://CRAN.R-project.org/package=chron. Accessed 23 January 2018. S original by David James, Report by Kurt Hornik.

Jordan, A., Krueger, F., & Lerch, S. (2017). *scoringRules: Scoring rules for parametric and simulated distribution forecasts. R package version 0.9.4*. https://CRAN.R-project.org/package=scoringRules. Accessed 23 January 2018.

Koenker, R. (2017). *quantreg: Quantile regression. R package version 5.34*. https://CRAN.R-project.org/package=quantreg. Accessed 23 January 2018.

Messner, J. W. (2016). Heteroscedastic censored and truncated regression with crch. *The R Journal, 8*, 2073–4859. https://journal.r-project.org/archive/2016-1/messner-mayr-zeileis.pdf.

Messner, J. W. (2017). *ensemblepp: Ensemble postprocessing. R package version 0.1-0*. https://CRAN.R-project.org/package=ensemblepp. Accessed 23 January 2018.

Messner, J. W., Mayr, G. J., Wilks, D. S., & Zeileis, A. (2014). Extending extended logistic regression: extended vs. separate vs. ordered vs. censored. *Monthly Weather Review, 142*, 3003–3014.

Messner, J. W., Zeileis, A., Mayr, G. J., & Wilks, D. S. (2014). Heteroscedastic extended logistic regression for post-processing of ensemble guidance. *Monthly Weather Review, 142*, 448–456.

MeteoSwiss. (2017). *easyverification: Ensemble forecast verification for large data sets. R package version 0.4.4*. https://CRAN.R-project.org/package=easyVerification. Accessed 23 January 2018.

Michna, P., & Woods, M. (2016). *RNetCDF: Interface to NetCDF Datasets. R package version 1.9-1*. https://CRAN.R-project.org/package=RNetCDF. Accessed 23 January 2018.

Möller, A., & Groß, J. (2016). Probabilistic temperature forecasting based on an ensemble autoregressive modification. *Quarterly Journal of the Royal Meteorological Society, 142*, 1385–1394.

Möller, A., Lenkoski, A., & Thorarinsdottir, T. L. (2013). Multivariate probabilistic forecasting using ensemble Bayesian model averaging and copulas. *Quarterly Journal of the Royal Meteorological Society, 139*, 982–991.

NCAR—Research Applications Laboratory. (2017). *verification: Weather forecast verification utilities. R package version 1.42*. https://CRAN.R-project.org/package=verification. Accessed 23 January 2018.

Nelder, J. A., & Wedderburn, R. W. M. (1972). Generalized linear models. *Journal of the Royal Statistical Society. Series A (General), 135*, 370–384.

Raftery, A. E., Gneiting, T., Balabdaoui, F., & Polakowski, M. (2005). Using Bayesian model averaging to calibrate forecast ensembles. *Monthly Weather Review, 133*, 1155–1174.

R Development Core Team. (2017). *An introduction to* R. Vienna, Austria. http://www.R-project.org/. Accessed 23 January 2018.

Ribatet, M. (2018). *Spatialextremes: Modelling spatial extremes. R package version 2.0-6*. https://CRAN.R-project.org/package=SpatialExtremes. Accessed 23 January 2018.

Rigby, R. A., & Stasinopoulos, D. M. (2005). Generalized additive models for location, scale and shape (with discussion). *Applied Statistics, 54*, 507–554.

Ryan, J. A., & Ulrich, J. M. (2017). *xts: eXtensible Time Series. R package version 0.10-1*. https://CRAN.R-project.org/package=xts. Accessed 23 January 2018.

Schefzik, R., & Möller, A. (2018). Multivariate ensemble post-processing. In S. Vannitsem, D. S. Wilks, & J. W. Messner (Eds.), *Statistical postprocessing of ensemble forecasts*. Elsevier.

Schefzik, R., Thorarinsdottir, T. L., & Gneiting, T. (2013). Uncertainty quantification in complex simulation models using ensemble copula coupling. *Statistical Science, 28*, 616–640.

Scheuerer, M. (2014). Probabilistic quantitative precipitation forecasting using ensemble model output statistics. *Quarterly Journal of the Royal Meteorological Society, 140*, 1086–1096.

Scheuerer, M., & Hamill, T. M. (2015a). Statistical postprocessing of ensemble precipitation forecasts by fitting censored, shifted gamma distributions. *Monthly Weather Review, 143*, 4578–4596.

Scheuerer, M., & Hamill, T. M. (2015b). Variogram-based proper scoring rules for probabilistic forecasts of multivariate quantities. *Monthly Weather Review, 143*, 1321–1334.

Schlather, M., Malinowski, A., Menck, P. J., Oesting, M., & Strokorb, K. (2015). Analysis, simulation and prediction of multivariate random fields with package RandomFields. *Journal of Statistical Software, 63*, 1–25.

Schmeits, M. J., & Kok, K. J. (2010). A comparison between raw ensemble output, (modified) Bayesian model averaging, and extended logistic regression using ECMWF ensemble precipitation reforecasts. *Monthly Weather Review, 138*, 4199–4211.

Siegert, S. (2017). *Specsverification: Forecast verification routines for ensemble forecasts of weather and climate. R package version 0.5-2.* https://CRAN.R-project.org/package=SpecsVerification. Accessed 23 January 2018.

Sloughter, J. M. L., Raftery, A. E., Gneiting, T., & Fraley, C. (2007). Probabilistic quantitative precipitation forecasting using Bayesian model averaging. *Monthly Weather Review, 135*, 3209–3220.

Stauffer, R., Umlauf, N., Messner, J. W., Mayr, G. J., & Zeileis, A. (2017). Ensemble post-processing of daily precipitation sums over complex terrain using censored high-resolution standardized anomalies. *Monthly Weather Review, 145*, 955–969.

Thorarinsdottir, T., & Schuhen, N. (2018). Verification: assessment of calibration and accuracy. In S. Vannitsem, D. S. Wilks, & J. W. Messner (Eds.), *Statistical Postprocessing of Ensemble Forecasts.* Elsevier.

Thorarinsdottir, T. L., Scheuerer, M., & Heinz, C. (2016). Assessing the calibration of high-dimensional ensemble forecasts using rank histograms. *Journal of Computational and Graphical Statistics, 25*, 105–122.

Trapletti, A., & Hornik, K. (2017). *tseries: Time series analysis and computational finance. R package version 0.10-42.* https://CRAN.R-project.org/package=tseries. Accessed 23 January 2018.

Umlauf, N., Klein, N., & Zeileis, A. (2017). BAMLSS: Bayesian additive models for location, scale and shape (and beyond). *Journal of Computational and Graphical Statistics*, in press. https://doi.org/10.1080/10618600.2017.1407325.

Wang, X., & Bishop, C. H. (2005). Improvement of ensemble reliability with a new dressing kernel. *Quarterly Journal of the Royal Meteorological Society, 131*, 965–986.

Wilks, D. S. (2009). Extending logistic regression to provide full-probability-distribution MOS forecasts. *Meteorological Applications, 368*, 361–368.

Wilks, D. S. (2011). *Statistical Methods in the Atmospheric Sciences.* London: Academic Press.

Wilks, D. S. (2018). Univariate ensemble post-processing. In S. Vannitsem, D. S. Wilks, & J. W. Messner (Eds.), *Statistical Postprocessing of Ensemble Forecasts.* Elsevier.

Wilks, D. S., & Hamill, T. M. (2007). Comparison of ensemble-MOS methods using GFS reforecasts. *Monthly Weather Review, 135*, 2379–2390.

Yuen, R., Gneiting, T., Thorarinsdottir, T., & Fraley, C. (2017). *ensembleMOS: Ensemble model output statistics. R package version 0.8.1.* https://CRAN.R-project.org/package=ensembleMOS. Accessed 23 January 2018.

Zeileis, A., & Croissant, Y. (2010). Extended model formulas in R: multiple parts and multiple responses. *Journal of Statistical Software, 34*, 1–13.

Zeileis, A., & Grothendieck, G. (2005). zoo: S3 infrastructure for regular and irregular time series. *Journal of Statistical Software, 14*, 1–27.

Zeileis, A., Koenker, R., & Doebler, P. (2015). *glmx: Generalized linear models extended. R package version 0.1-1.* https://CRAN.R-project.org/package=glmx. Accessed 23 January 2018.

Author Index

Note: Page numbers followed by *f* indicate figures and *t* indicate tables.

Subject Index

Note: Page numbers followed by *f* indicate figures and *t* indicate tables.

341

Printed in the United States
By Bookmasters